256
Topics in Current Chemistry

Topics in Current Chemistry

Recently Published and Forthcoming Volumes

Prebiotic Chemistry
Volume Editor: Walde, P.
Vol. 259, 2005

Supramolecular Dye Chemistry
Volume Editor: Wuerthner, F.
Vol. 258, 2005

Molecular Wires
From Design to Properties
Volume Editor: De Cola, L.
Vol. 257, 2005

Low Molecular Mass Gelators
Design, Self-Assembly, Function
Volume Editor: Fages, F.
Vol. 256, 2005

Anion Sensing
Volume Editor: Stibor, I.
Vol. 255, 2005

Organic Solid State Reactions
Volume Editor: Toda, F.
Vol. 254, 2005

DNA Binders and Related Subjects
Volume Editors:Waring, M.J., Chaires, J.B.
Vol. 253, 2005

Contrast Agents III
Volume Editor: Krause,W.
Vol. 252, 2005

Chalcogenocarboxylic Acid Derivatives
Volume Editor: Kato, S.
Vol. 251, 2005

New Aspects in Phosphorus Chemistry V
Volume Editor:Majoral, J.-P.
Vol. 250, 2005

Templates in Chemistry II
Volume Editors: Schalley, C.A.,Vögtle, F., Dötz, K.H.
Vol. 249, 2005

Templates in Chemistry I
Volume Editors: Schalley, C.A.,Vögtle, F., Dötz, K.H.
Vol. 248, 2004

Collagen
Volume Editors: Brinckmann, J., Notbohm, H., Müller, P.K.
Vol. 247, 2005

New Techniques in Solid-State NMR
Volume Editor: Klinowski, J.
Vol. 246, 2005

Functional Molecular Nanostructures
Volume Editor: Schlüter,A.D.
Vol. 245, 2005

Natural Product Synthesis II
Volume Editor:Mulzer, J.
Vol. 244, 2005

Natural Product Synthesis I
Volume Editor:Mulzer, J.
Vol. 243, 2005

Immobilized Catalysts
Volume Editor: Kirschning, A.
Vol. 242, 2004

Transition Metal and Rare Earth Compounds III
Volume Editor: Yersin, H.
Vol. 241, 2004

The Chemistry of Pheromones and Other Semiochemicals II
Volume Editor: Schulz, S.
Vol. 240, 2005

The Chemistry of Pheromones and Other Semiochemicals I
Volume Editor: Schulz, S.
Vol. 239, 2004

Orotidine Monophosphate Decarboxylase
Volume Editors: Lee, J.K., Tantillo,D.J.
Vol. 238, 2004

Low Molecular Mass Gelators

Design, Self-Assembly, Function

Volume Editor: Frédéric Fages

With contributions by
K. Araki · A. Brizard · F. Fages · A. R. Hirst · I. Huc · T. Kato
T. Kitamura · X.Y. Liu · N. Mizoshita · M. Moriyama · R. Oda
D. K. Smith · F. Vögtle · I. Yoshikawa · M. Žinić

Springer

The series *Topics in Current Chemistry* presents critical reviews of the present and future trends in modern chemical research. The scope of coverage includes all areas of chemical science including the interfaces with related disciplines such as biology, medicine and materials science. The goal of each thematic volume is to give the nonspecialist reader, whether at the university or in industry, a comprehensive overview of an area where new insights are emerging that are of interest to a larger scientific audience.
As a rule, contributions are specially commissioned. The editors and publishers will, however, always be pleased to receive suggestions and supplementary information. Papers are accepted for *Topics in Current Chemistry* in English.
In references *Topics in Current Chemistry* is abbreviated Top Curr Chem and is cited as a journal.

Visit the TCC content at springerlink.com

Library of Congress Control Number: 2005926237

ISSN 0340-1022
ISBN-10 3-540-25321-1 **Springer Berlin Heidelberg New York**
ISBN-13 978-3-540-25321-1 **Springer Berlin Heidelberg New York**
DOI 10.1007/b105250

Springer is a part of Springer Science+Business Media
springeronline.com

Printed in Germany

Cover design: KünkelLopka GmbH, Heidelberg; *Design & Production* GmbH, Heidelberg
Typesetting and Production: LE-TeX Jelonek, Schmidt & Vöckler GbR, Leipzig

Printed on acid-free paper 02/3141xv – 5 4 3 2 1 0

Volume Editor

Professor Frédéric Fages
Faculté des Sciences de Luminy
Université de la Méditerranée
Case 901
13288 Marseille Cedex 9,
France
fages@luminy.univ-mrs.fr

Editorial Board

Topics in Current Chemistry also Available Electronically

For all customers who have a standing order to Topics in Current Chemistry, we offer the electronic version via SpringerLink free of charge. Please contact your librarian who can receive a password for free access to the full articles by registration at:

springerlink.com

If you do not have a subscription, you can still view the tables of contents of the volumes and the abstract of each article by going to the SpringerLink Homepage,clicking on "Browse by Online Libraries", then "Chemical Sciences", and finally choose Topics in Current Chemistry.

You will find information about the

- Editorial Board
- Aims and Scope
- Instructions for Authors
- Sample Contribution

at springeronline.com using the search function.

Preface

There are certainly plenty of reasons accounting for the fascination exerted during the last several years by low molecular mass gelators (LMGs). At least two of them merit particular attention. On one hand, gelation of organic fluids or water with LMGs represents an extraordinary macroscopic expression of supramolecular self-assembly. It is indeed fascinating to realize how recognition events at the molecular level can lead so efficiently to the generation of three-dimensional continuous networks spanning whole sample volumes. Remarkably, the resulting self-assembled gel-phase materials, obtained at amazingly low concentration of a LMG, are fairly stiff, often at high temperature, and can retain their macroscopic shape, a property characteristic of the solid state of matter. On the other hand, gels are doubtlessly unique materials. They have been known since ancient times – their origin can be traced back to at least Neolithic times – and, ever since, they have played a crucial role in many decisive advances of technology, art and medicine. Gels continue to hold the front page as they not only remain irreplaceable materials in daily life, but are still considered to be one of the most promising materials in the 21th century.

Gels are especially complex systems and, in spite of the huge number of investigations, there is no simple definition of the gel state. It is well known that many polymers, synthetic and natural, form gels. This property largely stems the propensity of long-chain macromolecules to give rise to networks that immobilize the majority liquid component by surface tension. By contrast, LMGs are clearly defined molecular structures and it is their unidirectional self-assembly that serves to build thermoreversible networks of entangled fiber-like aggregates. A major attraction of the supramolecular approach toward gel-phase materials is the possibility to exquisitely control network properties and morphology by precise variations of the LMG chemical structure. As the spectrum of LMGs has considerably enlarged over the last decade, it is thus possible, via rational synthesis, to access a wide diversity of tunable functional materials for applications in separation technologies, medicine, biology, electronics, photonics, templated material synthesis, etc.

This book is intended to provide a comprehensive overview of some of the most exciting chemical and physical aspects in the field of low molecular weight organo- and hydrogelators. The contributions also illustrate the need for a multidisciplinary approach between synthetic, physical and biological chemistry, physics and material science. Chapter 1 presents the physical principles of the growth mechanism of fiber and fiber network with LMGs, as treated on the basis of the heterogeneous nucleation model. It also demonstrates that, beside chemical approaches, physical factors can be elegantly exploited to control and manipulate the morphology of self-assembled nanostructures in order to produce materials with desired rheological properties. The systematic synthesis and gelation ability of LMGs containing cholesterol and amide self-assembling motifs, two major classes of versatile gelators, are discussed in Chaps. 2 and 3, respectively. These chapters are intended to outline useful synthetic guidelines for the generation of an ever-increasing variety of molecular architectures within these two families of gelators. Recent developments in the chemistry of nucleobase-containing LMGs are described in Chap. 4. Hydrogen-bonding within these molecular systems involves complementary base pair formation, a process relevant to DNA double-helix formation. As such, their self-assembled gels have emerged recently as a very promising class of soft materials with biomimetic functional features. The self-assembly of chiral organo- or hydrogelators is the subject of Chap. 5. In many cases gelation of water or organic liquids with chiral LMGs is observed to lead to chiral supramolecular aggregates that exhibit a distinct helical or twisted structure, a feature reminiscent of biological systems. Liquid crystalline physical gels that result from the orthogonal self-assembly of liquid crystals and LMGs are presented in Chap. 6. The growth of self-assembled solid fibers in thermotropic liquid crystals leads to the formation of highly anisotropic composite materials with unique potential for the fabrication of optical, electrical, and photofunctional devices. The volume concludes with Chap. 7, a review of the emerging field of dendritic gels. Strictly speaking, dendrimers are not low molecular weight compounds. Yet, in contrast to the case of polymers, they have well-defined structures. In this respect, dendrimer gelators bridge the gap between LMGs and polymers and as such do offer exciting future directions to explore.

Of the many exciting achievements of supramolecular chemistry, it is arguably the control of multiple, specific recognition events at the molecular level that allows the construction of nanoscale architectures of increasing structural or topological complexity. As such, supramolecular synthesis represents a powerful bottom-up fabrication approach that allows one to generate not only novel, beautiful structures, but also uniquely functioning supramolecular devices and highly tunable materials. Clearly the field of low molecular mass organo- and hydrogelators has evolved into a sophisticated science at the frontiers of supramolecular chemistry.

Marseille, January 2005 Frédéric Fages

Contents

Contents of Volume 248

Templates in Chemistry I

Volume Editors: Schalley, Christoph A.; Vögtle, Fritz; Dötz, Karl H.
ISBN: 3-540-22547-1

Contents of Volume 249

Templates in Chemistry II

Volume Editors: Schalley, Christoph A.; Vögtle, Fritz; Dötz, Karl H.
ISBN: 3-540-23087-4

Top Curr Chem (2005) 256: 1–37
DOI 10.1007/b107170

Published online: 1 July 2005

Gelation with Small Molecules: from Formation Mechanism to Nanostructure Architecture

Xiang Y. Liu

Department of Physics, Faculty of Science, National University of Singapore, 2 Science Drive 3, 117542 Singapore, Singapore
phyliuxy@nus.edu.sg

Abstract The mechanism of fiber and fiber network formation of small molecular gelling agents is treated on the basis of a generic heterogeneous nucleation model. The formation of a crystallite fiber network can take place via the so-called crystallographic mismatch branching. At very low supersaturations, unbranched fibers form predominantly. As supersaturation increases, small-angle crystallographic mismatch branching occurs at the side face of growth fibers. At very high supersaturations, the so-called wide-angle crystallographic mismatch branching becomes kinetically favorable. Both give rise to the formation of fiber networks, but of different types. Controlling the branching of the nanofibers of small molecular gelatins allows us to achieve the micro/nanostructure architecture of networks having the desired rheological properties. In this regard, the engineering of supramolecular functional materials can be achieved by constructing and manipulating the micro/nanostructure in terms of a "branching creator", or by tuning processing conditions.

Keywords Nanofiber · Nucleation · Branching · Fiber network · Additive

Abbreviations

a	Activity
C	Concentration
D_f	Fractal or Hausdorff–Besicovitch dimension of a pattern
d	Diameter of an object
DIOP	Di-(2-ethylhexyl phthalate) $(C_8H_{17}COO)_2$ (C_6H_4)
EVACP	Ethylene/vinyl acetate copolymer
$f(m)$	Interfacial correlation function
G^*	Complex modulus
G'	Storage modulus
G''	Loss modulus
GP-1	*N*-lauroyl-L-glutamic acid di-*n*-butylamide
ΔG	Gibbs free-energy barrier
h	Height of step of crystal surface
Δh_m	Enthalpy of melting per molecule
ISA	Isostearyl alcohol
L-DHL	Lanosta-8,24-dien-3β-ol:24,25-dihydrolanosterol = 56:44
J	Nucleation rate
k	Boltzmann constant
m	Interfacial matching parameter
N	Number of particles or segments
N_g	Number of crystals per volume
P	Pressure
r_c	Radius of curvature of critical nucleus
R	Radius of gyration of a pattern
R_g	Growth rate of fiber along the axial direction
SA-CMB	Small-angle crystallographic mismatch branching
SEM	Scanning electron microscopy
T	Temperature
t_s	Nucleation induction time
t	Time
WA-CMB	Wide-angle crystallographic mismatch branching
v_g	Growth rate of bulk crystals
X	Crystallinity of a system
φ	Volume fraction of crystal materials
γ	Interfacial free energy
γ_{step}	Step free energy of crystal surface
μ	Chemical potential
η	Viscosity
θ	Contact angle
Ω	Volume per structural unit
τ	Induction time for the nucleation of new fibers on the host fibers
σ	Supersaturation
ω	Angular frequency
ξ	Branching distance

1 Introduction

Supramolecular functional materials having 3D fibrous network structures formed by, for instance, dilute solutions of polymers, proteins, and inorganic substances like silica or clays in water and organic solvents have been well studied. In recent years there has been rapidly growing interest in such materials, which is motivated by the many potential applications in photographic, cosmetics, food, and petroleum industries, drug delivery, lithography, catalyst supporters, scaffolds for tissue engineering, the novel separation for macromolecules, etc. [1–14]. Supramolecular functional materials with 3D fibrous network structures can be employed as a vehicle for drug delivery and controlled release. The mesh size of 3D fiber networks will determine the rate of drug release.

Macroscopic properties, in particular, the rheological properties of supramolecular functional materials are determined by the micro/nanostructure of fiber networks. These materials have continuous 3D entangled networks in the liquid, thereby preventing the liquid from flowing owing to the capillary force.

Among these materials, those formed from small organic molecules are a special class. In contrast to their macromolecular and inorganic counterparts, it is believed that the network structure formed by low molecular weight organogelators is held together solely by noncovalent forces, including hydrogen bonding, stacking, and solvophobic effects.

Fibrous networks with permanent interconnections will effectively entrap and immobilize liquid in the meshes, and possess both the elastic properties of ideal solids and the viscosity properties of Newtonian liquids. Consequently, self-supporting supramolecular materials will be obtained [8–10, 15–21]. In contrast, systems consisting of nonpermanent/or transient interconnecting (or entangled) fibers or needles can only form weak and viscous paste at low concentrations [6, 7].

Although the formation of supramolecular functional materials from small molecules is an excellent example of a supramolecular self-organization process, most such materials have been found by serendipity rather than design, and many aspects of supramolecular functional materials are still poorly understood. The control of gelation phenomena induced by small molecules and the design of new gelling agents are therefore challenging goals leading to a new area of fascinating organic materials, and it is only recently that a number of successes have been reported.

It was believed [6, 9] that the formation of interconnecting fiber networks, which leads to the formation of supramolecular materials, takes place via molecular self-assembly of nanofibers. Nevertheless, the latest research indicates that the 3D self-organized micro/nanostructure of supramolecular functional materials is controlled by a so-called crystallographic mismatch branching. This is essentially a special case of heterogeneous nucleation. This implies that even for areas such as supramolecular functional materials where conventionally crystallization was regarded unimportant, knowledge of nucleation is also very crucial.

It is the purpose of this chapter to analyze the kinetics of nucleation under the influence of substrates and additives from the point of view of the solid/fluid structure. On the basis of the knowledge acquired, the principles and strategies for the engineering of micro/nanostructures of various systems, in particular supramolecular functional materials, will be examined.

2 Crystallization of Nanofibers

As can be seen in the following sections, nucleation is the initial step in the formation of crystalline materials. It is also very crucial in determination of the structural synergy between crystals and the substrate. It will be shown that the formation of a fibrous structure of some supramolecular materials is actually controlled by special type of nucleation—crystallographic mismatch nucleation, on the growing tips of fibers. Therefore, a decent understanding of nucleation is very important.

2.1 Thermodynamic Driving Force

Nanofibers which form self-organized fibrous networks in organogels are sometimes found to have a 3D crystalline order [15–18, 21]. The formation of fibers, therefore, takes place in most cases via a crystallization process [15–18, 21], including *nucleation* and *growth* [22–32].

Crystallization is the process that the first-order phase transitions begin with. The driving force for the formation of new phases (e.g., crystals) is $\Delta\mu$, which is defined as the difference between the chemical potentials μ_{mother} and μ_{crystal} of the growth unit in the mother and the crystalline phases [22, 31, 34]:

$$\Delta\mu = \mu_{\text{mother}} - \mu_{\text{crystal}} \,. \tag{1}$$

When $\Delta\mu > 0$, it is said that the system is supersaturated. This is the thermodynamic precondition for nucleation and growth of the crystalline phase. Conversely, when $\Delta\mu < 0$, the system is undersaturated. Under such conditions, crystals will dissolve. In the case where $\Delta\mu = 0$, the mother phase is in equilibrium with the crystalline phase [20, 29, 32].This implies that under the given temperature T, pressure P, concentration C, etc., one always has

$$\mu_{\text{mother}}^{\text{eq}} = \mu_{\text{crystal}}\,, \tag{2}$$

where $\mu_{\text{mother}}^{\text{eq}}$ is the chemical potential of solute molecules in the phase equilibrium (or coexistence) between the mother and the crystalline phases. It follows that for a given condition, μ_{crystal} can be expressed by $\mu_{\text{mother}}^{\text{eq}}$. Therefore, in many cases of practical importance $\Delta\mu$ can be expressed as

$$\Delta\mu = \mu_{\text{mother}} - \mu_{\text{mother}}^{\text{eq}}\,. \tag{3}$$

For crystallization from solutions, the chemical potential of species i is given by [15, 23, 25]

$$\mu_i = \mu_i^0 + kT\ln a_i \approx \mu_i^0 + kT\ln C_i\,, \tag{4}$$

where a_i, and C_i denote the activities and concentrations of solute, k is the Boltzmann constant, and T is the absolute temperature. μ_i^0 denotes the standard state ($a_i = 1$) of the solute chemical potential. This then gives rise to the dimensionless thermodynamic driving force:

$$\frac{\Delta\mu}{kT} = \ln\frac{a_i}{a_i^{\text{eq}}} \approx \ln\frac{C_i}{C_i^{\text{eq}}}\,, \tag{5}$$

where a_{eq} and C_{eq} are, respectively, the equilibrium activities and concentrations of the solute.

Notice that the thermodynamic driving force for crystallization is often expressed in terms of supersaturation. If we define supersaturation as

$$\sigma = \left(a_i - a_i^{\text{eq}}\right)/a_i^{\text{eq}} \approx \left(C_i - C_i^{\text{eq}}\right)/C_i^{\text{eq}} \tag{6}$$

Eq. 5 can then be rewritten as

$$\frac{\Delta\mu}{kT} = \ln(1+\sigma)\,. \tag{7}$$

In the case of $\sigma < 1$, Eq. 7 can be approximated, after the Taylor series expansion, as

$$\Delta\mu/kT = \ln(1+\sigma) \cong \sigma\,. \tag{8}$$

For crystallization from melts at temperatures not far below the melting or equilibrium temperature, we have the thermodynamic driving force by applying similar thermodynamic principles as [31, 35]

$$\frac{\Delta\mu}{kT} = \Delta h_{\text{m}}\Delta T/kTT_{\text{e}}\,, \tag{9}$$

$$\Delta T = (T_{\text{e}} - T)\,, \tag{10}$$

where Δh_{m} is the enthalpy of melting per molecule, T_{e} is the equilibrium temperature, and ΔT is supercooling.

2.2
Fiber Formation

2.2.1
Nucleation of Fiber

The formation of fibers, in particular, crystalline fibers takes place in terms of nucleation. In distinction from the branching process, the nucleation associated with the generation of new fibers is hereafter referred to as *primary nucleation*. A characteristic feature of the nucleation process is that the substance with the properties of the new phase is fluctuating and localized in nanoscale small spatial regions. These are occupied by atoms or molecules of various numbers which constitute the so-called clusters. The clusters staying in equilibrium with the surrounding mother phase are the critical nuclei, and the smaller or the larger clusters are the subnuclei or the supernuclei, respectively. Only the supernuclei are the clusters that can grow spontaneously to reach macroscopic sizes. For simplicity, we call hereafter the subnuclei "clusters", and the supernuclei "nuclei".

The nucleation rate J describing the number of nuclei successfully generated from the population of clusters per unit time unit volume is determined by the height of a free-energy barrier, the so-called nucleation barrier. The occurrence of a nucleation barrier is attributed to the following two contradictory effects:

1. Since the crystalline phase is a stable phase, the occurrence of the new phase from the mother phase will lead to the lowering of the (Gibbs) free energy of the system.
2. Owing to the interfacial (or surface) free energy, the increase in the size of the crystalline new phase leads to the increase of the area of the interface (or surface), and consequently the interfacial (or surface) free energy. This will cause the increase of the Gibbs free energy of the system.

The combination of these two effects gives rise to the formation of the nucleation barrier.

The occurrence of a substrate in the system normally reduces the interfacial (or surface) free energy; therefore it will also lower the nucleation barrier. Note that the substrate can be any foreign body, such as dust particles, air bobbles, or the wall of a gelation container during the formation of new fibers. In the following sections, the so-called crystallographic mismatch branching, which is one of the key processes responsible for the fiber network formation, will be treated. In this case, the branching is a special case of heterogeneous nucleation, where the substrate can be parent fibers. Let ΔG^*_{homo} be the homogeneous nucleation barrier, and $\Delta G^*_{\text{hetero}}$ be the heterogeneous nucleation barrier (the nucleation barrier in the presence of the substrate.) The nucleation barrier can be given as

$$\Delta G^*_{\text{hetero}} = f \Delta G^*_{\text{homo}} , \tag{11}$$

$$\Delta G^*_{\text{homo}} = \frac{16\pi\gamma_{\text{cf}}^3 \Omega^2}{3\left[kT \ln(1+\sigma)\right]^2} , \tag{12}$$

and, f, a factor describing the lowering of the energy barrier of nucleation owing to the occurrence of substrate, can be expressed as a function of m as [22, 31, 34]

$$f(m) = \frac{1}{4}(2 - 3m + m^3) , \tag{13}$$

$$\text{with} \quad m = (\gamma_{\text{sf}} - \gamma_{\text{sc}})/\gamma_{\text{cf}} \approx \cos\theta , \quad (-1 \leq m \leq 1) , \tag{14}$$

where γ_{ij} is the surface free energy per unit surface between phases i and j and Ω is the volume per structural unit, the mother phase is represented by subscript f, the cluster of the crystalline phase by c and the foreign body by s.

In association with the nucleation barrier, the radius of curvature r_c of critical nuclei is determined by γ_{cf} and the driving force $\Delta\mu$ [22, 28, 30, 31], and can be given by

$$r_c = 2\Omega\gamma_{\text{cf}}/\Delta\mu = 2\Omega\gamma_{\text{cf}}/kT \ln(1+\sigma) . \tag{15}$$

Note that the factor $f(m)$ changes from 1 to 0 as m varies from –1 to 1. Obviously, this factor plays an important role in the determination of the heterogeneous nucleation barrier $\Delta G^*_{\text{hetero}}$. One can see from Eq. 11 that the influence of substrates on the nucleation barrier can be fully characterized by this factor.

Let us repeat the picture of the heterogeneous nucleation. On the substrate surface, some molecular processes occur owing to transient visiting molecules which adsorb, form short-lived unions, break up, desorb, etc. An instantaneous census would show some distributions of subcritical nuclei (or clusters) with 1,2,3, ...molecules per cluster.

Nucleation begins with the formation of a cluster of size r_c, with n^* molecules. r_c is given by Eq. 15.

The widely accepted kinetic model of nucleation (within the cluster approach) was used first by Farkas [35] in 1927. It is based on the scheme of successive "chain reaction" between monomer molecules and n-sized clusters:

$$\text{Monomer} \rightleftharpoons \text{dimer}\ldots \rightleftharpoons (n-1)\text{mer} \rightleftharpoons n\text{-mer} \rightleftharpoons (n+1)\text{mer}\ldots \tag{16}$$

Taking into account the effect of the substrate on both the nucleation barrier and the transport process, and the effect of the density and the size of foreign

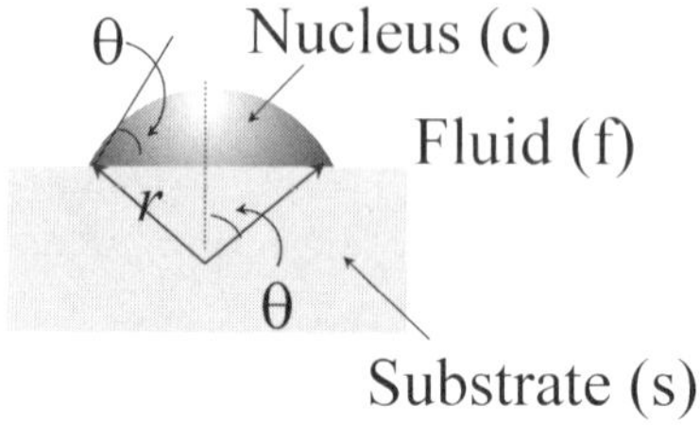

Fig. 1 A generic picture of nucleation on the foreign particle. θ is the virtual contact angle

particles occurring in the system, we obtain the nucleation rate as [31–34]

$$J = Bf''(m)\left[f(m)\right]^{1/2} \times \exp\left[-\frac{16\pi\gamma_{\mathrm{cf}}^{3}\Omega^{2}}{3kT\left[kT\ln(1+\sigma)\right]^{2}} f(m)\right], \tag{17}$$

$$\text{with} \quad f(m) = \frac{1}{2}(1-m), \tag{18}$$

where B is the kinetic constant. The growth of nuclei is subject to the effective collision and the incorporation of growth units onto the surfaces of the nuclei (Fig. 1). In the case of homogeneous nucleation, growth units can be incorporated into nuclei in all directions; however, in the case of heterogeneous nucleation, the presence of substrates will block the collision of growth units with the surfaces of these nuclei (Fig. 1). $f'(m)$ in the preexponential term describes the ratio between the average effective collision in the presence of substrates and that of homogeneous nucleation (no substrate).

Both $f(m)$ and $f''(m)$ are functions of m. When $m \to -1$, $f(m)f''(m) = 1$. This is equivalent to the case of homogeneous nucleation. In the case where $m \to 1$, one has $f(m)f''(m) = 0$. Normally, heterogeneous nucleation occurs in the range of m between 1 and – 1, or $f(m)$ between 0 and 1, depending on the interfacial structural match between the nucleating phase and the substrate, and supersaturation.

For a given system, the induction time t_{s} of nucleation can be associated with J by $t_{\mathrm{s}} = 1/J$. It follows from Eq. 17 that

$$\ln t_{\mathrm{s}} = \frac{\rho}{\left[\ln(1+\sigma)\right]^{2}} f(m) - \ln\left\{Bf''(m)\left[f(m)\right]^{1/2}\right\}, \tag{19}$$

$$\text{with} \quad \rho = \frac{16\pi\gamma_{\mathrm{cf}}^{3}\Omega^{2}}{3\left(kT\right)^{3}}. \tag{20}$$

According to Eq. 15, the plot of $\ln(t_{\mathrm{s}})$ against $1/\left[\ln(1+\sigma)\right]^{2}$ should give rise to a straight line whose slope is determined by ρ and $f(m)$. Obviously, for a given system ($\rho, B = \text{const.}$), the slope of the straight line will change accordingly to $f(m)$. In this sense, the slope of the $\ln(t_{\mathrm{s}})$ vs. $1/\left[\ln(1+\sigma)\right]^{2}$ plot gives the relative $f(m)$ for the given system. One can analyze the change of the correlation between the substrate and the crystalline phase in terms of the variation of the slope.

As given by Eq. 14, m is directly associated with γ_{cs}, which is determined by the interaction and/or structural match between the crystalline phase and the substrate. For a given crystalline phase and a substrate, the optimal structural match is the crystallographic orientation $\{hkl\}$ corresponding to the strongest *average* interaction between the crystalline phase and the substrate. This orientation corresponds to the (minimal) cusp in the γ-plot.

Evidently, an excellent structural match $[\gamma_{cs}(\alpha) \to 0$, at $\alpha \to 0]$ between the nucleating phase and the substrate leads to $m \to \gamma_{sf}/\gamma_{cf}$ [34]. (α is the mismatch angle between the nucleating phase and the substrate.) In the case where $\gamma_{sf} \approx \gamma_{cf}$, one has then $m \to 1$, and $f(m) \to 0$ (cf. Eq. 13). As mentioned before, this implies that $\Delta G^*_{\text{hetero}}$ almost vanishes completely (cf. Eq. 11.) It occurs only when the growth of the crystals is well orientated and ordered with respect to the structure of the substrate. In this case, the excellent epitaxial relation arises.

As the structural match varies from a perfect to a poor match, m decreases from 1 to 0, – 1. The extreme case will be $m \to -1$, corresponding to the situation where there is no correlation between the nucleating phase and the substrate. This is the case where the substrates exert almost no influence on nucleation, which is equivalent to homogeneous nucleation. Nuclei emerging in this case are completely disordered, bearing no correlation to the substrate. One has then $f(m) = 1$.

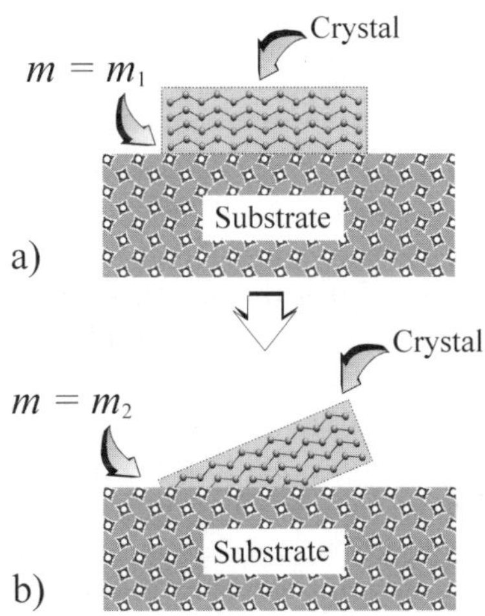

Fig. 2 The structural match between a nucleus and the substrate and the corresponding m. (**a**) Good interfacial structural match between the nucleus and the substrate. m_1. (**b**) Poor interfacial structural match between the nucleus and the substrate. $m_2 \cdot m_1 > m_2$, and $f(m_1) < f(m_2)$

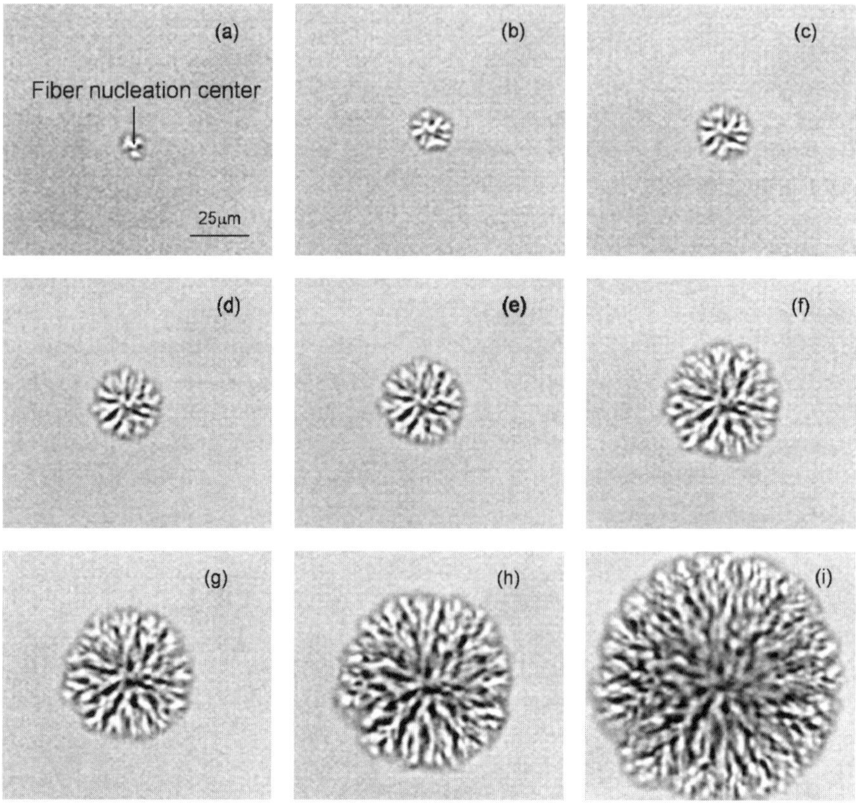

Fig. 3 In situ observation on the formation of an *N*-lauroyl-L-glutamic acid di-*n*-butylamide (*GP-1*) fiber network. (**a**) The formation of primary fibers initiates from a nucleation center. The formation of GP-1 fibers and the branching process is shown by (**a**)–(**h**), in which the time interval between two neighboring photographs is of 0.2 s. Solvent 1,2-propanediol; $\sigma = 6.92$; $T = 330$ K

Note that the primary nucleation of fibers is normally governed by heterogeneous nucleation. Figure 3 shows the initiation of an *N*-lauroyl-L-glutamic acid di-*n*-butylamide (GP-1) a fiber network (spherulite) from a nucleation center.

2.2.2 Growth of Crystalline Fibers

The process following nucleation is growth [22, 23]. The growth of crystals is actually a process of delivering growth units from the bulk to a crystal surface and incorporating them into the kinks at the surface (Fig. 4) [22, 23]. In the case of the faceted growth, the crystal face is atomically smooth and the kinks occur only at the steps. In this case, the steps can be regarded as "sinks" for

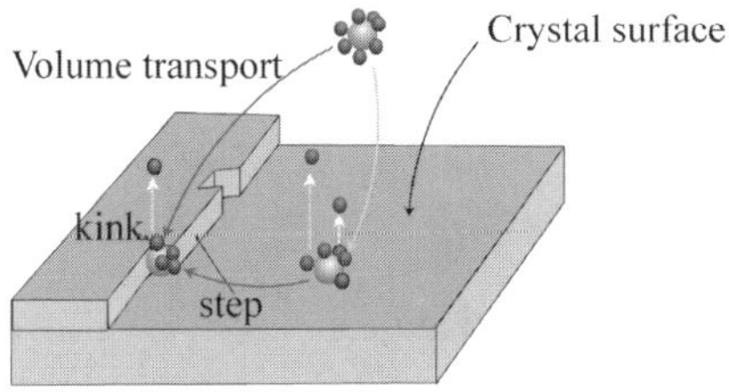

Fig. 4 Schematic illustration of the growth process occurring at the surface of a crystal

growth units to enter the crystals [22, 23, 37–39]. As shown by Fig. 4, each advancing step will be eliminated when it spreads over the surface and reaches the edge of the surface. Therefore, the step source or the creation of new steps will determine the growth rate of the crystal surface. Owing to the step free energy, the creation of a new layer on the existing layer of the crystal surface requires overcoming a free-energy barrier, the so-called 2D nucleation barrier [22, 23, 37–39]. Normally, for the growth of flat or faceted crystal surfaces, the screw dislocations occurring at the surface will provide uninterrupted step sources for the layer-by-layer growth [22, 23, 37]. In this case, the growth is controlled by the screw dislocation mechanisms [22, 23, 37].

If the crystals are free of screw dislocations, their growth is then governed by the mechanism of 2D nucleation [22, 23, 38, 39]. This implies that the growth of crystal faces takes place by growing crystal layers one on top of the other, and the occurrence of a new layer on the existing layer is via 2D nucleation [38, 39]. According to this model, the growth rate of the fibers can be expressed as [38, 39]

$$R_g = C_1 \left(\frac{\Delta\mu}{kT}\right)^{5/6} \exp\left(\frac{\Omega\gamma_{\text{step}}^2 \pi h}{3(kT)^2(\Delta\mu/kT)}\right), \tag{21}$$

where h and γ_{step} denote the height of the steps and the step free energy of the growing surface, respectively. C_1 is a coefficient associated with volume transport, and should be a constant for a given condition [38, 39].

In the case of the tips of fibrous or needle crystals, the growth of the crystal surfaces is then likely to occur in a rough mode. In this case, the 2D nucleation energy barrier is zero or very small [22, 23, 37–39]. The creation of new layers or steps will not cost any energy. Therefore, the overall crystallographic orientation of the crystal surface will not be maintained under this mode of growth [40]. The growth rate of rough (or normal) growth is expressed [22, 41] as

$$R_g = A_1 \beta_{\text{st}} \left(\frac{\Delta\mu}{kT}\right), \tag{22}$$

where A_1 is a coefficient associated with volume transport, and should be a constant for a given condition [22, 41].

3 Crystallographic Mismatch Branching

3.1 Templating and Shadow Effects

Kinetically, the occurrence of substrates will, on one hand, lower the nucleation barrier, leading to an increase in the nucleation rate; on the other hand, it will exert a negative impact on the surface integration. Nucleation on a substrate will reduce the effective collisions of structural units with the surface of clusters (Fig. 4), where the structural units are incorporated into the crystal phase. This will slow down the nucleation kinetics, in contrast to the effect of lowering the nucleation barrier. The latter, called the "shadow effect" of the substrate, is described by $f''(m)$ and $f(m)$ appearing in the preexponential term of Eq. 17.

These two contradictory effects play different roles in different regimes. At low supersaturations, the nucleation barrier is very high (Eqs. 11, 12). The nucleation rate will be substantially enhanced if the nucleation barrier is suppressed effectively $[f(m) \to 0]$. Therefore, the heterogeneous nucleation with a strong interaction and an optimal structural match (or the optimal epitaxial relation) between the substrate and the nucleating phase will be kinetically favored. In this case, the nucleation of crystalline materials will be best templated by substrates having excellent structural correlation with the crystalline phase. The structural synergy between the nucleating phase and the substrate will be optimal under this condition.

On the other hand, the epitaxial and templating relationship between the substrate and the nucleating phase cannot be maintained even for substrates having the optimal structural match with crystalline materials if supersaturation is too high. At higher supersaturations, the exponential term associated with the nucleation barrier becomes less important. Instead, the shadow effect of the substrate, described by the preexponential factors $f(m)$ and $f'(m)$, will govern the kinetics (Fig. 4). Nucleation on substrates having larger $f(m)$ and $f''(m)$ (or $m \to 0, -1$) corresponds to a lower degree of restriction from the substrate, and a higher degree of orientational freedom (or a larger entropy). This will reduce the shadow effect of the substrate, therefore become kinetically more favorable.

If σ progressively increases from low supersaturations to high supersaturations, nucleation will be governed by a sequence of progressive heterogeneous processes associated with increasing $f(m)$.

Notice that the templating and the supersaturation driven interfacial structural mismatch are two contradictory effects. By carefully adjusting these two effects, we should be able to engineer and fabricate complex structures at the micro/nanostructural level.

3.2 Crystallographic Mismatch Branching

3.2.1 General Patterns

Crystalline materials reveal a variety of patterns when they occur at different conditions. Pattern formation is the subject of significant fundamental and practical interest, and receives broad attention across various areas from materials science, biology, health, mineralogy, etc. [42–45]. This is because these patterns, which are most likely to capture liquid between the branches, determine the rheological and other physical properties of the system.

One of the most common patterns we encounter in our daily life is dendrite. When a liquid is supersaturated and begins to crystallize, the crystals grow and penetrate into the metastable liquid phase. The resulting crystallites often appear as needlelike dendrites [46, 47]. A common example of dendritic growth is a snowflake.

Dendrites are a common morphology for diffusion-controlled crystal growth in the present of anisotropy [44, 45, 48–50]. For the growth of crystals, some quantities, such as impurities and/or the latent head of crystallization, that are generated at the interface between the crystal and the melt during freezing, for example, must diffuse away from that interface. In the case where a very low or negative temperature gradient occurs on the liquid side of the interface, like fins on a thermal radiator, the sharp tip of a dendrite may promote diffusion of the head away from itself, allowing the crystal to grow rapidly in the direction in which the tip is pointing. When the ratio between the temperature gradient in the liquid phase G and the growth velocity of the crystal surface R_g is lower than a certain value, i.e.,

$$G/R_g < C, \tag{23}$$

surface instability will occur [22, 46, 47]. (Here C is a constant for a given crystallization system.) In general, the instability of growing crystal faces is due to a certain type of perturbation at the growing front, which triggers a local rapid growth compared with that for the rest of the crystal surface.

The formation of certain patterns of crystalline materials can also occur via an aggregation process. It is believed [43–45, 51] that the formation of a fractal pattern occurs via a diffusion-limited aggregation, a reaction control, or a cluster–cluster aggregation process.

Among the two pattern formation mechanisms, the anisotropy in the interfacial development plays an important role in this transformation [43–45, 51]. The microscopic interfacial anisotropy is essential to stabilize dendritic growth. Insufficient anisotropy encourages the formation of fractal patterns [33–35, 41–44].

The pattern formation of fiber networks to be discussed as follows is controlled by a completely new mechanism, the so-called crystallographic mismatch branching mechanism. The patterns produced by this mechanism can have both the microscopic interfacial anisotropy and the characteristic of fractals. In particular, the evolution of the pattern is supersaturation- and impurity-dependent.

3.2.2 Crystallographic Mismatch Nucleation

In Sect. 3.1, we showed that as supersaturation increases, the interface structural match between the substrate and the nucleating phase will deviate from

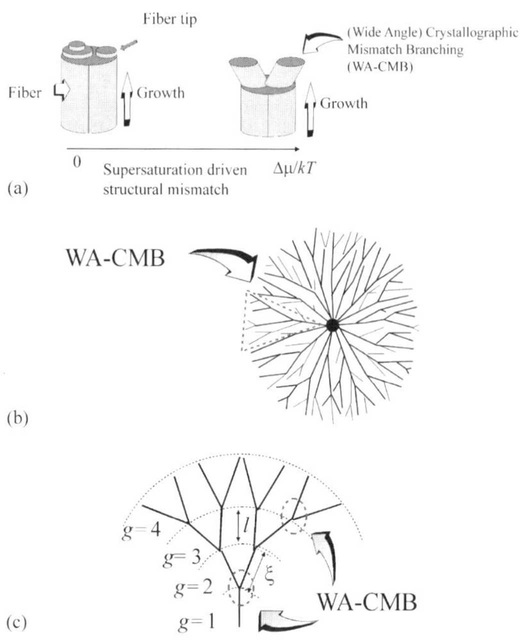

Fig. 5 (**a**) Supersaturation-driven structural mismatch at the tip of fiber crystals leads to wide-angle crystallographic mismatch branching (*WA-CMB*). This occurs at relatively high supersaturations, or in the presence of additives (including self-impurities). (**b**) Spherulitic pattern resulting from WA-CMB [10–12, 47, 48]. (**c**) Open interconnecting fiber network resulting from WA-CMB. In comparison with the WA-CMB spherulitic pattern, the open network structure shown in (**c**) occurs at higher supersaturations

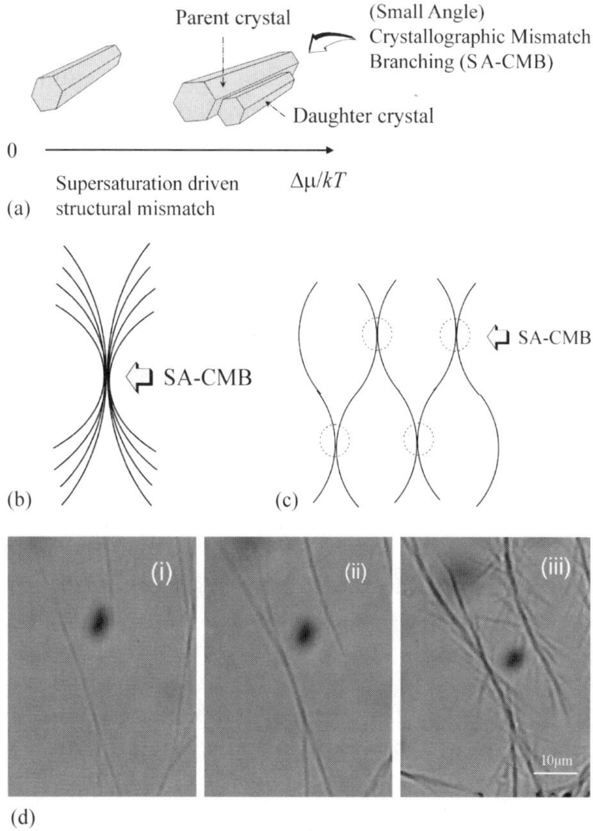

Fig. 6 Crystallographic mismatch nucleation and growth: the daughter crystal differs from the parent crystal by crystallographic mismatch (self-epitaxial) nucleation and growth. (**a**) Crystallographic mismatch nucleation and growth at a side face of a needle crystal, resulting in the small-angle crystallographic mismatch branching (*SA-CMB*). This normally occurs at relatively low supersaturations. (**b**) Spherulitic pattern resulting from SA-CMB [46]. (**c**) Open interconnecting fiber network resulting from SA-CMB. In comparison with the SA-CMB spherulitic pattern, the open network structure shown in (**d**) occurs at higher supersaturations. (**e**) In situ observation of the formation of a GP-1 fiber network in terms of SA-CMB. The formation of GP-1 fibers and the branching process is shown by panels i–iii, in which the time interval between two neighboring photographs is of 0.2 s. Solvent 1,2-propanediol; $\sigma = 2.17$; $T = 333$ K

the optimal structural match position. If the substrate is a growing crystal itself, the optimal structural match means that the new layer on the surface of the growing crystal matches exactly the crystallographic orientation of the parent crystal [15–18, 55, 56]. This corresponds to the growth of crystalline fibers in common sense. Once the crystallographic mismatch nucleation and growth takes place, a new crystalline domain will then occur on the surface of the parent crystal in different orientations (Figs. 5a, 6a).

Actually, during the growth of crystals, there is a tendency for the occurrence of the mismatch of new layers with respect to the crystal surface structure of the parent crystals at high supersaturations owing to the supersaturation-driven mismatch nucleation and growth. Similar to normal 3D nucleation, the mismatched domain should first nucleate on the growing crystal surface. If the energy cost to create a mismatch domain per area on the parent crystals is defined as the specific mismatch free energy γ_{mis}, the barrier of mismatch nucleation is determined by the surface supersaturation and the specific mismatch free energy γ_{mis} (cf. Eq. 12). Obviously, the crystallographic mismatch nucleation is a special case of heterogeneous nucleation, where $\gamma_{\text{mis}} = \gamma_{\text{cs}}$. If the mismatch growth does not deviate from much from the orientation of the parent crystal, we can in principle have $\gamma_{\text{cf}} \sim \gamma_{\text{sf}}$. It follows then that

$$m = 1 - \gamma_{\text{mis}}/\gamma_{\text{cf}} \,. \tag{24}$$

Likewise, the nucleation barrier and the rate of crystallographic mismatch nucleation is given by Eqs. 11–17. As indicated by Eqs. 11 and 12, ΔG^*_{mis} decreases as the surface supersaturation increases (see Eqs. 11, 12). Notice that the only difference between normal heterogeneous nucleation and the crystallographic mismatch nucleation is that in the crystallographic mismatch nucleation, the substrate is the growing crystal surface, whereas in normal heterogeneous nucleation, substrates are foreign bodies.

In the following discussion, we will examine briefly some key factors controlling the crystallographic mismatch nucleation and growth:

1. Supersaturation. Similar to normal nucleation and growth, the kinetics of crystallographic mismatch nucleation and growth also depends on supersaturation. At low supersaturations, ΔG^*_{mis} is very high (Eqs. 11, 12); therefore, the crystallographic mismatch nucleation is difficult at low supersaturations. As supersaturation increases, ΔG^*_{mis} will drop rapidly (Eq. 12). It follows that interfacial mismatch nucleation can occur much more easily in this case.
2. Impurities. Adsorbed impurities may disturb the interfacial structural match between nucleating layers and the parent crystal surfaces. This gives rise to the lowering of m; therefore, it will promote crystallographic mismatch nucleation (Sect. 4.1).
3. Slow surface integration. The crystallographic mismatch nucleation occurs at the surface of growing crystals; therefore, it is governed by surface supersaturation. The orientations with slow surface integration kinetics will therefore lead to higher surface supersaturations (much closer to the bulk supersaturation). It follows that at low supersaturations, the crystal faces with slow surface integration kinetics can take advantage of the highest possible supersaturation—the bulk supersaturation of the system, and will induce much more easily the crystallographic mismatch nucleation.

4. Low specific mismatch free energy. According to Eq. 12, a low specific mismatch free energy γ_{mis} corresponds to a low ΔG_{mis}. This implies that the crystallographic mismatch nucleation can occur much more easily in crystal surfaces with low γ_{mis}s. Notice that normally crystal surfaces with low γ_{mis}s often coincide with those with slow surface integration kinetics. Therefore, criteria 3 and 4 may be very likely applied to the same crystallographic orientation for a given crystalline material.

On the basis of criteria 1. and 3., we will expect that at low supersaturations the crystallographic mismatch nucleation takes place much more easily in slow growth crystallographic orientations (Fig. 6a), whereas at high supersaturations, the crystallographic mismatch nucleation may occur in faster growth crystallographic orientations (Fig. 6b). Note that another reason for the occurrence of crystallographic mismatch nucleation in faster growth crystallographic orientations at high supersaturations is that the faster growth crystallographic orientations can penetrate into the bulk easily, and "feel" much higher supersaturations in the bulk. This, according to Eq. 17, will trigger the crystallographic mismatch nucleation at the tips.

At relatively low supersaturations, owing to the large ΔG^*_{mis}, the crystallographic mismatch nucleation and growth will only occur at the side faces of needle crystals as the side faces are the slowest growing directions and the largest effective surface supersaturation, leading to "small-angle" crystallographic mismatch branching (SA-CMB) (Fig. 6). Figure 6d shows the formation of SA-CMB during GP-1 fiber network formation. The typical crystalline networks or patterns resulting from the SA-CMB are illustrated in Fig. 6a. At high supersaturations, the crystallographic mismatch nucleation and growth will take place at the tips, leading to "wide-angle" crystallographic mismatch branching (WA-CMB) (Fig. 5a–c). Figure 3 shows the formation of WA-CMB during GP-1 fiber network (or spherulite) formation

In general, the evolution of fibers to fiber networks of different types for a given system can be summarized as follows. At low supersaturations, single fibers form first without any branching, as the free-energy barrier ΔG^*_{mis} is very high. As supersaturation increases, the branching of fibers initiates from the side faces (SA-CMB). With further increasing supersaturation, ΔG^*_{mis} at the growing tip of fibers becomes very low. It follows that the WA-CMB becomes kinetically favorable and will take place at the tip of growing fibers. This evolution process is demonstrated in Fig. 7, in which the evolution of GP-1 fiber branching is shown.

As to be discussed in Sect. 4.1, the crystallographic mismatch branching can be utilized to construct self-organized interconnecting fiber network structures/patterns, so as to engineer supramolecular functional soft materials.

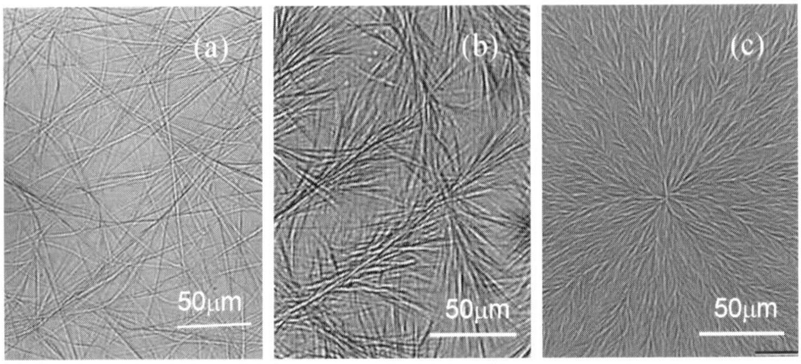

Fig. 7 Micrographs showing fibers and the fiber network evolution of GP-1 in a 1,2-propanediol solution with increasing supersaturation. (**a**) Unbranched GP-1 fibers; $\sigma = 0.188$, $T = 342$ K. (**b**) SA-CMB, and GP-1 fiber network $\sigma = 2.17$, $T = 333$ K. (**c**) WA-CMB, and GP-1 fiber network $\sigma = 4.76$, $T = 323$ K

3.2.3
Fibrous Network Formation

In this section, we will focus on the formation of fibrous networks/patterns via the (wide angle) crystallographic mismatch branching. Actually, the typical patterns that occurred owing to the (wide angle) crystallographic mismatch branching are shown in Fig. 5b and c.

As shown in Fig. 5b and c, these patterns consist of radius arms initiating from a core. The radius aims are often found to be branched with the Cayley tree structure (Fig. 5b, c) [51, 59, 60]. The patterns are found to initiate by 3D nucleation [15–18]. This 3D nucleation is a process to create the radial aims from the cores (Fig. 5a).

If we take into account this fact and the structural characteristics of a Cayley tree of fibrous networks, the process for the network formation can be regarded as: *initial nucleation*—growth—branching—growth—branching... (Fig. 5c.) Obviously, one of the key steps in building up the Cayley tree is the branching at the tips of growing nanofibers. Unlike dendritic branching, the daughter branches of the fibers cannot be correlated strictly to the crystallographic orientation of their parent fibers; therefore, the branching is referred to as *crystallographic mismatch* (or *noncrystallographic*) branching.

3.2.4
Branching Kinetics

The key question to be addressed is why and how crystallographic mismatch branching takes place. Obviously, the crystallographic mismatch branching takes place via the supersaturation-driven interfacial structural mismatch or the crystallographic mismatch nucleation and growth (Fig. 5).

The occurrence of crystallographic mismatch branching is controlled by the following two steps: (1) the growth of the surface of parent crystals; (2) the crystallographic mismatch nucleation on the surface.

As mentioned in Sect. 4.1, the crystallographic mismatch nucleation can be regarded as a special case of supersaturation-driven interfacial structural mismatch in heterogeneous 3D nucleation. Therefore, the nucleation rate can also be described by Eq. 17 with m given by Eq. 24.

In practice, crystallographic mismatch nucleation and the growth of fibers can be approximately regarded as two independent physical processes. Designating the induction time for the nucleation of new fibers on the host fibers as τ ($\tau \sim 1/J$, where J is the rate of the crystallographic mismatch nucleation), and the growth rate of fibers in the fibril axis direction as R_g, it follows then that the average branching distance can be expressed as

$$\langle \xi \rangle \sim R_g \tau \sim R_g / J \,. \tag{25}$$

In the case of 2D nucleation controlled growth, substituting Eqs. 17 and 21 into Eq. 25, we obtain

$$\langle \xi \rangle \sim C_2 \frac{(\Delta\mu/kT)^{5/6}}{f'' f^{1/2}} \exp\left[\frac{16\pi f}{3(\Delta\mu/kT)^2} \left(\frac{\gamma_{cf} \Omega^{2/3}}{kT} \right)^3 \right] , \tag{26}$$

where $C_2 = C_1/B$. To check this mechanism, a set of experiments were performed at different supersaturations.

In the case of rough growth (such as the growth of the tips of fibers or needles), substituting Eqs. 17 and 22 into Eq. 25, we obtain

$$\langle \xi \rangle = A_2 \frac{(\Delta\mu/kT)}{f} f^{1/2} \exp\left[\frac{16\pi f}{3(\Delta\mu/kT)^2} \left(\frac{\gamma_{cf} \Omega^{2/3}}{kT} \right)^3 \right] , \tag{27}$$

where $A_2 = A_1 \beta_{st}/B$. Actually, Eqs. 26 and 27 are very similar apart from the kinetic coefficients C_2 and A_2. In addition, the power of supersaturation $\left(\frac{\Delta\mu}{kT}\right)$ occurring in the pre-exponential term is 5/6 for 2D nucleation growth and 1 for rough growth, which again are very close to each other. Such a similarity may be attributed to the fact that the 2D nucleation model can also be applied to describe rough growth.

The average length $\langle \xi \rangle$ of branches, which is proportional to the mesh size in the fiber networks, was examined by scanning electron microscopy (SEM). The linear relation between $\ln[\langle \xi \rangle (\Delta\mu/kT)^{-1}]$ and $1/(\Delta\mu/kT)^2$ obtained for a lanosta-8,24-dien-3β-ol:24,25-dihydrolanosterol = 56:44 (L-DHL)/di-(2-ethylhexyl phthalate) $(C_8H_{17}COO)_2(C_6H_4)$ (DIOP) system with 0.03% ethylene/vinyl acetate copolymer (EVACP), $(C_4H_6O_2)_x(C_2H_4)_y$, approximately $M_w \sim 100\,000$, verifies the branching mechanism described by Eq. 27.

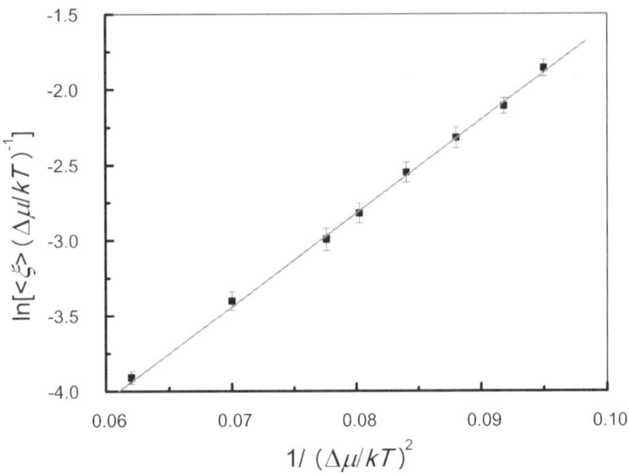

Fig. 8 The correlation between $\ln[\langle\xi\rangle(\Delta\mu/kT)^{-1}]$ and $1/(\Delta\mu/kT)^2$ for a lanosta-8,24-dien-3β-ol:24,25-dihydrolanosterol = 56:44 (*L-DHL*)/di-(2-ethylhexyl phthalate) $(C_8H_{17}COO)_2$ (C_6H_4) (*DIOP*) system with 0.01% ethylene/vinyl acetate copolymer (*EVACP*) at 20 °C. The linear relationship confirms the governing role of the crystallographic mismatch branching mechanism in the formation of organized interconnecting fiber networks. $T = 298.15$ K, $X = 0.026$. The supersaturation difference $\Delta\mu/kT$ is obtained by changing the molar fraction of the solute in the solutions [18]

According to this mechanism, the branching density $(1/\langle\xi\rangle)$ of fibers will increase with $\Delta\mu/kT$ (or ΔT) (cf. Eq. 27). For a given concentration, T_{eq} is constant. It follows from Eq. 9 that a lower T (or a large ΔT) corresponds to a larger $\Delta\mu/kT$. This will result in a highly branched and more open structure.

3.2.5 Structural Characteristics

Although the structure of fibrous networks of supramolecular materials is very important with respect to the macroscopic properties [1–9] and has the typical characteristic of the Cayley fractal tree, the measurement of the fractal dimension D_f remains as a challenging issue owing to the lack of an effective method. Since the average length of the branches is essentially above microns and there is a problem associated with multiple scattering, the fractal structure of this type of fibrous network is difficult to measure by the conventional light scattering method [59, 60]. Here, we introduce a new method to measure the in situ fractal growth of fibrous networks of supramolecular materials.

Fractal structures are self-similar in that the two-point density–density correlation function and their essential geometric properties are independent of the length scale [59, 61–63]. In d-dimensional space, they can be characterized by fractal or Hausdorff–Besicovitch dimension D_f [61, 63, 64]. The radius of gyration of such a pattern R is related to the number of particles or segments it contains N by [61]

$$N \sim R^{D_{\mathrm{f}}} . \tag{28}$$

A uniform object has $D_{\mathrm{f}} = d$, while more open structures in which the density decreases with distance from the center have $D_{\mathrm{f}} < d$.

Taking into account the fact that the formation of fibrous networks is controlled by nucleation and growth [15–19], the new method to be introduced is based on the famous Avrami equation [62], which was developed to describe the nucleation and growth of bulk crystals, taking into account the Poission distribution of crystal size [62]. It has the form

$$\ln[1 - X_{\mathrm{cr}}] =- k^0 t^d , \tag{29}$$

where k^0 is a constant; t is the time; X denotes the crystallinity of the system, equal to $\varphi(t)/\varphi(\infty)$ (where $\varphi(t)$ is the volume fraction of crystal material at t, and $\varphi(\infty)$ is φ (t) at t $\rightarrow \infty$). For the growth of spherical crystals, one has $- k^0 t^d = \frac{4}{3} N_{\mathrm{g}} \pi (\nu_{\mathrm{g}} t)^3$ (ν_{g} is the growth rate of bulk crystals; N_{g} is the number of crystals per volume) [62]. In the case of nonspherical crystals, this expression should be modified to $- k^0 t^d = K \frac{4}{3} N_{\mathrm{g}} \pi (\nu_{\mathrm{g}} t)^3$ (where K describes the geometrical deviation of the crystals from the spherical shape). In general, 1D or rodlike growth, 2D or platelike growth, and 3D growth will lead to d =1, 2, 3, respectively. In the case of fractal growth, one has growing fractal aggregates instead of uniform crystals. The length of bulk crystals (as a function of time) in the previous consideration should then be replaced by the radius of gyration of fibrous networks, and d by the fractal dimension D_{f}. This implies that the previous expression is given for fractal growth as

$$\ln[1 - X_{\mathrm{cr}}(t)] =- k^0 t^{D_{\mathrm{f}}} . \tag{30}$$

The key step in applying Eq. 30 is to measure X_{cr} as a function of t. This can be obtained from the correlation between φ (t) (the crystallinity) and properties such as the viscosity of the system [63, 64]. Einstein's relation has been applied to examine the dependence of $\varphi(t)$ on the viscosity of suspended particles and of polymer networks [12, 64]. According to this relation, the volume fraction of fibrous networks can be correlated to the specific viscosity η_{sp} as

$$\eta_{\mathrm{sp}} \approx F\varphi , \tag{31}$$

with $$\eta_{\mathrm{sp}} = \frac{\eta^* - \eta_0}{\eta_0} , \tag{32}$$

where η^* and η_0 are the complex viscosity of the system and the viscosity of the solvent, respectively, and F denotes the factor determined by the shape of te particles [64]. The crystallinity can be given by

$$X_{\mathrm{cr}}(t) = \frac{\varphi(t)}{\varphi(\infty)} = \frac{\eta^*(t) - \eta_0}{\eta^*(\infty) - \eta_0} = \frac{G^*(t) - G_0^*}{G^*(\infty) - G_0^*} , \tag{33}$$

where $\eta_{sp}(\infty)$ and $\eta^*(\infty)$ denote η_{sp} and η^* at $t \to \infty$, respectively, $\eta^*(t) = G^*(t)/\omega$, ($\omega$ is the angular frequency), $\eta^*(\infty) = G^*(\max)/\omega$, and $\eta_0^* \sim G_0^*/\omega$. Figure 9a shows that the change of G^* as a function of time during the formation of GP-1 fibrous networks from an isostearyl alcohol (ISA) solution (6.7 wt %). From this graph, $X_{cr}(t)$ is obtained according to Eq. 33.

In Fig. 9b, $\ln\left\{-\ln[1 - X_{cr}(t)]\right\}$ is plotted versus $\ln(t)$. The curve at the beginning of the network formation shows a certain scattering behavior and

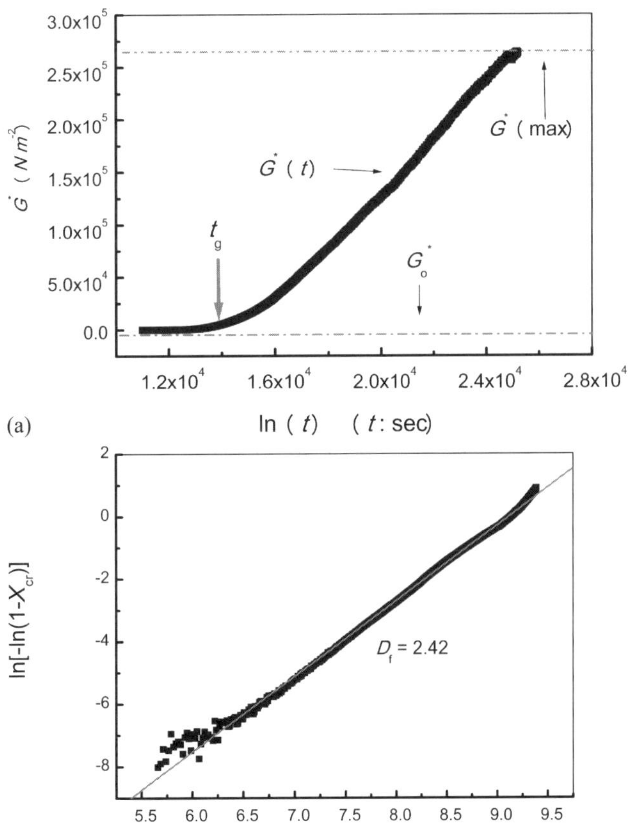

Fig. 9 (**a**) The variation of the complex modulus G^* before and after fibrous network formation. The induction time t_g of fiber formation is identified from the change of G^*. $T = 40\,°C$, $C = 6.7$ wt % [21]. For rheology, the concentration was focused on the storage modulus G' (describing the elastic property) and the loss modulus G'' (describing the viscosity property), and the complex modulus $G^*(= [(G')^2 + (G')^2]^{1/2})$ [13–16]. Obviously, the onset of such changes in G', G'', or G^* should correspond to the initiation of such networks. (**b**) Dependence of $\ln\left\{-\ln[1 - X_{cr}(t)]\right\}$ on $\ln(t)$. The fractal dimension is given by the slope ($D_f = 2.42$) [19]

Table 1 Dependence of fractal dimension of *N*-lauroyl-L-glutamic acid di-*n*-butylamide fibrous networks on temperature

	20 °C	30 °C	20 °C
D_f	2.10	2.40	2.42
$G'(\max)$ ($N\,m^{-2}$)	8.97×10^5	3.62×10^5	2.63×10^5
$G''(\max)$ ($N\,m^{-2}$)	6.53×10^4	2.79×10^4	2.65×10^4

a slight deviation from the straight line. This is due to the 3D nucleation (Fig. 9b). As the networks grow, the power-law domain invades the entire observable regime. This indicates that the essential geometrical properties are independent of the size increase of the gyration of fibrous networks during the growth, which demonstrates the self-similarity of the fiber fractal patterns [61, 64, 65]. According to Eq. 30, the slope is then legitimately interpreted as the fractal dimension D_f.

A fractal dimension of 2.42 was observed, which is close to diffusion-limited aggregation [61, 64, 65], for which $D_f \sim 2.5$. Nevertheless this does not imply that the formation of GP-l networks is controlled by the diffusion-limited aggregation mechanism. It follows from our experiments that D_f varies with experimental temperature T. The change of D_f with temperature within the range 20–40 °C is given in Table 1.

As shown, D_f changes from 2.42 to 2.10 as T decreases from 40 to 20 °C. Since a denser structure adopts a higher Hausdorff dimension, this result implies that the fibrous networks will adopt a more open structure at lower temperatures [8].

We notice that some rheological approaches have also been applied to examine the fractal structure of networks of polymers and suspended particles [52–54, 66]. For instance, Winter and his colleagues [54, 66] utilized the correlation between the frequency and the viscosity $\eta^* \sim (\omega)^{-2/(D_f+2)}$ to measure the fractal dimension of cross-linking polymers. However, these approaches are very unlikely to be applied to the point where a gel occurs. As pointed out, the molecular weight of growing polymer cluster diverges at the gel point, and the relationship does not hold. A similar problem is the fibrous networks of organic gels. It is worth noting that unlike most other methods which are only suitable for the measurement of fractal dimension, the approach developed here allows us measure in situ the fractal growth, which turns out to be very important in studying the kinetics of fractal network formation.

4 Supramolecular Materials Engineering

4.1 Fibrous Network Materials

As mentioned at the beginning, supramolecular materials have a variety of applications. In a recent paper in *Science*, Shi et al. [67] reported on new microcellular organic materials which were prepared by supercritical drying of organogels in CO_2. The most remarkable feature of this system is that except for the organogelator and CO_2, no other compounds or cosolvents are involved. This results in a single-step and environmentally friendly method to create these materials. The essence of this work is that the authors succeeded in the design of organic gelling agents for CO_2, that is, compounds that strongly associate and form an entangled network structure at lower temperature or pressure, but that would dissolve upon increasing the temperature or pressure. For this purpose gelling agents were developed in which urea groups, to achieve association through multiple hydrogen bonding, are combined with groups showing affinity for CO_2 to increase the solubility.

The results of this very basic design are remarkable. The compounds dissolve very well in supercritical CO_2 at higher temperatures and pressures, but upon cooling the solution turns into an opaque gel. After removal of the solvent by simply lowering the pressure, the network structure is preserved and a highly porous and self-supporting monolithic foam is obtained. The foams combine an extraordinary low density (less than 5% of the bulk density of the organogelators) with a submicrometer cellular size, properties which could not be obtained with traditional polymers in combination with CO_2 as a foaming agent. These organic analogues of silicate aero gels are therefore highly interesting materials for application in separation processes as well as for use as novel low dielectric or insulating materials and as supports for catalysts.

Other research groups including those of Terech, Hanabusa, Weiss, Shinkai, and Boden all made significant contributions towards the design of novel gelling agents and the understanding of the gelation phenomena [6, 7, 11]. Important guidelines for the design of novel gelling agents that emerged from these studies are:

1. The presence of strong self-complementary and unidirectional intermolecular interactions to enforce one-dimensional self-assembly
2. Control of fiber–solvent interfacial energy to control solubility and to prevent crystallization
3. Some factor to induce fiber cross-linking for network formation

Most new gelling agents that resulted from these studies are based on peptides or carbohydrates because of the strength and high directionality of hydrogen-bond interactions. Especially cyclic dipeptides and bisurea-based gelling agents appeared to be very efficient gelling agents and combine easy accessibility with a wide solvent scope, and in the case of the bisurea gelling agents it was shown that the supramolecular architecture can be controlled by the spacer between the urea groups [71–77]. The hydrogen bonding core of these cyclic dipeptides and bisureas provide a "gelating scaffold" which can easily be functionalized without loss of gelation ability. For example, it was reported some years ago that the polymerizable bisurea is a potent gelling agent for apolar solvents [77]. After photopolymerization of benzene gels and removal of the solvent by freeze-drying, an organic aerogel was obtained with very similar characteristics with respect to density and cellular size as that reported by Shi et al. [67].

Another example is the application of organogelators as a template for the preparation of nanostructured materials. First, Möller, Weiss, and later Nolte [57–59] prepared membranes with man-sized pores by the gel-template leaching process. This process takes advantage of the distinct feature of organogelators to *reversibly* form elongated fibers with well-defined dimensions and geometry. In the gel-template leaching process, gels of various gelling agents are prepared in polymerizable solvents like methacrylates or styrene in the presence of a cross-linking agent. After polymerization of the matrix, the organic gelling agent is removed again by extraction with a proper solvent. In this way the organogel fiber network is "imprinted" in the cross-linked polymer matrix, resulting in porous membranes with channels of submicrometer and in some cases even nanometer dimensions.

Apart from this, highly porous and self-supporting monolithic foams, which are found to have many important applications, including novel bioseparation [67], can be obtained from the previously mentioned materials after applying the supercritical CO_2 extraction.

A completely different but very appealing issue in gel research is the development of "smart" or "responsive" gel systems [79, 80]. Pioneering work on smart polymer gels which could reversibly swell or shrink upon exposure to some chemical or physical triggering event has been carried out by Tanaka [81]. Recently, the group of Tanaka developed a novel gel system which can bind guest molecules by multiple-point noncovalent interactions [82]. Most interestingly, swelling of the gel induced by a temperature change increases the distance between the individual interaction points. As a result the binding changes from a multiple-point to a single-point interaction, and the affinity decreases by more than 1 order of magnitude. It is foreseen that this principle can be applied for controlled drug-release systems, and gels can be designed to respond to other triggers, like solvent polarity or pH value, as well.

It should be emphasized that organogels are very attractive systems for the design of novel responsive gels, because their formation is completely reversible. Here the challenge is to couple the self-assembly process to a chemical or physical trigger. Having the feature of reversibility in mind, Aggeli et al. [71] designed various peptide-based gelling agents which self-assemble into extended P-sheet tapes and which, depending on the polarity of the solvents, entangle to form gels. To make gel systems that are responsive to pH some basic amino acids were incorporated in the peptide, and indeed it was found that a change in pH triggers the reversible transition from the gel state to a solution. Gels of other peptides respond to shear by a change of their viscoelastic properties. Shinkai and coworkers demonstrated some years ago by using an azobenzenesteroid-based gel system that it is possible to switch by irradiation with UV and visible light from solution to the gel state and back [70], and two other recent publications indicate that also host–guest interactions can be employed to control the structure or stability of organogels [81–85].

4.2 Micro/Nanonetwork Architecture

Nowadays, one of the key challenges to obtain supramolecular functional materials is the capability of designing and producing specific materials with desired structures and physical properties [8–10]. Owing to the lack of sufficient understanding, significant effort is required in order to obtain *appropriate functional materials* with the desirable properties. This includes the screening of a large number of compounds and solvent molecules, as well as suitable solvent capable of forming these structures [8, 10, 13–19]. In some cases, owing to the limitation in the choice of materials, screening becomes impossible. On the other hand, if interconnecting 3D micro/nanofiber networks with the required organization can be constructed, new functional materials with the required functionalities can be produced. Obviously, this is a completely new route in producing new functional materials. In this section, we will focus on a completely new approach to produce new macromolecular functional materials.

The strategy of this approach is to construct 3D permanent interconnecting nanocrystal fibrous networks from a system consisting of separate crystal needles using a tailor-made micro/nanostructure (Fig. 10a), and to engineer the materials with required micro/nanostructure (Fig. 10b), based on the CMB mechanism.

A multiphase system or a supramolecular material was obtained when L-DHL precipitated out from its DIOP solution at room temperature. The L-DHL powders used in our experiments were in a crystalline state [17, 18].

During the process of natural cooling of the aforementioned system (concentration of L-DHL of 10 wt %) from 120 °C to room temperature (approximately 20 °C), an opaque and viscous paste was obtained (Fig. 11a, inset). It can be seen that the system consists of only separated L-DHL *needles* (short

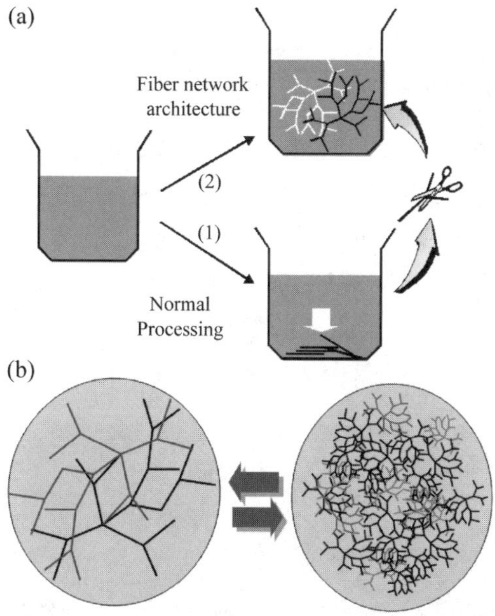

Fig. 10 Micro/nanostructure engineering of functional materials. (**a**) Network architecture (route *2*) with respect to normal fiber formation. (**b**) Modification of micro/nanostructure of 3D interconnecting fiber network

and thick fibers), which are in temporary contact with each other (Fig. 11a). This refers to the "normal processing" illustrated by Fig. 10a.

In order to create networks with permanent interlinking from such a system as illustrated by route 2 in Fig. 10a, an additive, a so-called "branching creator", EVACP (SP^2 Scientific Polymer Products) was introduced to the identical L-DHL/DIOP solution as described before at 120 °C. EVACP has the molecular structure given in Scheme 1.

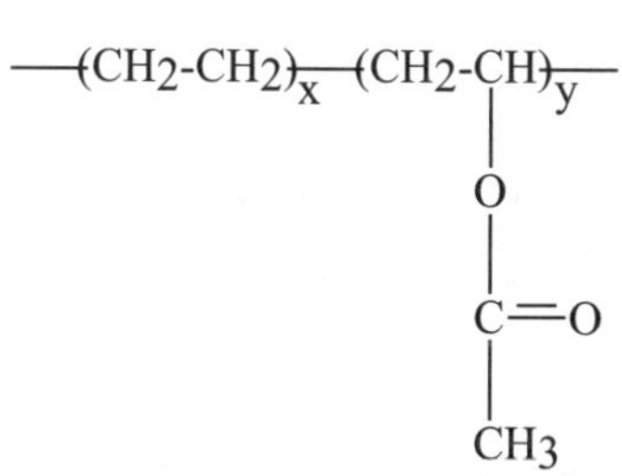

Scheme 1

The addition of a tiny amount of EVACP (0.004 wt %) is sufficient to create completely different micro/nanonetworks of interconnecting self-organized L-DHL nanocrystal fibers (Fig. 11b) at the experimental temperature. This

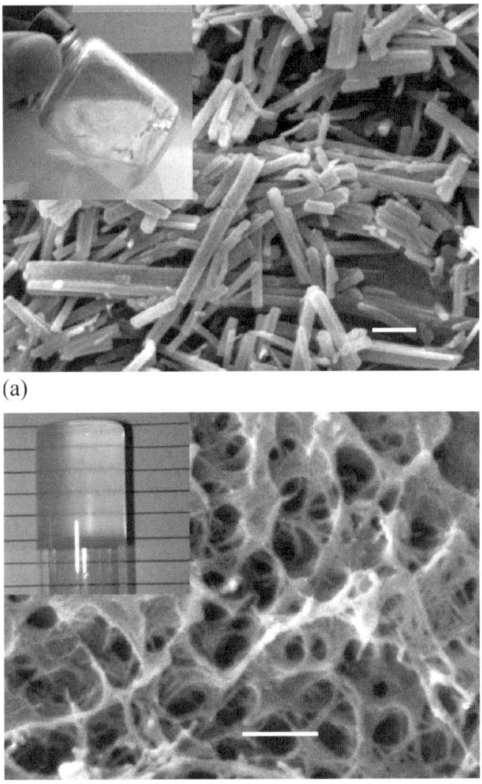

(a)

(b)

Fig. 11 (**a**) Separate fibers occurring in the 10 wt % L-DHL/DIOP system. This system gives rise to an opaque paste as shown in the *inset*. *Scale bar* 1 μm. (**b**) Interconnected fiber networks in the 10 wt % L-DHL/DIOP system after adding 0.004 wt % EVACP. *Scale bar* 1 μm. This system gives rise to a clear and tough gel as shown in the *inset*. The materials to be examined were obtained using L-DHL (Sigma) in DIOP (99%, Cognis) at approximately 120 °C, to form an L-DHL/DIOP solution, and lowering the sample temperature to a lower temperature (around room temperature), called the experimental temperature hereafter [19, 21, 83]. In order to examine the nano/microstructure of fibrous networks in supramolecular materials scanning electron microscopy coupled with a CO_2 supercritical fluid extraction technique (Thar Design), was applied. The purpose of implementing the latter was to remove the liquid captured in the networks, without disturbing the essential structure of the networks [15–18].

completely changes the rheological properties of the materials, and leads to the formation of a self-supporting soft, solidlike supramolecular material (Fig. 11b, inset).

Figure 12 shows the change of G^* as a function of time t for the L-DHL/DIOP system. G^* increases abruptly at $t \geq t_g$, where (t_g) is the gelation time, indicating the formation of L-DHL crystalline fibers. G^* tends to its

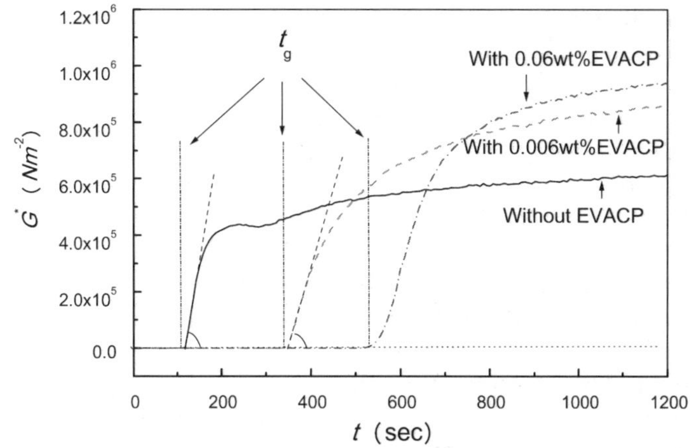

Fig. 12 The changes of the rheological properties of the 10 wt% L-DHL/DIOP system in the absence and presence of EVACP as functions of time or strain. The rheological properties of the system were measured by an advanced rheological expansion system (This advanced rheometer allowed us to control the temperature from 600.0 to – 150.0 ± 0.1 °C.) Dependence of the complex modulus G^* on time. The maximal G^* is almost doubled after adding EVACP.

maximum value G^*_{max} as the formation of the fiber networks approaches completion at $t \to \infty$. In comparison with the control sample, G^*_{max} is almost doubled when a tiny amount of EVACP is added. (Similar changes were also obtained for G' and G'', respectively.)

Figure 12 shows that a new macromolecular material with significantly modified macroscopic properties forms after the addition of EVACP. To the best of our knowledge, such a novel approach of creating new supramolecular materials by an additive has never been reported before.

4.3 Network Formation Mechanism

To explore the network promotion mechanism of EVACP, an adequate understanding on the cross-linking of fiber networks is required. The key issue to be addressed is whether or not the formation of a 3D interconnecting network of the small molecular organic gelling agents, such as L-DHL, is controlled by molecule self-assembly as suggested by most people [9, 10]. It is extremely unlikely that the occurrence of a trace amount of EVACP will cause the lash-shape of L-DHL crystal needles, as shown in Fig. 11a, to be assembled into the self-organized and regularly branched nanofiber networks, as shown in Fig. 11b.

X-ray diffraction analyses have shown that L-DHL powders, needles and 3D interconnecting fibers have identical crystalline structures [31, 32]. This implies that the formation of L-DHL needles and 3D interconnecting fibers is essentially controlled by a new kinetics associated with the nucleation–growth process. For the primary fiber formation, it should be controlled as for most crystalline materials by a nucleation process, which can be verified by the linear relationship between $\ln t_g$ and $1/(\Delta\mu/kT)^2$ [22, 30, 31].

According to the 3D nucleation model, we should then have a linear relationship between $\ln(t_g)$ and $1/(\Delta\mu/kT)^2$ (cf. Eq. 19) for the nucleation associated with a given f (including crystallographic mismatch nucleation). In Fig. 13, the linear fits between $\ln(t_g)$ and $1/(\Delta\mu/kT)^2$ obtained for both systems without and with EVACP indeed verify the nucleation control at the beginning of fiber network formation.

Referring to Fig. 10b, we can see that one of the key steps in building up the self-organized interconnecting fiber networks is the crystallographic mismatch branching at the growing tips of nanofibers (Fig. 5). The occurrence of the mismatch nucleation, leading to the occurrence of new fiber branches (Fig. 5), depends on both the structural match between the substrate and the nucleating phase, and on supersaturation [86–88]. For the growth of L-DHL needles, the deposition of new layers on the existing surface of growing tips requires a perfect structural match [86–88]. Since EVACP is an agent that will selectively adsorb on certain surfaces of organic crystals [40], the adsorbed molecules will disrupt structural match between new layers and the tip surface of parent fibers.

For the growth of L-DHL fiber tips (10 wt % L-DHL in DIOP solutions without additive and with 0.05 wt % EVACP), it is found [17, 18] (Eq. 1) that the growth rate of L-DHL fibers increases drastically with the supersaturation at the beginning, then gradually levels off as supersaturation increases further (Eq. 2). The introduction of EVACP causes a reduction in the growth rate of L-DHL fibers (only a half in comparison with that in the solution without EVACP) Since only a tiny amount of EVACP was added, the mutual diffusivity may not be changed much; therefore, the growth rate reduction is due to the adsorption on the growth tips of the fibers, which slows down the surface kinetics of the tip growth [22, 89–94].

The structural mismatch caused by the adsorption of EVACP has been verified by the enhancement of f in heterogeneous nucleation experiments. The structural match between newly created crystal layers and the tip surface can be described by f (and f') $(0 < f, f'' \leq 1)$ in Eq. 1 (and Eq. 5 given later). According to the heterogeneous nucleation models [29–31], the slope of the $\ln(t_g)$ vs. $1/(\Delta\mu/kT)^2$ plot corresponds to the specific f [29–31]. In the case of a perfect structural match (such as crystal growth), one has $f, f = 0$, while for completely structural mismatch, one has $f, f' = 1$ [31–33]. If the adsorption of EVACP on any substrate (including the tips of growing fibers) is capable of disrupting the structural match, one should observe a significant enhancement of f in the nucleation experiments [31–33]. This turns out to be

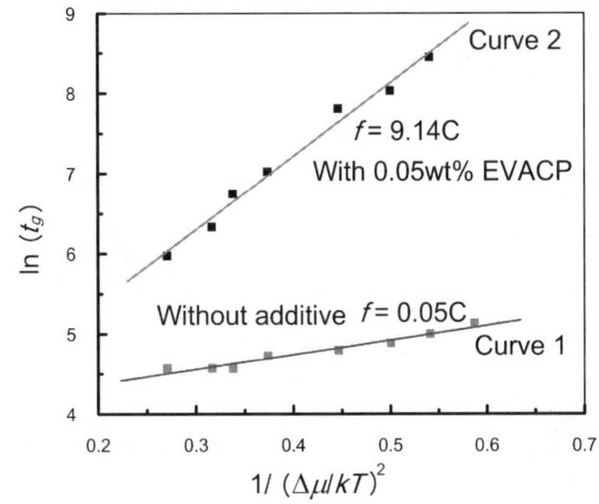

Fig. 13 $\ln t_g$ vs. $l/(\Delta\mu/kT)^2$ plots for the L-DHL/DIOP system in the absence and presence of EVACP (0.03 wt %). Both cases give rise to a linear relation, indicating the control of nucleation during the fiber formation [29–31]. The slope of the straight lines represents the specific f [29–31]. After introducing EVACP, f is raised by more than 183 times, confirming the significant mismatch between the nucleating fibers and the substrate caused by the additive.

the case for the L-DHL/DIOP system. Figure 13 shows that f indeed is raised by a factor of more than 183 after introducing EVACP.

Both the nucleation and the growth kinetics of L-DHL fibers in DIOP confirm that the significant structural mismatch is caused by the adsorption of EVACP on the growth tips of L-DHL fibers. This slows down the surface integration during the growth of fibers, and causes the interfacial structural mismatch nucleation and growth. Consequently, the (wide-angle) crystallographic mismatch branching is promoted (Fig. 5b).

In order to verify quantitatively the previously described mechanism, we will examine the correlation between the average branching distance $\langle\xi\rangle$ and supersaturation. The average length $\langle\xi\rangle$ of the branches, which is proportional to the mesh size in the fiber networks, was examined by SEM. The linear relation between $\ln[\langle\xi\rangle(\Delta\mu/kT)^{-1}]$ and $1/(\Delta\mu/kT)^2$ obtained for the aforementioned L-DHL/DIOP system with 0.03% EVACP (Fig. 8) verifies the branching mechanism described by Eq. 21.

4.4 Engineering Strategy

As illustrated by Fig. 10a and b, one aspect of the micro/nanostructure engineering is the architecture of a self-organized 3D interconnecting nanofiber structure (Fig. 10a), and the other aspect is to tune the micro/nanostructure in a predictive way (Fig. 10b). It was shown in the previous sections that the variations in the micro/nanostructure of functional materials will exert a dir-

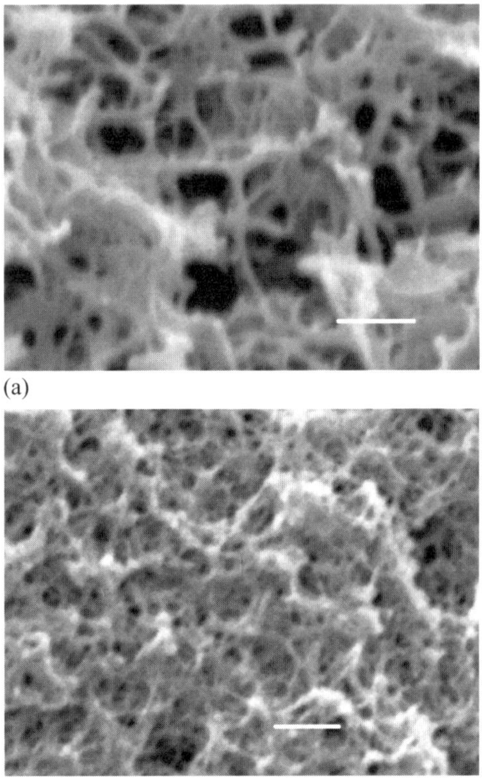

(a)

(b)

Fig. 14 Effects of EVACP concentration on the micro/nanofiber structure and rheological properties of L-DHL. The increase of EVACP concentration C_{EVACP} will give rise to a reduction of the mesh size of the L-DHL network ((**a**), (**b**)) and the rise of G^*. (**a**) Fiber network of L-DHL obtained from 10 wt %L-DHL in DIOP with 0.01 wt % EVACP. (**b**) Fiber network of L-DHL obtained from 10 wt %L-DHL in DIOP with 0.1 wt % EVACP

ect impact on the rheological or other physical properties. This implies that it is possible to control the mesh size and the corresponding fiber network structure so as to obtain functional materials with some desired micro/nanostructure and rheological properties. In other words, the materials with the required properties can be obtained by altering the experimental conditions.

The implication of Fig. 8 is that the micro/nanostructure can be modified by changing supersaturation according to Eq. 21. The decrease of the mesh size of the 3D interconnecting networks of the materials will lead to an increase of G^*, G', etc. Therefore, the change of supersaturation should enable us to alter the rheological or other physical properties of the materials.

Under identical conditions, the increase of the concentration of the branching promoter (additive) will also cause a reduction of the mesh size of 3D interconnecting networks of the materials. Figure 14a and b shows the 3D fiber network structures of L-DHL in 10 wt % L-DHL/DIOP gels with 0.01 and 0.1 wt % EVACP, respectively. Raising the EVACP concentration C_{EVACP} causes a reduction of the mesh size of the networks. This consequently will give rise to the increase of G^*, G', etc. [18]

Since the network can be created and controlled by the previously described approaches, we should be able to engineer directly the microstructure and macroscopic properties of the aforementioned materials in a controlled manner.

Notice that to engineer supramolecular functional materials by additives, an effective branching creator is one of the key steps. The criteria for the selection or the design of such molecules should take energetic and entropic factors into account. Normally, those relatively large molecules with a relatively rigid backbone and multifunctional groups capable of strong binding to the specific crystal surface are potential branching promoters. For more details, see Ref. [18].

5
Summary and Conclusions

In this chapter, the formation of fiber and fiber network was treated on the basis of a generic mechanism of heterogeneous nucleation. In terms of a so-called interface correlation factor $f(m)$ and the entropic barriers associated with kink integration, both classic and nonclassic interfacial effects and the particle size effect during fiber and fiber network formation were analyzed quantitatively. With the increase of supersaturation, instead of a single process, the heterogeneous nucleation occurring during the fiber and fiber network formation is a sequence of progressive processes which reveal a wide spectrum of heterogeneous characteristics described by $f(m)$ and $f''(m)$. At low supersaturations, nucleation will be governed by the process with an optimal structural match and a strong interaction between the substrate and the nucleating phase ($f(m) \to 0$); as supersaturation increases, the nucleation process will be controlled by the nucleation with the poor structural match and more orientational freedom ($f(m) \to 0$).

Supersaturation-driven interfacial structural mismatch is a new effect identified from the previously described model. This effect is directly responsible for the fiber network formation of many organic gels. With increasing supersaturation, the kinetics will be in favor of some random orientations other than the orientation with the optimal structural match between the nucleating phase and the substrate. This implies that the epitaxial relationship will break down at high supersaturations.

On the basis of fact that the surface of growing crystals can also serve as a substrate for the nucleation and growth of new crystals of the same structure, an important phenomenon arising from the supersaturation-driven interfacial structural mismatch is the so-called *crystallographic mismatch nucleation and growth*: new crystalline layers occurring at the surface of growing fibers can adopt a mismatch orientation with respect to the crystallographic orientation of the parent crystals. This process takes place via the crystallographic mismatch nucleation and growth process. For needlelike crystals, this leads to the so-called *crystallographic mismatch branching*—the branching controlling the topological structure of the fiber network and the formation of other crystalline patterns.

At very low supersaturations, unbranched fibers form predominantly. As supersaturation increases, SA-CMB occurs at the side face of growth fibers. At very high supersaturations, the so-called WA-CMB becomes kinetically favorable.

In many cases, the patterns created via WA-CMB have the characteristics of a Cayley tree structure (a type of fractal structure). The structure of the crystal networks can be then correlated to the rheological properties in terms of fractal dimension.

Supramolecular materials consist of a self-organized interconnecting fiber network, which can be created by the branching of fibrous crystals. Control of the branching of nanofibers allows us to achieve the micro/nanostructure architecture of networks. As the physical properties, in particular, the rheological properties of supramolecular functional materials, are determined by the micro/nanostructure architecture of the network, the engineering of supramolecular functional materials can be achieved by constructing and manipulating the micro/nanostructure. In this regard, we presented a typical example of a so-called "branching creator", EVACP, to promote the branching of L-DHL needlelike crystals to generate self-organized interconnecting fibrous networks and new supramolecular materials. The branching is due to the adsorption of the branching creator molecules at the growing tips of L-DHL needlelike crystals. This disturbs the normal growth of the needle crystals, leading to the (wide-angle) crystallographic mismatch branching. As indicated before, another effective way of manipulating the micro/nanostructure can also be achieved by changing supersaturation.

Acknowledgements The author would like to acknowledge the Science and Engineering Research Council, Singapore for the support of this research (project no. 022 101 0036). He also would like to thank R.Y. Wang for providing some of the pictures.

References

1. Mio MJ, Moore JS (2000) MRS Bull 25:6
2. Derossi D, Kajiwara K, Osada Y, Yamauchi A (1991) Polymer Gels – Fundamentals and Biomedical Applications. Plenum Press, New York, p 206

3. Guenet JM (1992) Thermoreversible gelation of polymers and biopolymers. Academic, London
4. Corriu RJP, Leclercq D (1996) Angew Chem 108:l524
5. Reetz MI (1997) Adv Mater 9:943
6. Terech P, Weiss RG (1997) Chem Rev 97:3133
7. Beginn U, Keinath S, Möller M (1998) Macromol Chem Phys 199:2379
8. Brunsveld L, Folmer BJB, Meijer EW (2000) MRS Bull 25:49
9. van Esth JH, Feringa BL (2000) Angew Chem Int Ed Engl 39:2263
10. Oya T, Enoki T, Grosberg AU, Masamune S, Sakiyma T, Tekeoka Y, Tanaka K, Wang G, Tilmaz Y, Feid MS, Dasari R, Tanaka T (1999) Science 286:1543
11. Terech P, Rodrigez V, Barnes JD, Mckenna GD (1994) Langmuir 10:3406
12. Hiemenz PC, Rajagopalan R (1997) Principles of colloid and surface chemistry. Dekker, New York, p 145
13. Lescanne, M, Grondin P, d'Ale'o, A, Fages F, Pozzo J-L, Mondain-Monval O, Reinheimer P, Colin A (2004) Langmuir 20:3032
14. Lescanne M, Colin A, Mondain-Monval O, Fages F, Pozzo J-L (2003) Langmuir 19:2013
15. Liu XY, Sawant PD (2002) Adv Mater 14:421
16. Liu XY, Sawant PD (2001) Appl Phys Lett 79:3518
17. Liu XY, Sawant PD (2002) Angew Chem Int Ed Engl 41:3641
18. Liu XY, Sawant PD, Tan WB, Noor IBM, Pramesti C, Chen BH (2002) J Am Chem Soc 124:15055
19. Liu XY, Sawant PD (2002) Chem Phys Chem 3:374
20. Liu XY, Bennema P (1994) J Cryst Growth 135:209
21. Liu XY, Bennema P (1993) J Appl Crystallogr 26:229
22. Chernov AA (1984) Modern crystallography III. Crystal growth. Springer, Berlin Heidelberg New York
23. Bennema P, Gilmer GH (1973) In: Hartman P (ed) Crystal growth—an introduction. North-Holland, Amsterdam, p 263
24. Kashchiev D (1969) Surf Sci 14:209
25. Shi G, Seinfeld JH, Okuyarna K (1990) Phys Rev A 41:2101
26. Wu DT (1992) J Chem Phys 97:2644
27. Shneidman VA, Weinberg, MC (1992) J Chem Phys 97:621
28. Kashchiev D (1995) In: van der Eerden JP, Bruinsma OSL (eds) Science and technology of crystal growth. Kluwer, Dordrecht, p 53
29. Liu XY (2001) Appl Phys Lett 79:39; Liu XY (2001) J Phys Chem B 105:11550
30. Ning D, Liu XY (2002) Appl Phys Lett 81:445
31. Liu XY (2001) In: Sato K, Nakajima K, Furukawa Y (eds) Advances in crystal growth research. Elsevier, Amsterdam, p 42
32. Liu XY (1999) J Chem Phys 111:1628
33. Liu XY (2000) J Chem Phys 112:9949
34. Fowler R, Giggenhein EA (1960) Statistical thermodynamics. Cambridge University Press, London
35. Farkas L (1927) Z Phys Chem 125:236
36. Gebhardt M (1972) In: Hartman P (ed) Crystal growth: an introduction. North-Holland, Amsterdam, p 105
37. Liu XY (2001) Surf Rev Lett 8:423
38. Liu XY, Maiwa K, Tsukamoto K (1997) J Chem Phys 106:1870
39. Liu XY, van den Berg EPG, Zauner ARA, Bennema P (2000) J Phys Chem B 104:7211942

40. Bennema P, Liu XY, Lewtas K, Tack RD, Rijpkema JJM, Roberts KJ (1992) J Cryst Growth 121:679
41. Cahn JW, Hillig WB, Sears GW (1964) Acta Metall 12:1421
42. Vicsek T (1992) Fractal growth phenomena, 2nd edn. World Scientific, Singapore
43. Barabasi AL, Stanley HE (1995) Fractal concepts in surface growth. Cambridge University Press, Cambridge
44. Langer JS (1989) Science 243:1150
45. Ben Jacob E, Garik P (1990) Nature 343:523
46. Huang SC, Glicksman ME (1981) Acta Metall 29:717
47. Langer JS (1980) Rev Mod Phys 52:1
48. Kessler DA, Koplik J, Levine H (1988) Adv Phys 37:255
49. Brener EA, Melnikov VI (1991) Adv Phys 40:53
50. Ben-Jocob E (1993) Contemp Phys 34:247
51. Kepler J (1966) De nive sexangula transl Hardie. Clarendon, Oxford
52. Vandewalle N, Ausloos M (1997) Phys Rev E55:94
53. Havlin S, Nossal R, Trus B (1985) Phys Rev A32:3829
54. Muthukumar M, Winter HH (1986) Macromolecules 19:1284
55. Liu XY, Strom CS (2000) J Chem Phys 112:4408
56. Strom CS, Liu XY, Wang M (2000) J Phys Chem B104:9638
57. Phillips PJ (1994) In: Hurle DTJ (ed) Handbook of crystal growth. North-Holland, Amsterdam, p 1167
58. Keith HD, Padden FJ (1963) J Appl Phys 34:2409
59. Cayley A (1858) Philos Mag 28:374
60. Vandewalle N, Ausloos M (1997) Phys Rev E55:94
61. Barabasi L, Stanley HE (1995) Fractal concepts in surface growth. Cambridge University Press, Cambridge
62. Avrami M (1939) J Chem Phys 7:1103
63. Witten TA, Sander LM (1981) Phys Rev Lett 47:1400
64. Hiemenz PC, Rajagopalan R (1997) Principles of colloid and surface chemistry. Dekker, New York, p 145
65. Vicsek T (1992) Fractal growth phenomena, 2nd edn. World Scientific, Singapore
66. te Nijenhuis K, Winter HH (1989) Macromolecules 22:411
67. Shi C, Huang Z, Kilic S, Xu J, Enick RM, Beckman EJ, Carr AJ, Melendez RE, Hamilton RD (1999) Science 286:1540
68. Hanabusa K, Yamada M, Kimura M, Shirai H (1996) Angew Chem 108:2086; (1996) Angew Chem Int Ed Engl 35:1949
69. Lin Y, Kachar B, Weiss RG (1989) J Am Chem Soc 111:5542
70. Yoza K, Amanokura N, Ono Y, Aoka T, Shinmori H, Takeuchi M, Shinkai S, Reinhoudt DN (1999) Chem Eur J 5:2722
71. Aggeli A, Bell M, Boden N, Keen IN, Knowles RF, McLeish TCB, Pitkeathly M, Radford SE (1997) Nature 386:259
72. Yan Esch J, Schoonbeek F, de Loos M, Kooiman H, Spek AL, Kellogg RM, Feringa BL (1999) Chem Eur J 5:937
73. (a) Oda R, Hue I, Candau SJ (1998) Angew Chem 110:2835; (b) Oda R, Hue I, Candau S (1998) Angew Chem Int Ed Engl 37:2689
74. Hafkamp RJH, Feiters MC, Nolte RJM (1999) J Org Chem 64:412
75. Hanabusa K, Matsumoto Y, Miki T, Koyama T, Shirai H (1994) J Chem Soc Chem Commun 1401
76. Hanabusa K, Shimura K, Hirose K, Kimura M, Shirai H (1996) Chem Lett 885

77. van Esch J, De Feyter S, Kellogg RM, De Schryver E, Feringa BL (1997) Chem Eur J 3:1238
78. de Loos M, van Esch J, Stokroos I, Kellogg RM, Feringa BL (1997) J Am Chem Soc 112:12675
79. Osada Y, Gong JP (1998) Adv Mater 10:827
80. Flory RJ (1974) Faraday Discuss Chem Soc 57:8
81. Tanaka I (1981) Am Sci 244:110
82. Beginn U, Keinath S, Möller M (1998) Macromol Chem Phys 199:2379
83. Gu W, Lu L, Chapman GB, Weiss RG (1997) Chem Commun 543
84. Hafkamp RJ, Kokke BPA, Danke IM, Geurts HPM, Rowan AE, Feiters MC, Nolte R JM (1997) Chem Commun 545
85. Engelkamp H, Middelbeek S, Nolte RJM (1999) Science 284:785
86. Liu XY, Lim SW (2003) J Am Chem Soc 125:888
87. Liu XY, Tsukamoto K, Sorai M (2000) Langmuir 16:5499
88. Laaksonen A, Talanquer V, Oxtoby DW (1995) Annu Rev Phys Chem 46:489
89. Liu XY, Boek ES, Briels WJ, Bennema P (1995) Nature 374:342
90. Liu XY, Bennema P (1993) J Chem Phys 98:5863
91. Liu XY(1999) Phys Rev B 60:2810
92. Liu XY, Bennema P (1993) J Chem Phys 98:5863
93. Davey RJ, Garside J, Hilton AM, McEwan D, Morrison JW (1996) J Cryst Growth 166:971
94. da Silva Couto M, Liu XY, Meekes H, Bennema P (1994) J Appl Phys 75:627

Top Curr Chem (2005) 256: 39–76
DOI 10.1007/b107171

Published online: 1 July 2005

Cholesterol-Based Gelators

Mladen Žinić[1] · Fritz Vögtle[2] · Frédéric Fages[3] (✉)

[1]Rudjer Boskovic Institute, POB 180, 10002 Zagreb, Croatia

[2]Kekulé-Institute für Organische Chemie und Biochemie, Universität Bonn, Gehard-Domack-Str. 1, 53121 Bonn, Germany

[3]Faculté des Sciences de Luminy, Université de la Méditerranée, Case 901, 13288 Marseille Cedex 9, France
fages@luminy.univ-mrs.fr

Abstract This chapter is intended to provide a comprehensive overview of the chemistry and molecular structures of cholesterol-based low molecular mass organogelators and to describe the underlying principles that are responsible for self-assembly and, in turn, organogel formation within this important class of systems. In particular, the properties of the resulting organogels are discussed in terms of the relationship between the designed structure of the gelator, the nature of the gelled organic fluid component, the influence of molecular interactions (π–π stacking, Van der Waals, hydrogen bonding, ...), and chirality. Emphasis is given to the remarkably high synthetic tunability of the cholesterol-based organogelators, which enable access to a range of functional gels. The prospects for future applications of cholesteryl molecular gels will be illustrated and discussed at the end of the chapter.

Keywords Cholesterol · Steroids · Organogels · Self-assembly · Supramolecular chemistry

Abbreviations

A	Aromatic unit
AFM	Atomic force microscopy
Bn	Benzyl
CD	Circular dichroism
DCC	*N*,*N*-Dicyclohexylcarbodiimide
DEAD	Diethyl azodicarboxylate
DMAP	4-(Dimethylamino)pyridine
DMF	Dimethyl formamide
DMSO	Dimethyl sulfoxide
L	Linker unit
LMG	Low molecular mass gelator
LMOG	Low molecular mass organogelator
NMR	Nuclear magnetic resonance
S	Steroidal unit
SANS	Small-angle neutron scattering
SEM	Scanning electron microscopy
TEOS	Tetraethoxysilane
TEM	Transmission electron microscopy
T_{gel}	Gel-to-sol phase transition temperature
TPP	Tetraphenylporphyrin
UV	Ultraviolet
vol	Volume
wt	Weight

1 General Introduction

Although the gelation of water or organic fluids by low molecular weight molecules has been known for a long time [1,2], it is only very recently that the development of self-assembling hydro- and organogelators emerged as a new challenging area of supramolecular chemistry and materials sciences [3–5]. Most of the appeal of LMGs arises from their well-defined molecular structures, in contrast to the case of gelling polymers, and the flexibility with which such structures can be synthetically varied, which, in turn, enables to fine tune the properties of the gel-phase assemblies. In that connection, it is remarkable how the field of low molecular weight gelators (LMGs) progressed rapidly from a situation where serendipity played a decisive role to the one where it became possible to chemists to rationally design new molecular structures capable of self-assembly and gelation [6]. Over the last decade, LMGs did open access to a range of functional, nanostructured materials having potential applications in separation technologies, medicines, biomaterials, electronics, photonics etc.

The phenomenon of gelation with LMGs is recognized as stemming from preferential unidirectional self-assembly of molecular units into fiber-like superstructures. The solid-state fibers, displaying remarkably high aspect ratio (usually tens of nanometers in diameter and up to several micrometers in length), become entangled through the formation of junction zones (crosslinks), which results in the formation of a continuous, three-dimensional network. As a result of a large solid–liquid interfacial area, the major fluid component is entrapped by the minor network component, and, at the macroscopic level, the complete volume of solvent is immobilized. Usually, the sample-spanning gel material is obtained simply by dissolving a small amount of gelator compound in a hot solvent and cooling the solution below the phase transition temperature. Self-assembled gels, based on LMGs, are physical gels and, thus, thermoreversible. Heating the gel above the gel-to-sol phase transition temperature disrupts the gel superstructure and re-forms the isotropic solution. Upon cooling, the sol reverts to the gel; the process is reversible and can be repeated over a great number of cycles. Another remarkable feature pertaining to LMGs is that, very often, a very low concentration (0.1–5 wt %) of gelling agent is enough for gelation to occur, which is not the case for the polymeric counterparts. It can occur that extremely low concentrations (below 0.1 wt %) are even sufficient to obtain very stable gels, such exceptional LMGs fall into the category of the so-called "super-gelators".

Critical to the rational design of LMGs is the control of (i) the self-assembling mode at the molecular level, one-dimensional assembly being required for efficient fiber growth, and (ii) the balance between solubility and insolubility in a given solvent, which relies on specific interactions between solvent and gelator molecules. If the latter feature still remains difficult to predict, the former condition is easily achieved as long as strong and highly directional self-assembling groups are selected. Hydrogen-bonding functionalities belong to that class of building-blocks, and amides and urea represent to date the most developed class of LMGs capable of gelling water and/or organic fluids [3–7]. Very recently, Shinkai and coworkers elegantly demonstrated that sugar derivatives could also be conveniently used for the systematic design of hydrogen-bonding LMGs and for producing novel hydro-and organogels with tailored properties [8, 9].

Although several steroids have found outstanding applications in many areas of supramolecular chemistry [10], it is arguably cholesterol (Scheme 1) that offered itself as the most versatile unit on which to base the systematic design of functional LMGs for the gelation of organic fluids [11]. Functional

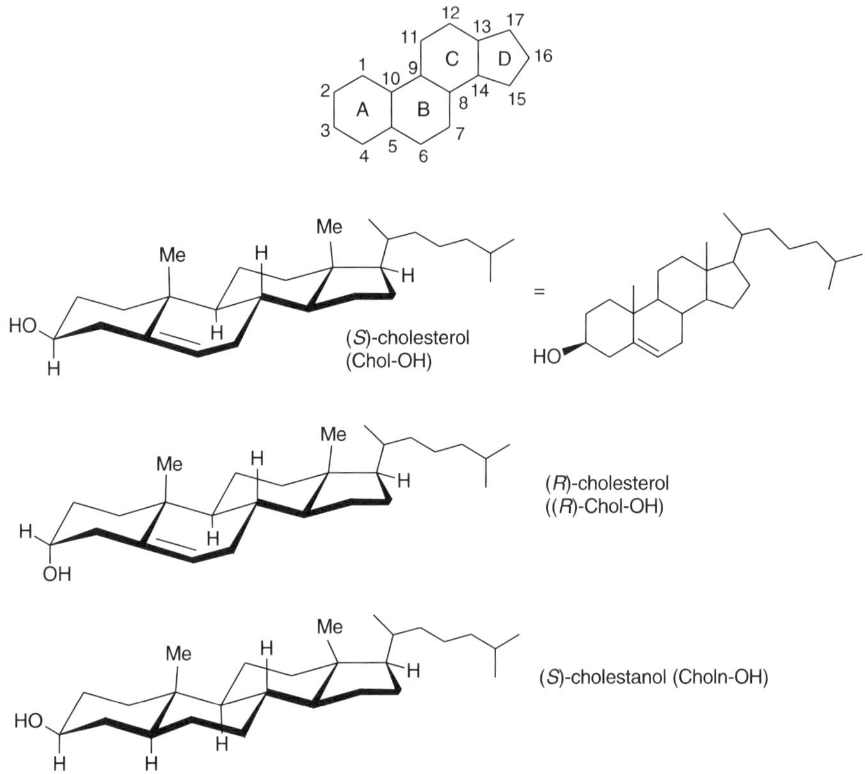

Scheme 1 Molecular structures of the cholesterol (α- and β- epimers) and cholestanol (β-epimer) building blocks

cholesterol-based LMOGs (O for organo) were discovered unexpectedly in 1987 by Weiss and coworkers and still remain a actively investigated class of compounds at the time of this writing [3, 4]. In cholesterol-containing LMOGs, it is the covalently bound cholesteryl group that induces directional self-association through weak Van der Waals interactions.

In this chapter, we hope to provide a comprehensive overview of the chemistry of cholesterol-based LMOGs and to describe the underlying principles that are responsible of self-assembly and, in turn, organogel formation. In particular, we will focus on the properties of the resulting organogels in terms of relationship between the designed structure of the gelator, the nature of the gelled organic fluid component, the influence of molecular interactions (π–π stacking, Van der Waals, hydrogen bonding, ...) and chirality. The prospects for future applications of cholesteryl molecular gels will be illustrated and discussed at the end of this chapter.

2 Families of Cholesterol-Based Organogelators

Cholesterol is essential to life. The name originates from the Greek *chole* (bile) and *stereos* (solid) as cholesterol was first isolated in the solid form from gallstones in 1784. More than 200 years later, it is now well established that important biological functions and solubility properties of cholesterol are intimately linked [12]. Indeed its insolubility in water explains why this molecule is an important component of the lipid membranes of animal cells. It is this very property that makes cholesterol deleterious, if not lethal. Indeed its deposition within the wall of an artery can lead to the formation of arteriosclerotic plaques. Remarkably these are also the unique self-aggregation properties of cholesterol derivatives that were at the heart of the discovery in 1888 of a new state of matter, the liquid crystal [13]. In many respects, cholesterol is a fascinating molecule; a unique molecular building-block that has been widely exploited in supramolecular and materials chemistry [14–17].

Cholesterol, as one of the steroids, has four rings fused together, three six-membered and one five-membered (Scheme 1). In cholesterol, the cyclohexane rings are *trans* fused, which implies that they are locked in the chair conformation. As a consequence, the cholesterol skeleton is rigid, flat, and straight. The methyl groups at C-10 and C-13 and the hydrogen atoms at C-8, C-9, and C-14 are axial. Those substituents that lie above the molecular plane, such as the two methyl groups and the hydrogen atom at C-8, are termed β, whereas those that lie below the plane are termed α. Cholesterol has a β aliphatic chain at C-17 of ring D, a double bond in ring B, and a hydroxy group on C-3. In the naturally occurring cholesterol molecule, the latter is equatorial and the stereochemistry at C-3 of the steroid moiety is thus β. In other words, the C-3 stereogenic center has the (*S*) configuration. 5α-Cholestanol (Scheme 1) presents a molecular structure very close to that of cholesterol and has been used as steroidal fragment of interest for the synthesis of LMOGs. As compared to cholesterol, 5α-cholestanol lacks the double bond in ring B, and the extra hydrogen atom at C-5 is axial (Scheme 1).

Although cholesterol possesses the typical features that characterize a surfactant molecule, e.g., a polar hydroxyl group and a long hydrocarbon skeleton, it does not behave as a surfactant in that it is not observed to form membrane assemblies by itself. However, it is the amphiphilic structure of cholesterol that explains its ready incorporation into cell membranes as a "plasticizer" agent. Cholesterol is also not a gelator molecule in its own right. Yet, simple synthetic derivatives of cholesterol or cholestanol are able to gel organic liquids, but they are very few (Scheme 2). For example, a long-chain ester derivative of cholesterol, cholesteryl laurate (**3**), was observed to form partial, fragile gels at 5 wt% in a silicone fluid [18]. Similarly 5α-

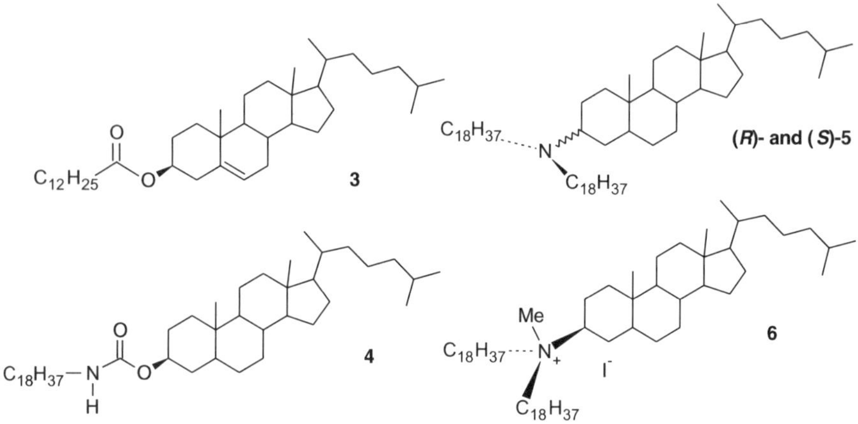

Scheme 2 LMOGs based on long-chain substituted derivatives of cholesterol and cholestanyldialkylamine

cholestan-3β-yl N-octadecylcarbamate, **4**, was shown to gelate benzyl alcohol and ethanol at ca. 0 °C [19]. Both α- and β-epimers of cholestanyldioctadecylamine, (R)-**5** and (S)-**5**, yielded gels with Dow-Corning 704 silicone oil [20], while several ammonium derivatives, such as **6**, were found to gel a variety of organic liquids at very low concentration (< 5 wt/vol %) [21]. The gelation of hydrocarbons was first reported for a d-homoandrostanyl derivative containing a hydroxyl group at C-3 and an amine (**1**) or nitroxide (**2**) functionality at the 17a-aza position (Scheme 3) [22]. The structures and properties of the organogels of **1** and **2** have been investigated in detail [23–25].

It is worth mentioning the use of naturally occurring bile acids, such as cholic acid derivatives, as steroidal building-blocks for the design of supramolecular receptors [10] and organogelators [26, 27]. Some of these simple steroid derivatives, such as the dihydroxy bile salt sodium deoxy-

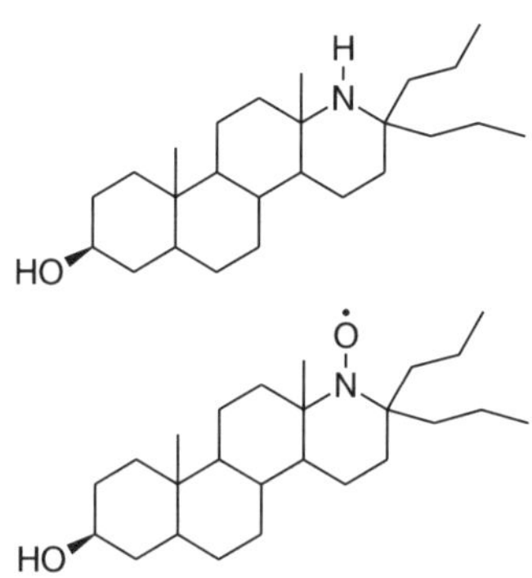

1

2

Scheme 3 The two d-homoandrostanyl derivatives act as gelators of hydrocarbons cholate [28] and dihydro-lanosterol [29], are known to gel aqueous solutions or organic liquids at low concentration.

Historically, the first rational synthesis of cholesterol-based LMOGs was proposed by RG Weiss and coworkers. They showed that molecular systems comprising an aromatic (A) moiety connected to a steroidal (S) group through a functionalized link (L) could display effective, and somewhat predictable, gelation ability (Scheme 4) [30, 31]. A great number of such ALS gelators, in which the nature of the aromatic and link components (A, L) varied, have been synthesized and investigated during the last decade. Variations of the S part are so far more limited as most of the known ALS structures are based on either the cholesterol or the 5α-cholestanol units. In particular, structure–property relationship studies have outlined the critical role of the aromatic unit. The latter is not only necessary for gelation to be observed, but it also allows a fine tuning of the molecular organizations within the self-assembled superstructures and, in turn, the self-assembly properties of the gel phase materials. Moreover, the presence of the π-delocalized unit incidentally endows the gels with electronic and optical properties, a feature that has been largely exploited for the generation of functional materials (vide infra). Recently, a new generation of steroidal gelators has appeared. They contain the same basic components but arranged in a dimer structure, $A(LS)_2$, in which the central aromatic fragment is symmetrically substituted with two cholesterol units (Scheme 4).

For a certain number of gelator molecules, especially those with the ALS architecture, the role of the configuration at C-3 of the cholesterol skeleton on the gelation ability has been investigated. In the following, when the properties of α- and β-epimers are compared, the number that designates the compound is preceded by (*R*)- or (*S*)- in order to distinguish between the two epimers. When no explicit mention of the configuration is given, this means the cholesterol possesses the naturally occurring β-stereochemistry, which is true for the vast majority of compounds reported so far. Moreover, for the sake of simplicity, a schematic representation of the cholesterol and cholestanol moieties (Scheme 1), in which only the configuration at C-3 is explicit, is used.

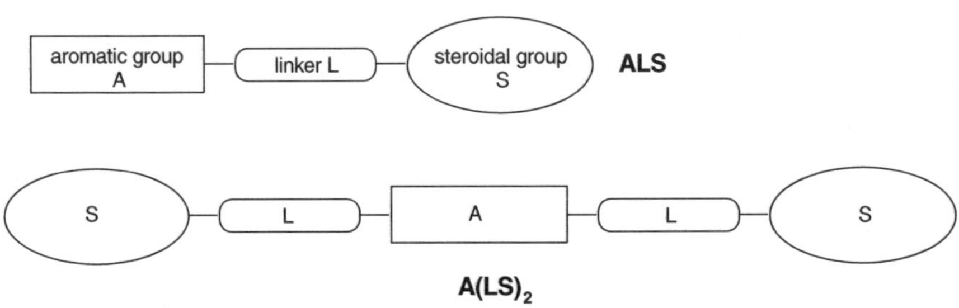

Scheme 4 ALS and $A(LS)_2$ architectures of cholesterol-based LMOGs

3 ALS Gelators

We have subdivided ALS organogelators into distinct classes according to the nature of the aromatic component.

3.1 Polycyclic Aromatic Hydrocarbons

3.1.1 Anthracenes

Cholesteryl 4-(2-anthryloxy)butanoate, **7**, was reported by Weiss and coworkers as the first member of the family of ALS gelators (Scheme 5) [30]. The longer homologue of **7**, containing the pentanoate linker (**8**), and the shorter one, **9**, were also shown to be effective gelators (Scheme 5) [31].

Thermally reversible organogels were also obtained with ALS derivatives (**10–12**) containing a nitrogen atom in the linker (Scheme 5) and 1- and 2-propanol, 1-butanol, 1-pentanol, and 1-octanol as solvents [19]. Within this series, **10** appeared as a very versatile and efficient gelator as it was observed to gel additionally *n*-dodecane and *n*-hexadecane solvents. Interestingly, compound **13**, containing the free N – H group, was designed as an aza-analogue of the excellent gelator **7** in order to investigate the potential influence of hydrogen bond formation. Actually **13** failed to gel any of the solvents examined in the study, which is in contrast to the *N*-protected aza-ALS **10–12** [19]. It was argued that hydrogen bonding was not a critical factor in the aggregate of anthracene-appended cholesterols.

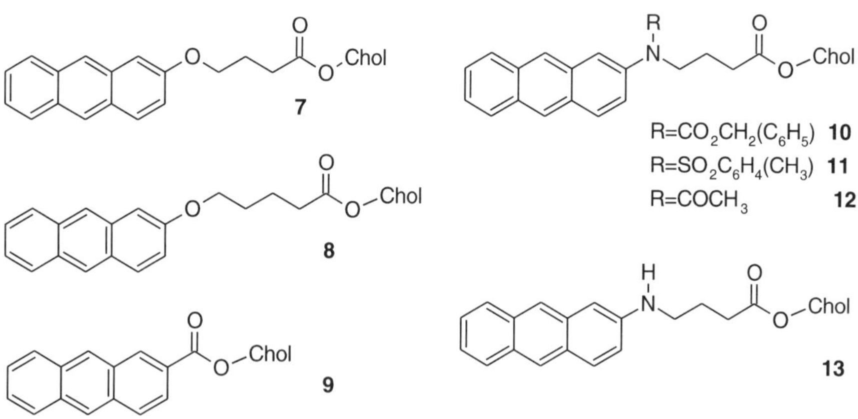

Scheme 5 Anthracene-appended cholesterol derivatives that act as organogelators

Among the anthracene-appended ALS organogelators, it is **7** that has been the subject of intense investigation. Owing to the well-known excited-state properties of anthracene, the anthracenyl chromophore in **7** could be conveniently used as a photochemical and photophysical probe to investigate the supramolecular organization at the molecular level and monitor gel formation. Electronic absorption, fluorescence emission, and CD spectroscopies proved useful to that end. The photophysical and photochemical results showed that the anthracene chromophores are overlapping within the aggregates, but not in a parallel fashion as no excimer emission could be detected. It was reported that small quantities of **7** (0.25–1.5 wt %) could gel reversibly with a wide variety of organic liquids, including hydrocarbons, alcohols, aldehydes, esters, amines etc [31]. In particular, the gelation properties of **7** were compared in two kinds of solvents, long-chain alcohols such as *n*-octanol, and *n*-alkanes such as dodecane [30–35]. Remarkably, the morphological features of the **7** gel superstructure were shown to be strongly influenced by the solvent nature. Although CD spectroscopy showed that **7** molecules are helically stacked in the dodecane gels, the observation on the mesoscopic level of twisted fibers was reported for the gels in *n*-octanol, but not in *n*-alkane solvents. Furthermore, a strong dependence of the morphology of aggregates on solvent was evidenced using TEM. Consistently, SANS profiles were also different in *n*-decane and *n*-butanol [33, 35]. In decane, the fiber-like aggregates are swollen with lyotropic organizations owing to the penetration of solvent molecules into the supramolecular structures. In alcohol solvents, the swelling process is not operative. In the latter solvents, the structural organization in the gel results from the formation of a more compact gel structure and is related to the hexagonal ordering found in the solid state. As a consequence of swelling, the size of the strand aggregates was one order of magnitude larger in *n*-alkanes than in *n*-alkanols. These results are consistent with the fact that T_{gel} values are higher in alcohols (63 °C in 1-octanol at 0.80 wt %) than in alkanes (40 °C in *n*-octane at 0.81 wt %). Indeed, the exclusion of alcohol molecules from the aggregate leads to the formation of somewhat "dry", cohesive junction zones. As a result, melting of the gel requires heating to higher temperatures in alcohols than in alkane solvents.

The synthesis of two compounds, **14** and **15**, containing the 2-anthracenyl group connected to the oxygen atom at C-3 of 5α-cholestanol (with the naturally occurring (*S*) configuration) via a carbamate linker was reported

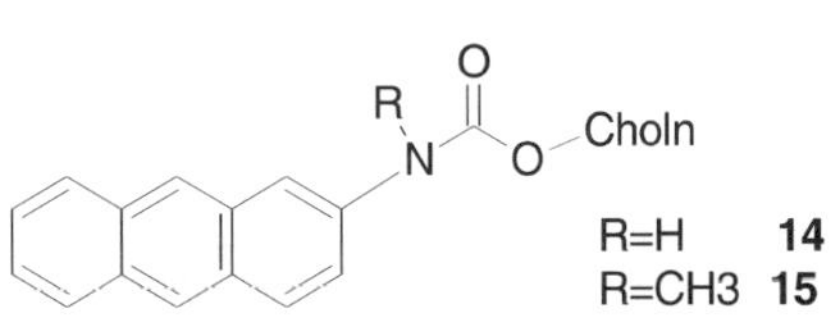

Scheme 6 Anthracene-appended ALS gelators based on a nitrogen atom-containing linker

by Weiss and coworkers (Scheme 6) [19]. They can be considered as aza-analogues of the gelator **9** (Scheme 5). **14** formed thermally stable and exceptionally long-lived (> 1 year) gels with several *n*-alkanes (T_{gel} ca. 100 °C at 1.0 wt % in *n*-dodecane) and linear alcohols (T_{gel} < 40 °C at 0.80 wt % in 1-pentanol), while **15** only gelated 1-butanol (at 2.0 wt %) and 1-pentanol (at 1.0 wt %).

The more effective gelation ability of **14** as compared to its analogue methylated at the carbamate nitrogen, **15**, was postulated to stem from the smaller size of the *N*-substituent rather than from hydrogen bonding. In contrast to **7**, **14** and to a lesser extent **15**, exhibited new electronic absorption and fluorescence emission bands in the gels, due to strong interaction between pendant anthryl groups. CD spectroscopy confirmed the presence of J-type aggregates with a helical arrangement of the molecular stacks within the gelled samples. The exceptional stability of some of the **14** gels at low gelation concentration (2 wt %) was thus related to the enhanced π–π interaction. Consistently, when the A part contained smaller aromatic rings, such as the 2-naphthyl moiety, gelation was still observed but the gels were less stable. In the case of the even smaller phenyl-based derivatives, the gelation property was lost.

3.1.2
Anthraquinones

Weiss and coworkers reported a series of anthraquinone-containing systems based on either the cholesterol or the 5α-cholestanol unit (Scheme 7) that were proven effective gelators of organic liquids. The behavior of cholesteryl anthraquinone-2-carboxylate, **16**, the anthraquinone-containing derivative of **9**, has been investigated in detail [31, 33, 36–38]. It was shown that the structure and properties of **16** gels in *n*-alkanes and long-chain alcohols, at

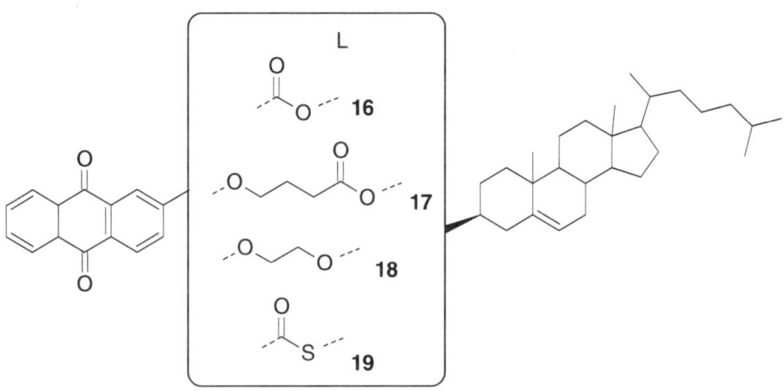

Scheme 7 Anthraquinonone-containing ALS organogelators

concentrations of 1–4 wt %, are very sensitive to the nature of the solvent, as observed in the case of the anthryl gelator **7** (Scheme 5) [31, 33]. Specific interactions between solvent and gelator molecules in the gel strands were invoked to account for the dependence of gel structure on solvent type, as confirmed by the use of neutron and X-ray scattering techniques [38]. A strong change in CD intensity accompanied the sol-to-gel phase transition in 1-octanol or hexadecane, which is in keeping with helical molecular association involving some overlap of the aromatic rings of neighboring molecules. Moreover, the CD spectra of the **16**-hexadecane gel were shown to depend on the cooling rate used for the gel preparation from the isostropic solution [38]. Other anthraquinone-containing compounds, such as **17**–**19**, also formed gels with a wide variety of linear and branched alcohols and alkanes, and other solvents at concentrations near 2 wt % [36]. This indicated that linkers (L) of different nature (ether, ester) and length could also impart the ALS with gelation ability. Interestingly, compound **17** formed *n*-alkane gels that exhibited gel-to-sol transition temperatures (59 °C in dodecane) higher than those recorded for alcohol gels (43 °C in 1-octanol) at the same concentration (2 wt %) [36], which was consistent with the behavior of **14** (Scheme 6) [19] but in contrast with that of the anthracenyl ALS **7** [38]. This effect was rationalized in terms of the better solubility of the compounds **14** and **17** in alcoholic solvents relative to **7**, as due to the presence of additional heteroatoms in the A or L units.

The grafting of the anthraquinone ring to the 5α-cholestanol moiety afforded an effective gelator of several alcohols (2.91 wt % in 4-heptanol) and alkanes (1.53 wt % in dodecane) in the case of the naturally occurring epimer (*S*)-**20** (Scheme 8) [31]. Interestingly, the α-epimer, (*R*)-**20**, failed to gel the same liquids under the same conditions [36], which illustrates the importance of the configuration at C-3 of the steroid skeleton. In general, an ALS with β-stereochemistry can preferentially adopt an extended conformation with

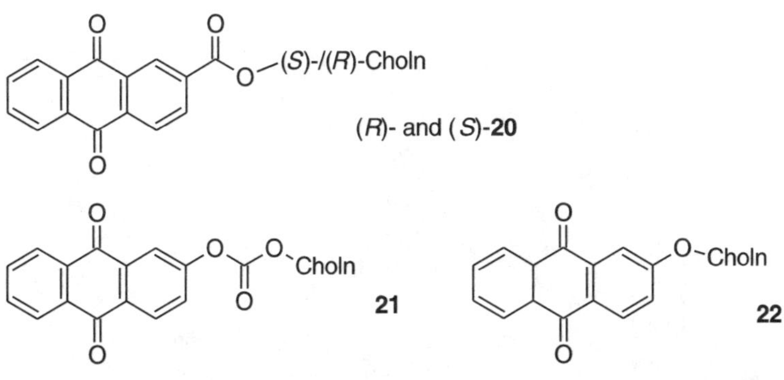

Scheme 8 The β-epimer of the cholestanol-based gelator (*S*)-**20** acts as an efficient organogelator of alcohols. The other anthraquinone-containing systems are non-gelators

a rod-like shape. In contrast, the α-epimer is likely to better experience a bent conformation, being thus less prone to one-dimensional stacking and to stable fiber formation. The compound with the carbonate link, **21**, and the one with the simple ether link, **22**, actually the ALS with the shortest link known to date, failed to gel a representative series of organic liquids.

3.1.3 Anthracene or Anthraquinone-Appended Systems Based on Non-Cholesteryl Steroidal Parts

Steroid units other than cholesterol or cholestanol have been used for the synthesis of ALS structures featuring anthraquinone or anthracene as A part (Scheme 9) [31, 36]. Particularly those compounds in which the steroidal unit lacks a C_8H_{17} chain at C-17 failed to produce gels in any of the solvents where **7** and **9**, for example, are effective [31]. Slight changes at the C-17 aliphatic chain, such as introduction of unsaturation (**23**) and/or ethyl group (**24**), were shown to have little impact on the gelation ability [36]. Compound **25**,

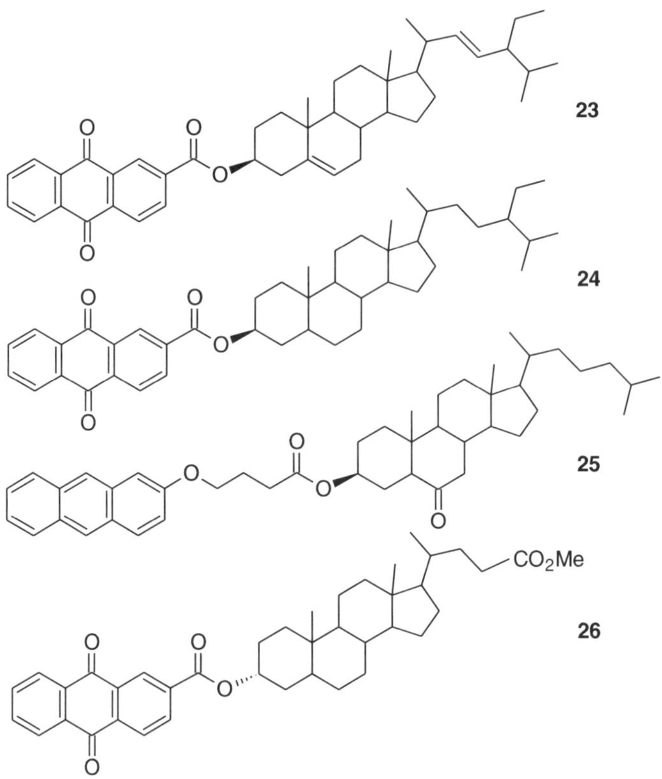

23

24

25

26

Scheme 9 Anthraquinone-appended systems based on non-cholesteryl steroidal parts containing the ketocholestanyl group was found effective in forming gels in dodecane and 1-octanol at low concentration (1.13 wt % in *n*-dodecane and 0.89 wt % in 1-octanol [31]. The fact that ALS **26** failed to produce gels is more likely due to the α-stereochemistry, as stated above, rather than to the structure of the steroidal part itself [36].

3.2
Conjugated Molecular Rods

Whitten and coworkers reported the synthesis and the study of both α- and β-epimers of two ALS gelators, **27** and **28** (Scheme 10), based on the *trans*-stilbene and the squaraine chromophores, respectively [39]. This group had previously investigated the prominent role of the squaraine and stilbene units in the aggregation properties of amphiphile derivatives containing these conjugated units [40]. Compounds **27** and **28** were thus designed to investigate the effect of two kinds of competing interactions, e.g., chromophore–chromophore and cholesterol–cholesterol interactions, on organogelation ability. Moreover, stilbenes and squaraines produce self-aggregated structures with characteristic spectroscopic signatures and represent valuable optical probes for the investigation of self-aggregation and organogelation phenomena [41]. For the squaraine-containing compound **28**, the link was chosen longer and more flexible than in compound **27**. For both ALS **27** and **28**, the effect of stereochemistry at C-3 on gelation ability has been investigated. Both epimers of the stilbene–cholesterol gelator, (*R*)-**27** and (*S*)-**27**, were found to be effective in the gelation of several solvents, including ben-

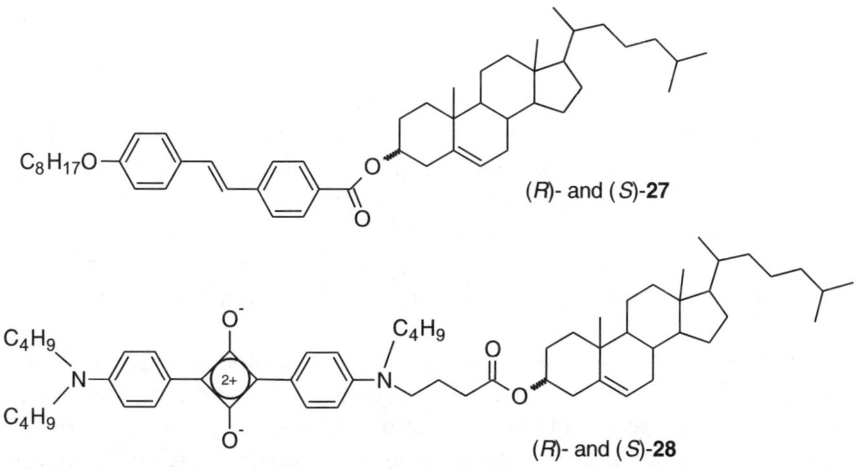

Scheme 10 α- and β-Epimers of two ALS gelators based on the *trans*-stilbene (**27**) and squaraine chromophores (**28**)

zene, nitrobenzene, toluene, haloalkanes, dioxane, acetonitrile, pyridine, and alcohols at concentrations below 1 wt %. However, the "unnatural" epimer, (*R*)-**27**, was observed to give more transparent and stable gels than the "natural" isomer, (*S*)-**27** [39]. Gelation of acetonitrile occurred at level as low as 0.1 wt % of the latter. The reverse situation was found in the case of the squaraine–cholesterol systems. Compound, (*S*)-**28**, with the natural cholesterol unit, could gel several alcohols at concentrations of 1 wt % or less, and was more effective than the α-epimer. The gels of (*R*)-**27** and (*S*)-**28** were investigated by a variety of techniques, including absorption, fluorescence and NMR spectrocopies, AFM and SEM [39, 42, 43]. These studies brought important insights into the molecular organization within the gel structures. Firstly, extensive aggregation of the aromatic chromophores was pro-active in organogel formation with both squaraine and stilbene derivatives, which resulted in fluorescence quenching and CD activity. The spectroscopic signature of the squaraine H-aggregate was recorded in the case of (*S*)-**28**. Secondly a significant amount of solvent molecules, up to 30% in the case of a gel of (*R*)-**27** in *n*-octanol, were shown to participate in the process of fiber formation and to penetrate into the fibers, being held in strong interaction with the gelator molecules [43]. This observation was consistent with the fact that the two epimers of the stilbene-containing gelator were observed to photoisomerize within the gel whereas *trans*-stilbene derivatives do not isomerize in the crystalline state. This is also in keeping with the behavior of anthracene- and anthraquinone–cholesterol gelators, of which the fibers were shown by Weiss and Terech to incorporate solvent molecules and to be swollen with lyotropic organizations [33, 35, 38]. Finally, CD spectra of a gel of (*R*)-**27** in octanol indicated the formation of a helical organization within the fibers [43].

3.3 Azobenzenes

Shinkai and coworkers reported the synthesis of a large number of cholesterol derivatives bearing a chromophoric azobenzene moiety [44–46]. These systems were primarily designed in order to investigate *cis*–*trans* photoisomerization in the azobenzene unit as a mean to reversibly control the characteristics of cholesteryl liquid crystals [47]. Among them, those bearing a *p*-alkoxyazobenzene fragment, (*S*)- and (*R*)-**29R** (Scheme 11), were shown to act as excellent gelators of a great variety of organic fluids at concentrations of 1–7 wt %.

The detailed investigation of these compounds allowed important conclusions to be drawn about the gelation ability of azobenzene–cholesterol gelators [46]. The gelation ability was found to increase with an increase in chain length. Compound **30** lacking a *para*-substituent (Scheme 11) indeed showed a rather weak gelation ability. Remarkably, (*R*)-**29Dec** gelated most of the 55 organic fluids and solvents tested in the study. Certain compounds

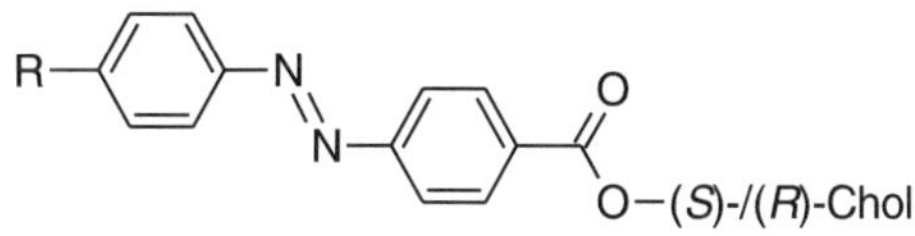

(*S*)-**29R**
R = Me, Et, Pr, Bu, Dec

(*R*)-**29R**
R = Me, Et, Pr, Bu, Pent, Dec

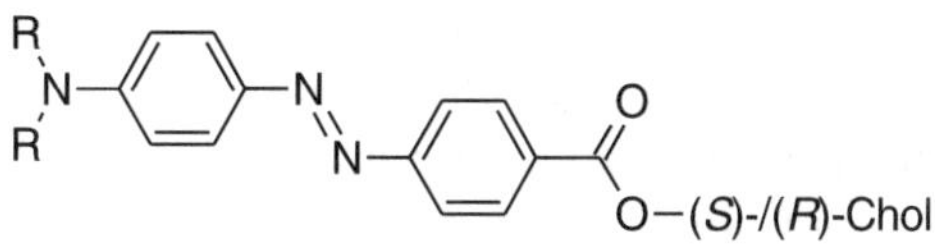

(*S*)-**31R**
R = Me, Et

(*R*)-**31R**
R = Me, Et, Pr

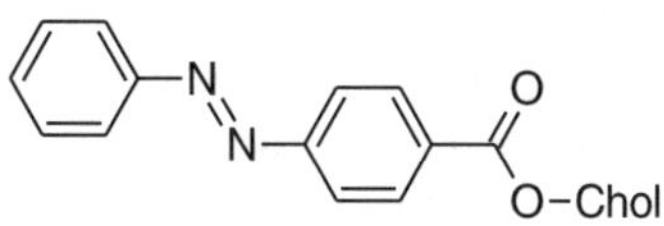

30

Scheme 11 Azobenzene-appended ALS gelators

gelated alcohol solvents at levels below 1 wt %. A strong effect of C-3 configuration was observed, as the gelation ability of the cholesteryl derivatives with the inverted (*R*)-chirality was found to differ from that of the cholesteryl compounds with the natural (*S*)-chirality. In general, (*S*)-**29R** were found more useful for gelation of apolar solvents whereas their diastereomers (*R*)-**29R** were superior in polar solvents. This behavior was rationalized in terms of the specific solubility properties of (*S*)-**29R** and their diastereomers (*R*)-**29R** in apolar and polar solvents. The high solubility of (*R*)-**29R** in aromatic hydrocarbons, halogen solvents, and ethers correlated well with their tendency to give solutions in such media and to gelate "poor" polar solvents preferentially, such as methanol, ethanol, acetic acid for example. This behavior was supported by the conformational properties of the two compounds in the solid state. Indeed, energy-minimized structures of (*S*)- and (*R*)-**13Me** showed that the natural epimer adopts an extended conformation, whereas the inverted epimer exits in a bent conformation with a L shape. Such geometries are likely to be a consequence of either the equatorial or axial orientation of the substituent at C-3 of the cholesterol skeleton. In the extended conformation (isomer β), the azobenzene unit is available for experiencing significant π-overlap with azobenzene chromophores of neighboring molecules, which can provide additional stabilization (and high crystallinity) to the aggregate. In contrast, the α-epimer, with the bent conformation, is not prone to allow face-to-face interactions between azobenzene chromophores.

Compounds with (*S*)-configuration consistently gave gels with higher sol–gel phase transition temperature values than those determined with the (*R*)-epimers. Moreover, the monolayer formation properties of (*S*)- and (*R*)-**13Me** were studied at the air–water interface and were found to be profoundly influenced by the configuration at C-3 [48]. Compared to the compounds containing a *p*-alkoxy substituent, those bearing a *p*-dialkylamino group ((*S*)-**31R** and (*R*)-**31R**, Scheme 11) displayed a poorer gelation ability owing to their higher solubility in organic solvents [46].

Drastic CD spectral changes were observed below and above the sol–gel phase transition [45, 46]. In the sol phase, the compounds (*R*/*S*)-**29R** and (*R*/*S*)-**31R** were totally CD-silent while, in the gel phase, they displayed CD spectra containing exciton coupling bands at the position of the azobenzene absorption band with a zero crossing close to the gel absorption maximum. This indicated that the ALS gelators self-associate in a specific chiral arrangement, forcing the azobenzene units to experience asymmetric interactions in the excited state. The either clockwise or anticlockwise orientation of the azobenzene dipoles, as deduced from the sign of the Cotton effect in the CD spectra, was found to depend not only on the nature of the solvent and the length of the alkyl substituent, but also on the gel preparation conditions. The latter feature was also noticed by Weiss and colleagues [38]. Compound (*R*)-**31Me** formed gels in cycloalkane solvents that displayed CD spectra with the negative exciton coupling (a sign for (*S*)-chirality of the aggregate, anticlockwise orientation of azobenzene dipoles) in the case of a slow cooling rate. Conversely, the positive exciton coupling (gel with the (*R*)-chirality) was recorded for a gel formed under a fast cooling regime. Remarkably, when the gel obtained with the (*R*)-chirality was slowly heated to temperatures not exceeding the T_{gel} value, the CD sign was observed to gradually invert from (*R*) to (*S*), which allowed determination of a critical inversion temperature for the gel material. These results indicated that the gel with (*R*)-chirality, formed by rapid cooling, is metastable. It is worth mentioning that no correlation was found to exist between the chirality of the C-3 stereogenic center in the ALS gelators (*R*/*S*)-**29R** and (*R*/*S*)-**31R** and that of the resulting gels. Besides, SEM results confirmed the formation of three-dimensional networks containing helical fibrils. The gels of (*R*)-**31Me** in cyclohexane with (*R*)-chirality in CD possessed a right-handed helix, whereas the opposite was found for the gels with (*S*)-chirality.

Finally it was observed that UV light irradiation of a (*S*)-**29Me** gel in 1-butanol induced the *trans*-to-*cis* isomerization in the azobenzene unit, thereby causing the gel to turn into the solution. The gel could be re-formed by visible light irradiation. The sol–gel phase transition could be controlled reversibly by light and monitored efficiently by CD spectroscopy [44, 46].

An important class of azobenzene-based ALS organogelators is that of cholesterol derivatives bearing a crown ether moiety at the azobenzene unit, such as compounds **32-2**, **33**, and **34X** (Scheme 12). Compound **32-2** with

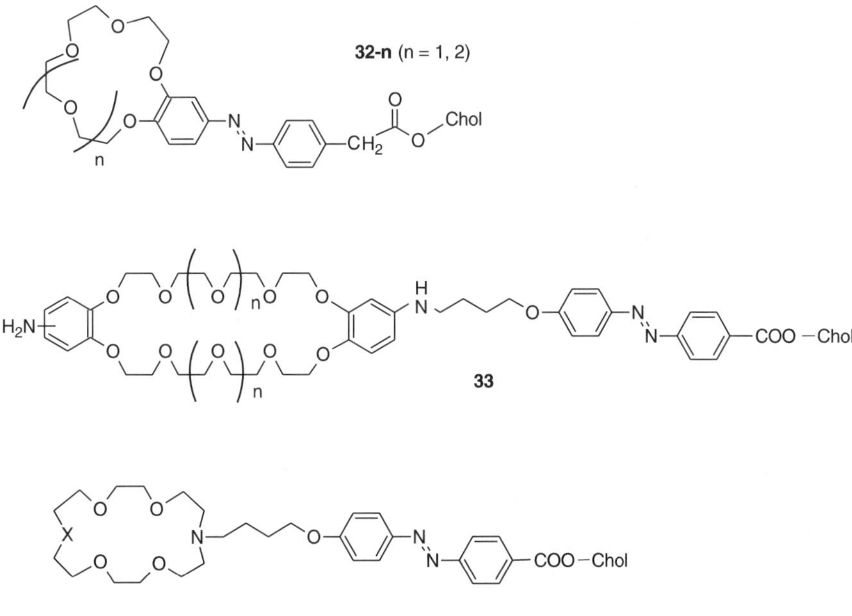

34X (X = O, NH, NBn)

Scheme 12 Azobenzene-appended cholesterol LMOGs bearing a crown ether moiety display metal-cation sensitive gelation ability

a benzo-18-crown-6 moiety acts as a gelator of several hydrocarbons and alcohols [44], whereas, under the same conditions, the smaller analogue **32-1** gives crystals only. Interestingly, the addition of alkali metal cations (Li^+, Na^+, K^+, Rb^+) induced an increase of the gel-to-sol phase transition temperature of gels of **32-2** in a mixture of methylcyclohexane and benzene as solvent [44]. In contrast, the opposite effect was observed with the larger Cs^+ cation, which indicates that the formation of the 1 : 2 metal-crown sandwich complex destabilizes the gel structure.

Compound **33** featuring a large dibenzo-[30]-crown-10 crown moiety (Scheme 12) gelated various organic solvents, such as 1-hexanol, DMSO, DMF, or propionic acid, under 1.0 wt % [49]. The propionic acid gel of **33** exhibited a negative sign for the first Cotton effect, which suggests that the azobenzene dipole moments interacts in an anticlockwise manner. Remarkably, **33** enabled observation of the formation of tubular-like assemblies with 45–75 nm wall thickness and 170–390 nm inside tube diameter. Three ALS gelators, **34X** (X = O, NH, NBn) bearing an aza-containing 18-crown-6 ether moiety were reported to display interesting gelation abilities (Scheme 12). Compound **34NBn** could gelate eight of the 17 solvents tested, including alcohols, DMSO, and amines [50, 51]. It was found to be particularly effective for alcohols. The CD properties of 1-butanol gels of **34NBn** in the absence and presence of $AgNO_3$ and $CsClO_4$ were investigated. Positive exciton bands clearly appeared in the CD spectra of **34NBn** in the presence of metal ions, while the effect was less pronounced in the case of the metal-free organogel. These observations indicated that the aggregates of **34NBn** are stabilized by intermolecular cholesterol–cholesterol and azobenzene–azobenzene interactions. Compound **34NH**, lacking the *N*-benzyl group (Scheme 12), also displayed a good gelation ability in alcohols and DMSO [52, 53]. The organogels of **34NH** were observed to be remarkably stabilized in the presence of mono- and diamines through the formation of hydrogen-bonded complexes between the amine guest and the crown ether moiety. Moreover, it was possible to discriminate the "chirality" of amines in the gelation process [52]. The monoaza-crown ether derivative **34O** behaved similarly.

Compound (*R*)-**35** containing an ammonium group presents the structural features of a conventional cationic surfactant (Scheme 13) [54, 55]. It was shown to gelate only polar solvents such as methanol, ethanol, acetic acid, and benzylamine, being thus less versatile than its neutral analogue (*R*)-**29Pent** (Scheme 11). Compounds **36-n** ($n = 2, 3, 4, 5$), containing a linear diamine unit (Scheme 13), were able to gelate various polar solvents such as alcohols, DMSO, and DMF. The morphology of the aggregate was remarkably sensitive to the number of methylene units bridging the two amino groups and, clearly, an odd–even relationship governs the aggregation modes of the diamine-containing ALS gelators [56]. The xerogels showed a lamellar structure for $n = 2$ and 4 and a fiber structure for $n = 3$ and 5.

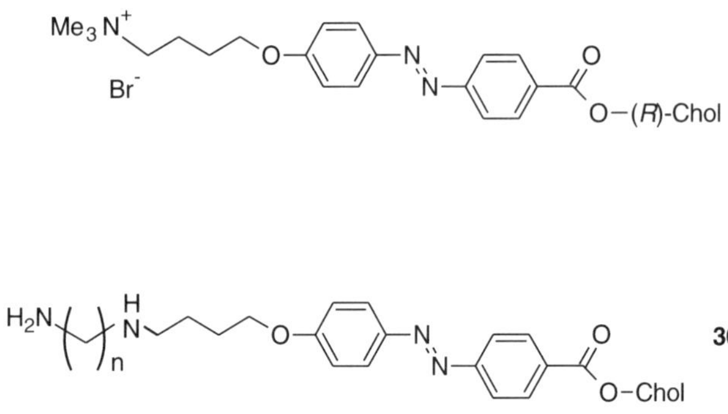

(*R*)-**35**

36-n (n = 2, 3, 4, 5)

Scheme 13 Azobenzene-appended cholesterol LMOGs bearing pendant ammonium or diamine units as effective gelators of polar organic solvents

3.4 Porphyrins

ALS compounds comprising a porphyrin group as the aromatic unit were shown to act as organogelators and to provide access to gel phase materials based on ordered chromophoric arrays with light-harvesting functions. Two epimeric compounds, (*S*)- and (*R*)-**37**, containing the free 5-(4-phenoxyacetic acid)-10,15,20-tritolylporphyrin unit connected to the cholesterol unit (Scheme 14), were tested as gelators in a wide range of organic solvents [57]. It was found that only cyclohexane and methylcyclohexane at 25 °C and diphenyl ether at 4 °C could be gelled by the natural (*S*)-epimer at 9 mM, whereas the inverted one gave isotropic solutions with these solvents. Although these compounds were not as versatile as the azobenzene gelators, their gelation ability showed a similar dependence on the configuration at C-3 of cholesterol [46]. From the CD studies, the cyclohexane gel of (*S*)-**37** gave rise to a strong exciton-coupling-type band and to a red-shift of the Soret ab-

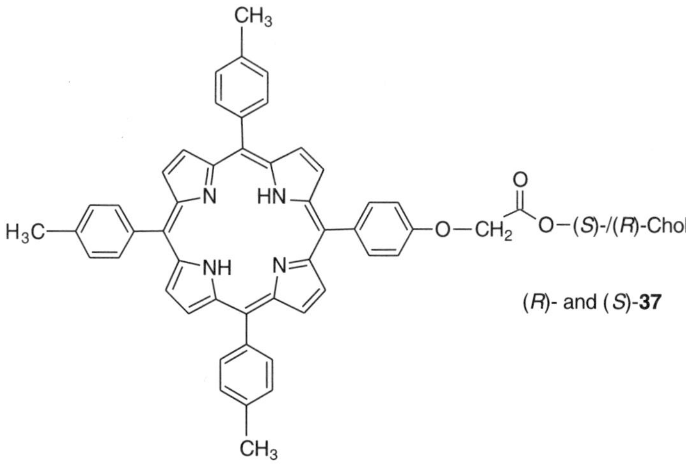

(*R*)- and (*S*)-**37**

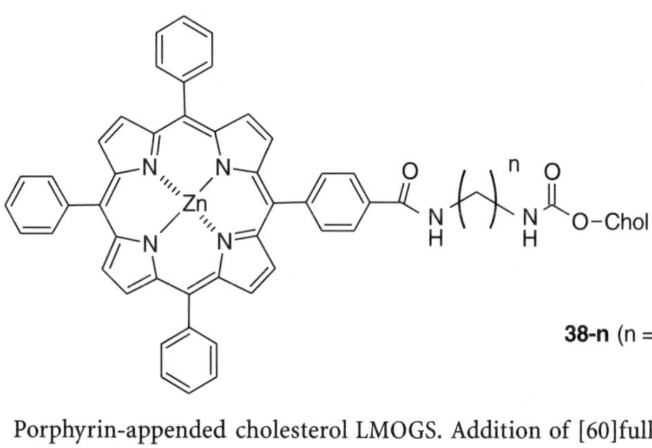

38-n (n = 2, 3, 4, 5)

Scheme 14 Porphyrin-appended cholesterol LMOGS. Addition of [60]fullerene reinforces the gelation ability of **38-n**

sorption band, which is indicative of the formation of porphyrin J-aggregates in the gel phase.

A series of ALS containing a Zn(II) porphyrin moiety appended to the (*S*)-cholesterol unit (**38-n**) via a diamine tether of variable length ($n = 2$, 3, 4, 5) were described and tested in various solvents at a concentration of 25.5 mM [58, 59] (Scheme 14). As observed in the case of the previous porphyrin-containing gelator, the gelation ability of **38-n** was rather weak. Indeed **38-4** failed to gel any of the solvents tested at room temperature whereas **38-3** and **38-5** formed turbid gels in cyclohexane and methylcyclohexane. At 5 °C **38-2** with a $(CH_2)_2$ spacer gelated aromatic hydrocarbons such as benzene, toluene, and *p*-xylene, but these gels where converted into sols at 20 °C. Remarkably, the gelation ability could be considerably increased when [60]fullerene was added to a solution of **38-2** in the aromatic solvents; the gels were then observed to be stable even at 20 °C. A similar stabilization effect was also observed with **38-4** under the same conditions. The reinforcement of the gel structure was rationalized in terms of the favorable intermolecular Zn(II) porphyrin–[60]fullerene interaction [60]. Sol–gel phase transition temperature measurements and spectroscopic studies supported the formation of 2 : 1 Zn(II) porphyrin/[60]fullerene sandwich complex only in the gel phase. CD spectroscopy was particularly useful for confirming the chiral orientation of **38-2** and **38-4** molecules in the gel phase as induced by the interaction with [60]fullerene. In contrast, the C60-assisted strengthening of the gel structure was not operative in the case of compounds **38-3** and **38-5** which have an odd number of methylenic units in their linker.

In the same connection, the gelation test of a [60]fullerene-appended cholesterol gelator, (*S*)-**39**, was performed in 28 solvents. (*S*)-**39** could only gelate dichloromethane at 25 °C at above the concentration of 16.2 mM, whereas the inverted epimer (*R*)-**39** was inactive in all solvents tested [61] (Scheme 15). CD spectroscopy studies indicated that the [60]fullerene moiety in (*S*)-**39** is chirally oriented in the gel state. The effect of zinc porphyrin/[60]fullerene interaction was investigated by adding Zn TPP into the gel of (*S*)-**39**. Interestingly, the addition of Zn TPP disrupted the gel, which is in contrast to the behavior of **38-n** in the presence of fullerene.

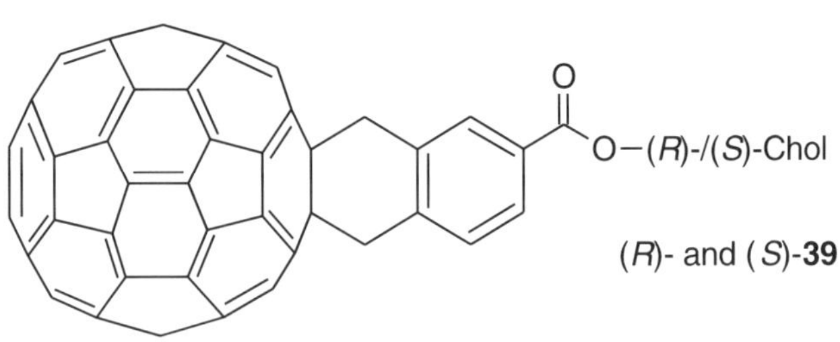

(*R*)- and (*S*)-**39**

Scheme 15 [60]Fullerene-based LMOGs

3.5
Small Rings

In general, cholesterol derivatives bearing a small aromatic ring or a heterocyclic nucleus display a weak gelation ability [31], as compared to their counterparts containing large π-delocalized systems. This is illustrated by the benzocrown ether derivative **40-2** that formed a gel in hexane only and at low temperature (– 6 °C at 2 wt %). Similarly, the shorter analogue **40-0** and related derivatives **41-n** ($n = 0$, 2) failed to gel any of the solvents investigated (Scheme 16) [44], which indicates that the size of the crown ether moiety is a particularly critical factor.

The gelation ability of a benzylamine-containing cholesterol **42** in alcoholic solvents, including *n*-butanol, was reported (Scheme 17) [62]. TEM photographs of a **42** organogel (5 wt % in *n*-butanol) confirmed the formation of a fibrous network. This study aimed at the use of the gel structure as a template for the benzylamine-catalyzed formation of silica. Although compound **43**, the trimethyl ammonium derivative of **42**, failed to gelate any of the solvents investigated, an equimolar mixture of **42** and **43** could gelate 1-butanol. Interestingly, proton NMR experiments ascertained that both compounds were present in the gel fibers. Pyridine-containing cholesterols recently appeared as structurally simple yet versatile gelators of organic solvents (Scheme 17) [63]. Especially, compound **44**, in which the pyridine is substituted at the *para* position, showed a superior gelation ability relative to its regioisomers **45** and **46**. Indeed it was observed to gelate various solvents ranging from apolar alkanes to polar solvents such as alcohols, ketones, acetonitrile, and DMSO at concentrations between 5 and 35 g cm^{-3} [63]. Re-

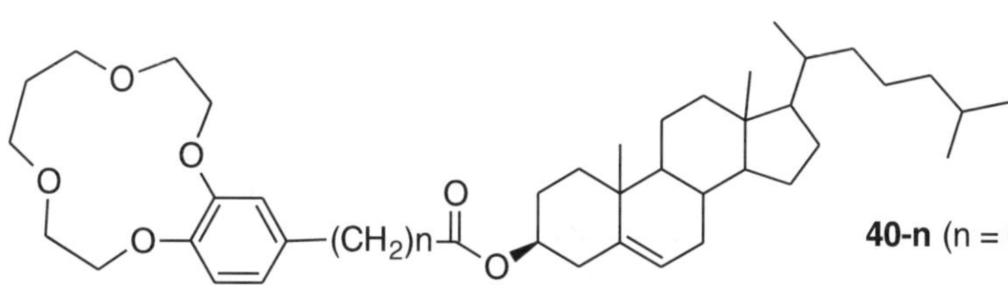

40-n (n = 0, 2)

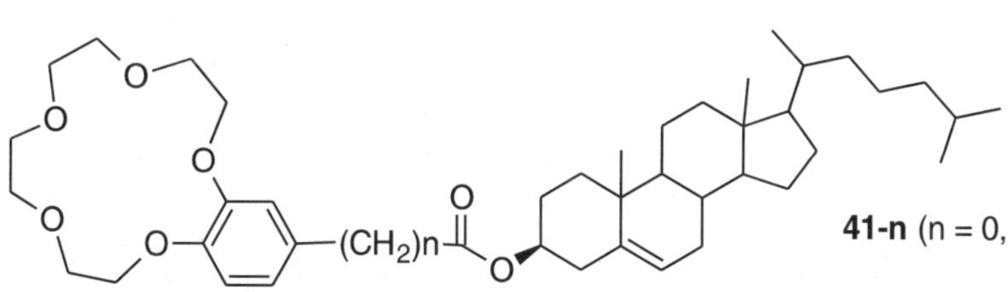

41-n (n = 0, 2)

Scheme 16 Cholesterol-based derivatives containing a benzo-crown ether moiety that behave as weak gelators

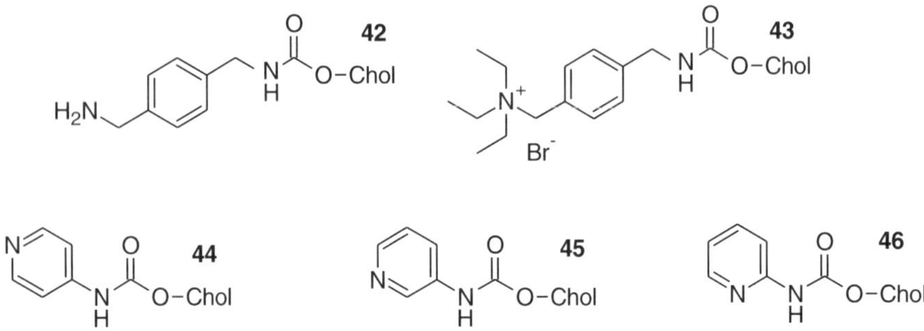

Scheme 17 Cholesterol derivatives that bear a benzylamine or a pyridine moiety are versatile organogelators. The gel of **45** in diphenyl ether is stabilized in the presence of silver(I) metal ions

markably, the gelation properties of compound **45** in diphenyl ether were found to be significantly improved in the presence of added silver(I) metal cations whereas they were not affected for **44** and **46**. This effect was rationalized in terms of a favorable orientation of two nitrogen atoms in the neighboring pyridyl groups for an efficient coordination of Ag(I) inside the gel aggregates.

Shinkai and coworkers reported the design and synthesis of cholesterol derivatives bearing heterocyclic units prone to hydrogen bond formation with the idea of investigating the cooperative participation of hydrogen-bonding and cholesterol–cholesterol interactions in the gelation process. As a first approach, the study of compounds **47** and **48**, containing an isocyanuric acid moiety and a triaminopyrimidine unit (Scheme 18), respectively, indicated that **47** forms a gel only in toluene whereas **48** precipitates from the solution. In contrast, using a 1 : 1 mixture (2 wt %) of **47** and **48** gelated methylcyclohexane, toluene, and 1-octanol the stability of the resulting gels was comparable to that of ALS containing an aromatic hydrocarbon group [64, 65]. Infra-red spectroscopy results showed that the reinforced gelation ability did not stem from "molecular tape" formation, as expected for the regular aggregation of the complementary hydrogen-bond-forming motifs [66], but rather involved complex hydrogen-bonded patterns.

This approach was further extended to the case of nucleobase-appended cholesterol derivatives, such as compounds **49R** (R = H, Me) functionalized by a uracil group (Scheme 18) [67, 68]. The fact that **49H** could form a very stable gel as compared to the methylated analogue, **49Me**, provided strong evidence for gel stabilization by hydrogen bond formation in addition to the cholesterol–cholesterol interaction. The latter was ascertained on the basis of CD spectroscopic results which confirmed the formation of helical columnar stacks as already noticed for previous ALS gelators. The addition of complementary nucleoside derivatives to **49H** allowed control of the stability and morphology of the gels [67, 68]. A remarkable behavior was observed when

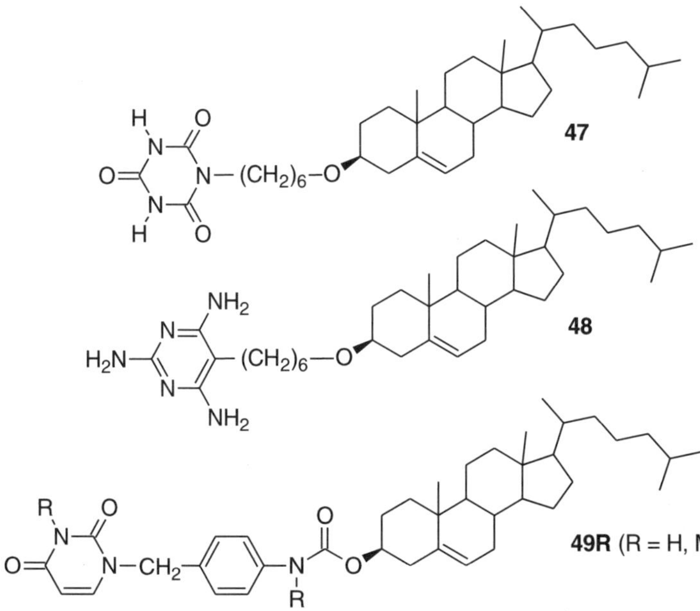

Scheme 18 Cholesterol-based LMOGs that served to investigate the influence of hydrogen bonding with complementary guests on gelation ability

49H was mixed with the complementary polynucleotide poly(A) – formation of a unique helical supramolecular structure [69].

The interplay between hydrogen-bonding and cholesterol–cholesterol interactions was also investigated in ALS compounds having a saccharide moiety tethered to the cholesterol unit. Sugar derivatives represent a recent class of effective gelators of water and organic solvents [8, 9]. Saccharide–cholesterol conjugates did behave as outstanding gelators [70] but the gelation ability was found to be strongly dependent on structure and absolute configuration of the sugar moiety. As an example, compound **50** (Scheme 19), containing the α-*D*-glucopyranosyl unit forms transparent gels with many solvents at a concentration of 4 wt/vol %, whereas its β-anomer (α/β refers in this particular case to the sugar moiety) is insoluble in most solvents owing to a more straight geometry that favors extensive face-to-face stacking. In some cases, such as the benzene gel of **50**, the stability of the gels was found to be excessively high and T_{gel} values were comparable with the boiling point of the solvent. This confirmed the cooperative participation of hydrogen bonds between sugar moieties, and van der Waal forces between cholesterol units. It is worth noting that the peracetylated β-anomer derivative of **50** could gel some organic solvents [70] and, especially, ionic liquids such as *N*,*N*′-dialkyl-substituted imidazolium and *N*-alkyl pyridinium ions [71]. In the latter fluids, the gels form at very low concentration (below 0.4 wt %)

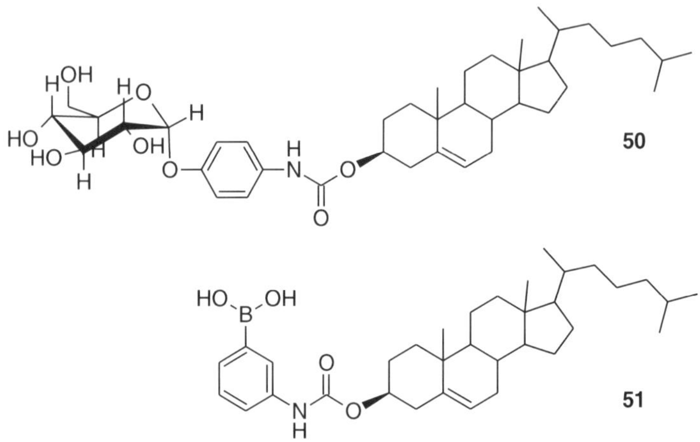

Scheme 19 Example of a saccharide–cholesterol (**50**) and a boronic acid–cholesterol (**51**) LMOG. Owing to cooperative sugar–sugar and cholesterol–cholesterol interaction, **50** forms remarkably stable gels

and are thermostable ($T_{gel} > 100\,°C$) above 0.4 wt %. When compound **50** was combined with boronic acid-appended poly(*L*-lysine) in a DMSO/water solvent mixture, a novel organopolymer gel formed, which consisted of vesicles of gelator **50** cross-linked by the polymer backbone [72]. Conversely complex formation between saccharide molecules and cholesterol boronic acid **51** (Scheme 19) also assisted organogelation [47].

3.6
ALS Organogelators: Main Outcomes

The use of the general ALS structure as design principle for novel organogelators enables access to a very broad family of cholesterol-based systems. Interestingly, such systems can gelate a range of solvent types, including polar and apolar, protic and aprotic solvents, depending on the nature of the A, L, and S components of the ALS structure. As gelation of cholesterol-based molecules is based primarily on weak forces (van der Waals, π-stacking), the solvent plays a critical role and has been shown to control gel stability, fiber helicity, and aggregate size. Swelling of the aggregates also appears to represent a common feature of the self-assembly of cholesterol-based LMOGs. Yet, the gelation ability of a designed ALS in a given fluid or even class of solvent still remains difficult to predict. This property results from a subtle balance between solubility and insolubility, which depends on the specific solvation of each A, L, and S parts. In this respect, the type of aromatic unit and the chemical nature, length, and flexibility of the linking group have a large impact. Furthermore, the stereochemistry at C-3 of ALS molecules determines the conformational preference of a ALS system, and, in turn, its solubility properties. Those compounds with an extended conformation, such as those based on the naturally occurring β-epimer, are generally less soluble and give better gels due to their higher propensity to stack relative to the α-epimer.

The structural feature common to effective ALS gelators is the presence of an aromatic unit tethered to the oxygen atom of the cholesteryl group. As a matter of fact, aromatic–aromatic and cholesterol–cholesterol interactions cooperatively act as driving forces in self-aggregation and gelation processes. It is likely that the additional self-association between aromatic groups assists the growth of unidirectional aggregates at the expense of the large three-dimensional structures that are usually observed with liquid crystalline or amphiphilic cholesterols [73]. In the case of anthracenyl–cholesterol gelators, such as **7**, it is the association of the aromatic parts that was believed to play a crucial role in the formation and ordering at the molecular level of the fiber aggregates [35]. This interpretation was based on the similar diffraction patterns in the solid-state of **7** and that of a long aliphatic chain-substituted anthracenyl organogelator lacking the cholesterol group [74, 75]. In contrast, for azobenzene- or *trans*-stilbene-appended cholesterol systems, the aggregation is more likely driven by the stacking of the cholesteryl moieties [42, 46]. In the primary unit aggregate, the stacking of the cholesteryl subunits is postulated to produce a one-dimensional helical array with the pendant aromatic moieties organized in a helical fashion at the periphery of the columnar core and sticking outwards (Scheme 20), consistent with AFM observations [40]. Such geometry of the unit aggregate is reminiscent of that of a spiral staircase [63]. In the case of azobenzene–cholesterol ALS having the β-configuration at C-3 (preferential extended conformation), and a long

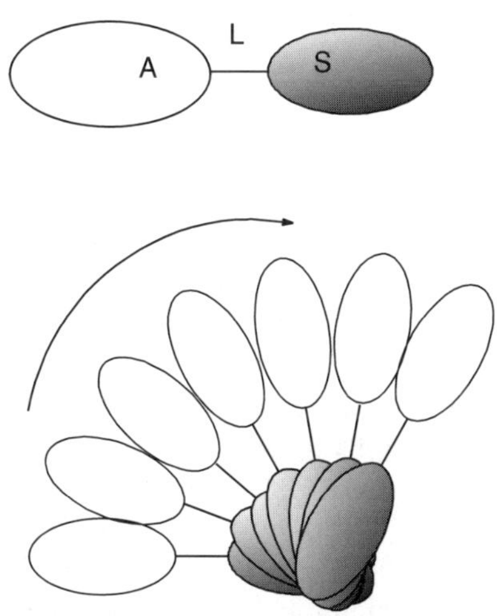

Scheme 20 Unidirectional self-assembly of LMOG molecules driven by cholesterol–cholesterol interactions forms a helical array as self-assembled unit fiber

tether (L), π–π interactions between the aromatic part of stacked molecules within the same unit column can account for the higher stability of the resulting gels, as compared to those obtained with the bent-shaped α-epimer [46].

Moreover, an additional association process, involving intercolumnar aromatic–aromatic interaction, is believed to account for the edge-to-edge association of adjacent unit fibers leading to the formation of larger fiber bundles which can link together to generate the gel network. As a result, the size and shape of the aromatic group A represent a key structural factor toward gelation. Generally, chromophores with a rod-like shape give better results, which is consistent with the aforementioned aggregation model. Those with larger rings (porphyrins) or smaller rings (benzene) showed instead weaker gelation ability, unless their aggregates are stabilized through supramolecular interactions with exogenous guests, such as metal ions, amines, nucleobases, or [60]fullerene.

4 $A(LS)_2$ Organogelators

Recently dimeric cholesterol-based derivatives were reported as a new class of organogelator architecture. The examples published so far display a $A(LS)_2$ structure in which a central unit A is "sandwiched" by two cholesterol moieties. Moreover, all such "gemini"-type cholesterol-based structures reported to date are based on the cholesterol unit having the natural (*S*) configuration at C-3. They are listed below according to the main functional characteristics of the central unit.

4.1 Chromophores

The synthesis and gelation properties of compounds **52–55**, containing aromatic chromophores, were reported (Scheme 21). The central unit A of gelator **52** is based on two azobenzene chromophores that are connected by a central flexible diamino chain [76]. This compound was observed to gelate various organic solvents, such as 1-hexanol, 1-octanol, toluene, *m*-xylene, and *p*-xylene, under 1.0–5.0 wt %. As observed in the case of ALS systems, CD spectroscopy pointed to the chiral orientation of **52** molecules within the gel aggregate and TEM photographs showed the occurrence of helical fibers.

The cholesterol organogelator **53** contains two benzene chromophores linked by a flexible chain combining both amide and ethylene glycol sequences [77]. Interestingly, both hydrogen-bond formation and π–π interaction play a key role in explaining fiber formation. Electronic absorption spectroscopy studies showed the formation of H-aggregates in which the ben-

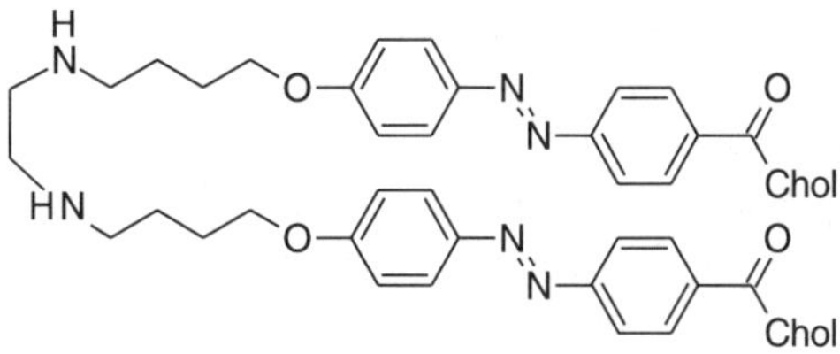

52

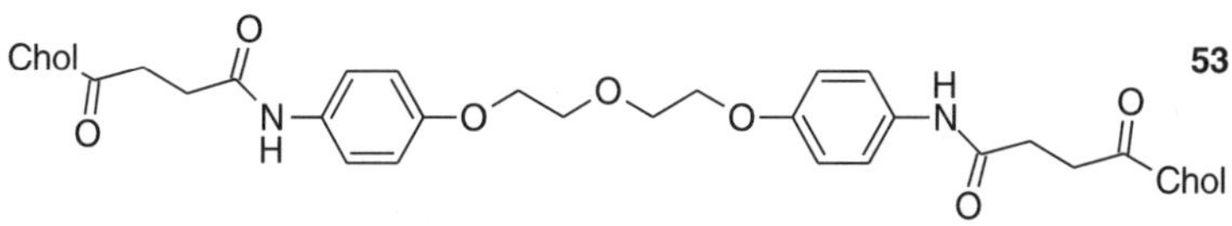

53

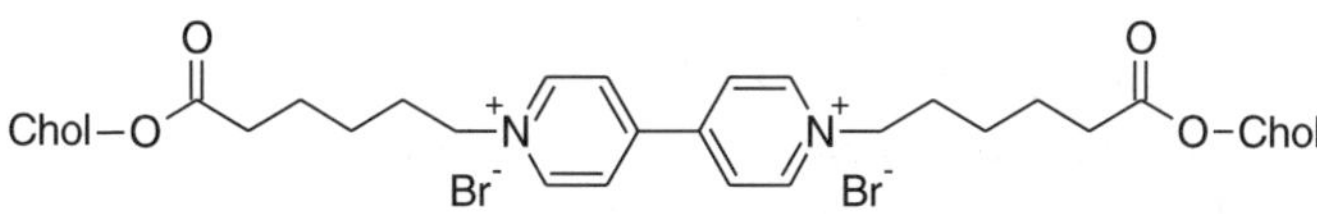

54

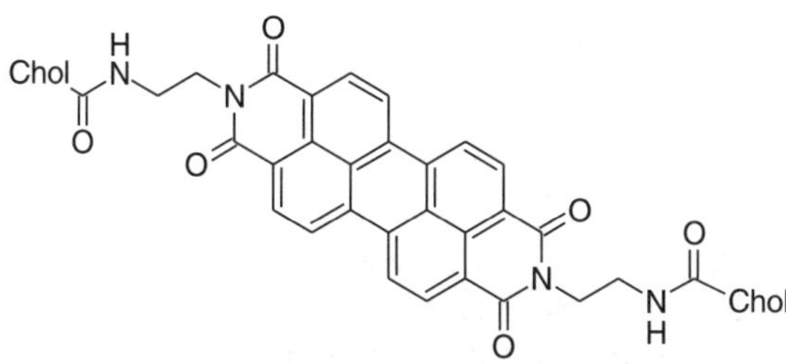

55

Scheme 21 $A(LS)_2$ LMOGs bearing a chromophoric unit

zene units are held in a cofacial fashion. These results together with those obtained from X-ray studies led the authors to propose that **53** could exist in a folded conformation in the gel state, leading to the formation of lamellar, bilayer structure. The preferential folded conformation of ALS **52** was also assumed by Shinkai and coworkers to prevail in the gel [76].

The dimeric $A(LS)_2$ structure was found remarkably effective even in the case where the unit A is a ionic moiety, such as viologen in **54** [78], or a strongly self-associating group, such as the *N*,*N*′-disubstituted 3,4,9,10-perylenetetracarboxylic diimide core in **55** [79]. It was found that **54** forms a stable gel only in 1-butanol at low concentration (below 1 wt/vol %) of which the CD spectrum displayed the features typical of a chiral, clockwise aggregation of the gelator molecules. X-ray diffraction patterns of the xerogel of **54** were similar to those of the neat crystal obtained from ethanol. They indicated the formation of a lamellar organization with an interlayer distance of 4.96 nm. The molecules **54** were proposed to exist in an extended conformation in the gel state and to be tilted relative to the normal of the layer plane [78]. A similar extended conformation was proposed to occur in the gels of **55**. The formation of one-dimensional stacks was driven by the extraordinary strong propensity of perylene derivatives to aggregate into columnar assemblies [79]. The latter property is at the origin of the known weak solubility of perylenetetracarboxylic diimide compounds. Transparent gels of **55**, suitable for spectroscopic experiments, were obtained in binary mixtures of a polar (alcohol, nitrobenzene) and an apolar (toluene, *o*-xylene, *p*-xylene) solvent. For example, the **55** gel formed at 3.8 mM in *p*-xylene/1-propanol (3 : 1 v/v) was thermally stable. CD spectroscopy confirmed the formation of chiral aggregates as due to the cholesterol–cholesterol interaction. Remarkably, confocal laser scanning microscopy could be used to image the fiber assemblies acting as photonic wires. Moreover, a series of analogues of **55**, in which the perylene nucleus is functionalized at the bay region (1 and 7-positions) with bulky groups, did not act as gelators under the same conditions. Nevertheless stable gels could be obtained in mixtures of the 1,7-disubstituted compounds containing up to 50% of **55**. Electronic energy transfer took place efficiently between the mixed chromophores incorporated into the gel fibers.

4.2 Ligands

A series of $A(LS)_2$ compounds in which the unit A is a metal ion chelating site were also described as versatile organogelators (Scheme 22). Two cholesterol units were introduced onto both extremities of benzo- or diaza-crown ether units via an azobenzene molecule, such as in compounds **56-n** and **57**, respectively [80]. Gelator **56-0** and **56-1**, containing a dibenzo-[24]-crown-8 or a dibenzo-[30]-crown-10 central unit, respectively, acted as a good gelator of organic solvents (propionic acid, acetic acid, 1-butanol, DMSO, DMF etc at 1.0 wt %) [49, 81]. The CD spectrum of the gel of **56-0** in acetic acid exhibited a negative sign for the first Cotton effect, which supported the anticlockwise orientation of the azobenzene dipole moments in the gel state, similar to that previously mentioned for the analogous ALS gelator **33** (Scheme 12) [49]. In contrast, the gel of **56-1** exhibited different CD features, which pointed to a peculiar organization of the molecules in the gel structure. To explain, in particular, the opposite sign for the first Cotton effect, the authors suggested that the latter compound could adopt a folded conformation in the gel that would favor efficient intramolecular cholesterol–cholesterol and azobenzene–azobenzene interactions [81]. Very interestingly, the xerogel of **56-1** consisted of both tubular structures with about 520 nm outer diameter, and helical ribbon structures with 1700 nm pitch, as observed by TEM and SEM. The diaza-[18]-crown-6-containing system **57** (Scheme 22) was found to act as a more effective organogelator than its mono-type cholesterol ALS analogue **34-NBn** (Scheme 12). Indeed it gelated a greater range of organic solvents at 5.0 wt %, including long-chain alcohols and amines [51]. The aggregation mode of the two compounds **57** and **34-NBn** was observed to be different, on the basis of electronic absorption and CD measurements. Particularly the azobenzene chromophores in the acetic gel of **57** were found to be oriented into the anticlockwise direction. Strikingly, SEM and TEM investigation of the acetic acid gel of **57** revealed the spherical morphology of the aggregates. Two different spherical structures, with 200 nm and 2500 nm diameter, were observed. The connection of the smaller particles to one another in a pearl necklace fashion resulted in three-dimensional network formation. Such gel structure is unprecedented and in stark contrast with the fibrillar structure usually observed in organogels. The TEM pictures established the multilayer

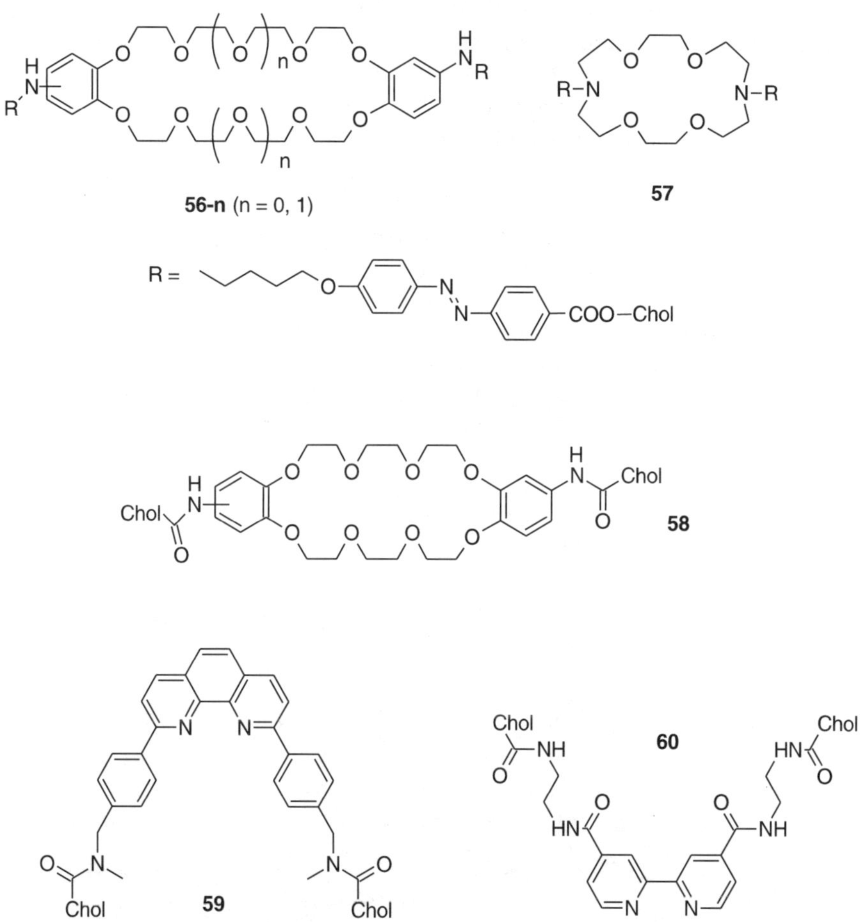

Scheme 22 Ligand-containing $A(LS)_2$ LMOGs

structure of the spherical vesicles. The value of ca. 5 nm measured for the layer thickness is consistent with a folded conformation of **57** [51].

Pseudo-rotaxane complex formation was used to reversibly trigger the conformational preference of the bis-cholesterol derivative **58** (Scheme 22). Compound **58** was observed to gel cycloalkane and diphenyl ether at 19 mM but failed to gelate aromatic solvents under the same conditions [82]. However, addition of a diammonium guest into a benzene sol of **58** induced gelation to occur. This effect was attributed to a change in host conformation. Indeed pseudo-rotaxane formation forced the molecule to adopt the extended geometry, which resulted in a more favorable situation for one-dimensional aggregation and, in turn, gel formation [82].

Dipyridine ligands, such as phenanthroline and 2,2′-bipyridine, were used for the construction of two $A(LS)_2$ organogelators, **59** and **60**, respectively (Scheme 22). Compound **59** is capable of gelating cycloakanes, alcohols, polar aprotic solvents (DMF and DMSO), and organic acids. In most cases the minimal gelation concentration at room temperature is below 1.0 wt %, indicating that compound **59** falls into the category of supergelators [83, 84]. The gels were remarkably stable, a T_{gel} value of 82–88 °C was determined for a 1-propanol gel at a concentration of 0.3 wt %. The fluorescence emission properties of the gels were sensitive to protonation at the phenanthroline nitrogen atoms. The 2,2′-bipyridine ligand **60** represented a unique example in that not only the free ligand but also its copper(I) complex, Cu(I)·**60**$_2$, acted as organogelators. The latter displayed gelation ability toward nitrile solvents at 8.7 mM [85]. As such, the copper(I) complex represents a first example of a new class of cholesterol-based LMOGs in which four cholesterol units are preorganized in a tetrahedral fashion around the cationic metallic center. TEM photographs of gels of **60** and of Cu(I)·**60**$_2$ showed that the network structure is composed of fibers with 13–100 nm and 40–100 nm, respectively.

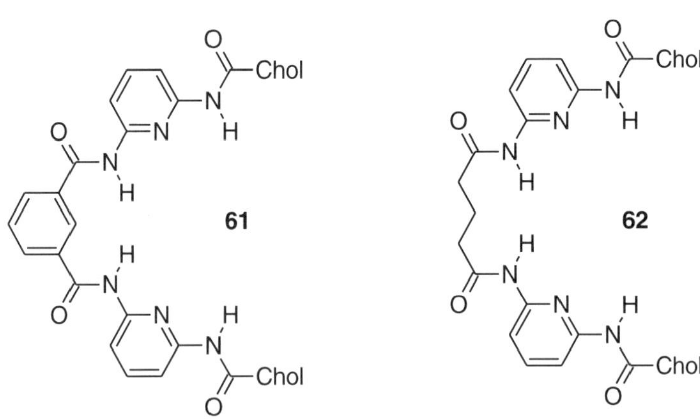

Scheme 23 Hydrogen-bonding receptors that act as $A(LS)_2$ organogelators

For the free ligand, the fibers were observed to have a left-handed helical shape. Remarkably, both color change and sol–gel phase transition could be reversibly controlled by the redox state of the metal in the complex. It is worth mentioning that metal coordination was also useful to control the self-assembling properties of cholic amide–phenanthroline gelating agents [86].

Bis-cholesterol hydrogen-bonding receptors **61** and **62** (Scheme 23) were found to act as gelators in organic solvents. Their gelation ability could be markedly modulated upon appropriate host–guest interaction [87].

4.3 Polymerizable Unit

Diacetylene cholesteryl esters having two urethane linkages **63X-n** formed organogels in cyclohexane and, for some derivatives, in hexane, diethyl ether, DMF, and ethanol (Scheme 24) [88, 89]. Cyclohexane as solvent was far superior as concentrations as low as 0.2–0.5 wt % of the LMOGs were sufficient to produce stable gels (T_{gel} = 70 °C for a cyclohexane gel of $63(CH_2)_3$-**4** at 0.2 wt %). Examination of the gelation properties of the model compounds **64-n** and **65** led to the important conclusion that the main driving force for

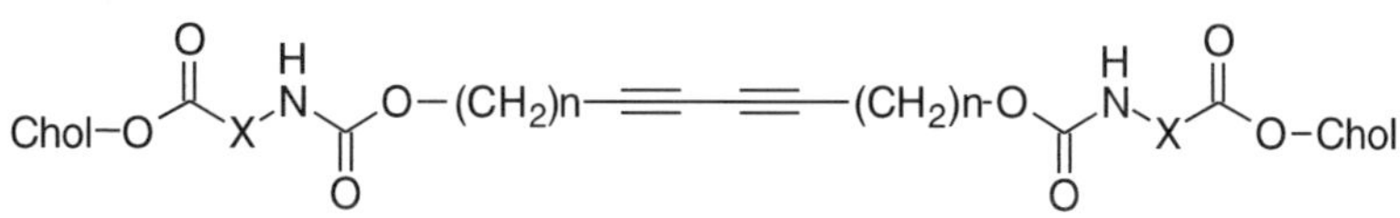

63X-n

X = CH_2, $(CH_2)_2$, $(CH_2)_3$,$CHCH_3$ (L-Ala, D-Ala), CHCH $_2CH(CH_3)_2$ (L-Leu)
n = 1, 2, 3, 4

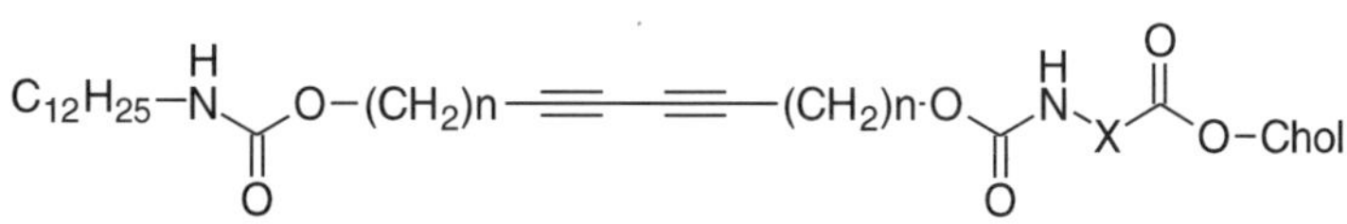

64X-n

X = CH_2, $(CH_2)_2$
n = 3, 4

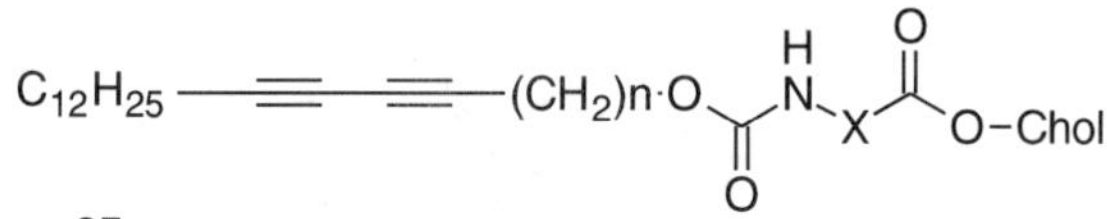

65

Scheme 24 $A(LS)_2$ LMOGs combining hydrogen-bonding and cholesterol units. Due to the presence of the photoreactive diacetylene moiety in the central linker moiety, topochemical photopolymerization in the gel state makes new stable polymeric materials one-dimensional self-assembly and gelation is the hydrogen bonding of the urethane groups.

4.4 $A(LS)_2$ Organogelators: Main Outcomes

$A(LS)_2$ structures do offer new potentialities for the construction of novel cholesterol-based organogelators. Indeed, from the examples listed above, ionic aromatic species or even a coordination complex moiety could be successfully incorporated into gel superstructures, thereby leading to stable organogels with unique functionalities. Clearly, the $A(LS)_2$ gelators have been so far the subject of less investigation than the ALS counterparts. Much work remains to be done in order to unravel the role of the two pendant cholesterol units in gel fiber structure and formation. To date, two different kinds of molecular organization have been postulated to occur within the gel unit aggregates. They involve either a folded or an extended conformation of the $A(LS)_2$ molecule (Scheme 25). In the first case, one-dimensional stacking is believed to result from both intramolecular and intermolecular cholesterol–cholesterol interactions. The aggregation mode is somewhat reminiscent of the spiral staircase arrangement described for ALS systems (Scheme 20) in that it leads to the formation of a central columnar array of cholesterol units surrounded by the folded AL_2 segments organized in a helical arrangement. Of course, this configuration requires the two linker units L to be long and flexible enough. In the second case, one-dimensional stacking of the fully

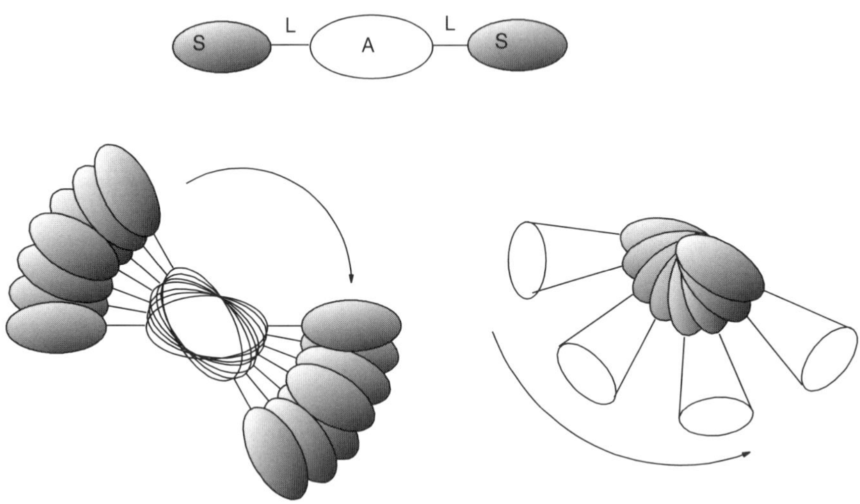

Scheme 25 Possible molecular packing for $A(LS)_2$ organogelators in either the folded (*right*) or extended (*left*) conformation

extended molecules could involve the formation of a twisted central column of stacked aromatic units stabilized at each extremities by cholesterol–cholesterol interactions (Scheme 25). As mentioned above, the influence of the configuration at C-3 has not been examined in the case of $A(LS)_2$ gelators as all the examples known to date have the (*S*,*S*) configuration.

5 Synthetic Aspects

Functionalization at the C-3 hydroxy group of the commercially available 3β-cholesterol and 3β-cholestanol enables the ready preparation of a variety of cholesteryl derivatives with the natural configuration. In conventional esterification procedures, either the acid chlorides or the carboxylic acids in the presence of DCC and DMAP react with the hydroxy function at C-3 to afford the desired compound in rather good yields. In such cases, the configuration at C-3 is retained as the naturally occurring configuration. In order to obtain the corresponding α-epimers, a condensation method was found particularly effective to induce inversion of C-3 [46] when carboxylic acids and cholesterol (cholestanol) are treated with triphenylphosphine and DEAD as a condensation reagent under Mitsunobu conditions. 3β-Cholesteryl chloroformiate is also commercially available and represents a versatile precursor for the straightforward synthesis of organogelators containing a urethane group in the linker part. Indeed reaction of the chloroformiate derivative with an amino derivative leads in one step and high yield to the corresponding ALS or $A(LS)_2$ compounds.

6 Applications

The applications of cholesterol-based gelators encompass a wide range of scientific and technological areas. Steroid-based ALS structures especially hold great promise for the generation of new functional materials [6]. A remarkable consequence of the required presence of an aromatic unit in the molecular structure is that the gel fibers consist of arrays of well-organized π-conjugated units, which thereby endows the material with optical and/or electronic properties [4]. Clearly the concepts of supramolecular chemistry as applied to gels offer elegant opportunities to create supramolecular photonics and electronics [90]. ALS that contain anthracene or perylene derivatives as chromophoric units, such as **7** [30] and **55** [79], respectively, produce highly luminescent gels that have great potential for the design of photonic devices, such as light-emitting diodes, fluorescent sensors, or artificial photosynthetic systems with light-harvesting properties. The use of molecular units with solid-state semiconducting units was reported in the case of a series of hydrogen-bonded gelators combining oligothiophene oligomers and urea functionalities [91]. Although there is no example at the time of this writing of such systems based on cholesterol, ALS gelators properly substituted with oxidizable/reducible groups might represent valuable attractive candidates for the generation of organic charge-transport devices [92]. In that connection, one-dimensional assembly of porphyrin-containing ALS represented a first approach towards the generation of photonic wires transporting charges and/or photons for electronic and photovoltaic applications [57–59, 93]. The [60]fullerene-porphyrin interaction has been exploited by Shinkai et al. to reinforce the organogel structure of porphyrin-appended cholesterol derivatives **38-n**. A [60]fullerene-containing cholesterol derivative (**39**) has also been reported to gelate dichloromethane [61]. Given the tremendous potential of [60]fullerene dyads for photovoltaic applications and new electronic devices [94], these examples might open the way to future developments. Careful design of organogelator molecules containing donor or acceptor groups may also provide a useful approach for the control of donor and acceptor units organization within gel fibers, which could be amenable to the generation of charge transfer interactions for electrical conductivity. In that respect ALS containing anthraquinone as an electron acceptor has been shown to form thixotropic gels in the presence of 20% of a non-gelator ALS containing the dimethoxyanthracene ring as donor. Charge-transfer interactions were indeed observed to occur in the gel state [4].

Another class of important photofunctional gels are those based on a photoisomerizable group, which may lead to the conception of light-driven devices that release or uptake ionic or molecular species under irradiation [95, 96]. Photoisomerization in the stilbene (**27**) or azobenzene (**29Me**) chromophore was shown to actively trigger the sol–gel phase transition [39, 44]. Recently, a gel-to-sol transition with spatial control was achieved with the used of a stilbene photosurfactant [97]. Host–guest interactions have also been shown to control, to a large extent, the stability and morphology of the gel superstructure [44, 50–52, 60, 63, 66, 69, 72, 82]. Such gels sensitive to external stimuli represent promising examples of transducers that convert a host–guest recognition event into a readable output [98].

Polymerization of self-assembled LMOG molecules within the fibers of a molecular gel locks the scaffold-like structures and can make new stable materials [99, 100]. Photopolymerization in the gels of diacetylenic cholesteryl derivatives, such as **63X-n** [88, 89], has been successfully employed to access organic nanowires with potential electrical conductivity, which could be used in molecular electronics. This strategy was based on the known propensity of diacetylenes to undergo topochemical solid-state 1,4-addition reactions leading to conjugated polymers [101].

The use of organogels as templates for the generation of inorganic structures and materials has received considerable attention [102]. Three-dimensional networks permit inorganic materials to be arranged on the microscopic level, leading to otherwise unattainable structures. Inorganic nanostructures are obtained by coating the fibers with the inorganic precursor followed by polycondensation or crystallization, and then calcination to remove the organic template. Remarkably, the process yields hollow fibers, whose morphology and chirality are directly related to that of the templating supramolecular assemblies. In that connection, cholesterol-based organogelators offer themselves as very versatile systems [102]. Indeed, their gels were found to be stable under the conditions required for the polycondensation of TEOS. For example, silica fibers could be generated with a right-handed helicity from the transcription of an acetic acid gel of a mixture of **35** and (*S*)-**29Bu** in the presence of TEOS and water [55]. In a similar way, a variety of silica nanostructures could be obtained, depending on the structure of the ALS employed in the transcription process [102]. Interestingly, transcription of the organogel of compound **34O** in the presence of $AgNO_3$ resulted in the incorporation of nanosized metal particles into silica [102]. Such metal-containing silica could present interesting catalytic properties. In ALS **42**, the benzyl amine moiety served as an immobilized catalyst for TEOS polycondensation reactions, which avoided the use of an additional catalyst [62].

The transcription of organogel superstructures into semiconducting inorganics holds promise for the development of the next generation of nanostructured materials in electronics [103]. In that connection, the organogel fibers of the $A(LS)_2$ derivative **53** served as nucleation sites for the mineralization of CdS, which led to the formation of CdS nanofibers [77].

7 Conclusion

This chapter illustrates the variety of structures of cholesterol-based LMOGs that have been reported over the last 15 years. Clearly the chemistry of such systems has progressed rapidly from unexpected discovery to rational synthesis. Cholesterol-based organogelators present many exciting attributes in terms of easy synthetic accessibility and tunability, chirality, and gelation ability. In particular, they display versatile solubility properties, which enable the gelation of a variety of solvents. They frequently exhibit low critical gelation concentration and the resulting gels are usually thermally stable. In the rational synthesis of novel organogelators, ALS architectures represent quite reliable targets. Recently $A(LS)_2$ systems appeared as promising alternative structures for the generation of three-dimensional fiber networks with enhanced gel properties. In the light of these recent advances, one can foresee that more complex, three-dimensional molecular structures, incorporating several cholesteryl units held together in a well-defined arrangement, could be amenable to unique self-assembly and gelation properties. The report on the gelation ability of a copper(I) complex (Cu(I)·$\mathbf{60}_2$), bearing four cholesterol groups oriented in a tetrahedral fashion, represents a first breakthrough. Cholesterol-based organogelators have been largely exploited to create self-assembled, nanostructured materials with fascinating functional features. It is clear that the properties of these soft materials can form the basis of far-reaching applications.

Acknowledgements Frédéric Fages thanks the Université de la Méditerranée, the CNRS, and La Région PACA for their valuable help. Fritz Vögtle is grateful for initial support by Deutsche Forschungsgemeinschaft. COST Chemistry is gratefully acknowledged (D-11 Supramolecular Chemistry Action, Project D11/0015/99).

References

1. Brenzinger K (1892) Z Physiol Chem 16:552
2. Fiero GW (1940) J Am Pharm Assoc 29:502
3. Terech P, Weiss RG (1997) Chem Rev 97:3133
4. Abdallah DJ, Weiss RG (2000) Adv Mater 12:1237
5. Estroff LA, Hamilton AD (2004) Chem Rev 104:1201
6. van Esch J, Feringa BL (2000) Angew Chem Int Ed 39:2263
7. Aggeli A, Bell M, Boden N, Keen JN, Knowles PF, McLeish TCB, Pitkeathly M, Radford SE (1997) Nature 386:259
8. Kiyonaka S, Shinkai S, Hamachi I (2003) Chem Eur J 9:976
9. Gronwald O, Shinkai S (2001) Chem Eur J 7:4329
10. Virtanen E, Kolehmainen E (2004) Eur J Org Chem 3385
11. James TD, Kawabata H, Ludwig R, Murata K, Shinkai S (1995) Tetrahedron 51:555
12. Brown MS, Goldstein JL (1985) Nobel Lecture. http://nobelprize.org. Cited 2005
13. Reinitzer F (1888) Monatsch Chem 9:421
14. Wallimann P, Marti T, Fürer A, Diederich F (1997) Chem Rev 97:1567
15. Chen SH, Mastrangelo JC, Shi H, Bashir-Haschemi A, Li J, Gelber N (1995) Macromolecules 28:7775
16. Klok HA, Hwang JJ, Iyer SN, Stupp SI (2002) Macromolecules 35:746
17. Weiss RG (1988) Tetrahedron 44:3413
18. Bujanowski VJ, Katsoulis DE, Ziemelis MJ (1994) J Mater Chem 8:1181
19. Lu L, Cocker M, Bachman RE, Weiss RG (2000) Langmuir 16:20
20. Lu L, Weiss RG (1996) Chem Commun 2029
21. Lu L, Weiss RG (1995) Langmuir 11:3630
22. Martin-Borret O, Ramasseul R, Rassat A (1979) Bull Soc Chim Fr 7–8:II-401
23. Terech P, Ramasseul R, Volino F (1985) J Phys France 46:895
24. Terech P, Ramasseul R, Volino F (1983) J Colloid Interface Sci 91:280
25. Terech P, Berthet C (1988) J Phys Chem 92:4269
26. Babu P, Sangeetha NM, Vijaykumar P, Maitra U, Rissanen K, Raju AR (2003) Chem Eur J 9:1922

27. Willemen HM, Vermonden T, Marcelis ATM, Sudhölter EJR (2001) Eur J Org Chem 2329
28. Jover A, Meijide F, Rodríguez Núñez E, Vázquez Tato J, Mosquera M, Rodríguez Prieto F (1996) Langmuir 12:1789, and references therein
29. Liu XY, Sawant PD (2002) Angew Chem Int Ed 41:3641
30. Lin Y, Weiss RG (1987) Macromolecules 20:414
31. Lin Y, Kachar B, Weiss RG (1989) J Am Chem Soc 111:5542
32. Furman I, Weiss RG (1993) Langmuir 9:2084.
33. Terech P, Furman I, Weiss RG (1995) J Phys Chem 99:9558
34. Lin Y, Weiss RG (1989) Liq Cryst 4:367
35. Terech P, Furman I, Weiss RG, Bouas-Laurent H, Desvergne JP, Ramasseul R (1995) Faraday Discuss 101:345
36. Mukkamala R, Weiss RG (1996) Langmuir 12:1474
37. Ostuni E, Kamaras P, Weiss RG (1996) Angew Chem Int Ed Engl 35:1324
38. Terech P, Ostuni E, Weiss RG (1996) J Phys Chem 100:3759
39. Geiger C, Stanescu M, Chen L, Whitten DG (1999) Langmuir 15:2241
40. Song X, Geiger C, Farahat M, Perlstein J, Whitten DG (1997) J Am Chem Soc 119:12481
41. Chen H, Law KY, Perlstein J, Whitten DG (1995) J Am Chem Soc 117:7257
42. Duncan DC, Whitten DG (2000) Langmuir 16:6445
43. Wang R, Geiger C, Chen L, Swanson B, Whitten DG (2000) J Am Chem Soc 122:2399
44. Murata K, Aoki M, Nishi T, Ikeda A, Shinkai S (1991) J Chem Soc Chem Commun 1715
45. Murata K, Aoki M, Shinkai S (1992) Chem Lett 21:739
46. Murata K, Aoki M, Suzuki T, Harada T, Kawabata H, Komori T, Ohseto F, Ueda K, Shinkai S (1994) J Am Chem Soc 116:6664
47. Shinkai S, Murata K (1998) J Mater Chem 8:485
48. Kawabata H, Murata K, Harada T, Shinkai S (1995) Langmuir 11:623
49. Jung JH, Lee SH, Yoo JS, Yoshida K, Shimizu T, Shinkai S (2003) Chem Eur J 9:5307
50. Jung JH, Ono Y, Shinkai S (2000) Angew Chem Int Ed 39:1862
51. Jung JH, Ono Y, Sakurai K, Sano M, Shinkai S (2000) J Am Chem Soc 122:8648
52. Jung JH, Ono Y, Shinkai S (1999) Tetrahedron Lett 40:8395
53. Jung JH, Ono Y, Shinkai S (2000) Langmuir 16:1643
54. Ono Y, Nakashima K, Sano M, Kanekiyo Y, Inoue K, Hojo J, Shinkai S (1998) Chem Commun 1477
55. Ono Y, Nakashima K, Sano M, Hojo J, Shinkai S (2001) J Mater Chem 11:2412
56. Jung JH, Shinkai S (2000) J Chem Soc, Perkin Trans 2:2393
57. Tian HJ, Inoue K, Yoza K, Ishi-i T, Shinkai S (1998) Chem Lett 27:871
58. Ishi-i T, Jung JH, Shinkai S (2000) J Mater Chem 10:2238
59. Ishi-i T, Iguchi R, Snip E, Ikeda M, Shinkai S (2001) Langmuir 17:5825
60. Shoji Y, Tashiro K, Aida T (2004) J Am Chem Soc 126:6570
61. Ishi-i T, Ono Y, Shinkai S (2000) Chem Lett 29:808
62. van Bommel KJC, Shinkai S (2002) Langmuir 18:4544
63. Kawano S, Fujita N, van Bommel KJC, Shinkai S (2003) Chem Lett 32:12
64. Jeong SW, Murata K, Shinkai S (1996) Supramolecular Sci 3:83
65. Jeong SW, Shinkai S (1997) Nanotechnology 8:179
66. Mathias JP, Simanek EE, Whitesides GM (1994) J Am Chem Soc 116:4326
67. Snip E, Shinkai S, Reinhoudt DN (2001) Tetrahedron Lett 42:2153
68. Snip E, Koumoto K, Shinkai S (2002) Tetrahedron 58:8863
69. Numata M, Shinkai S (2003) Chem Lett 32:308

70. Amaike M, Kobayashi H, Shinkai S (2000) Bull Chem Soc Jpn 73:2553
71. Ikeda A, Sonoda K, Ayabe M, Tamaru S, Nakashima T, Kimizuka N, Shinkai S (2001) Chem Lett 30:1154
72. KobayashiH, Amaike M, Jung JH, Friggeri A, Shinkai S, Reinhoudt DN (2001) Chem Commun 1038
73. Urata K, Takaishi N (2001) Eur J Lipid Sci Technol 103:29
74. Brotin T, Utermöhlen R, Fages F, Bouas-Laurent H, Desvergne JP (1991) J Chem Soc Chem Commun 416
75. Pozzo JL, Desvergne JP, Clavier GM, Bouas-Laurent H, Jones PG, Perlstein J (2001) J Chem Soc Perkin Trans 2:824
76. Jung JH, Shinkai S, Shimizu T (2003) Chem Mater 15:2141
77. Xue P, Lu R, Huang Y, Jin M, Tan C, Bao C, Wang Z, Zhao Y (2004) Langmuir 20:6470
78. Xue P, Lu R, Li D, Jin M, Bao C, Zhao C, Wang Z (2004) Chem Mater 16:3702
79. Sugiyasu K, Fujita N, Shinkai S (2004) Angew Chem Int Ed 43:1229
80. Jung JH, Shinkai S (2001) J Inclus Phenom Macro 41:53
81. Jung JH, Kobayashi H, Masuda M, Shimizu T, Shinkai S (2001) J Am Chem Soc 123:8785
82. Kawano S, Fujita N, Shinkai S (2003) Chem Commun 1352
83. Sugiyasu, K, Fujita N, Takeuchi M, Yamada S, Shinkai S (2003) Org Biomol Chem 1:895
84. Jung JH, Nakashima K, Shinkai S (2001) Nano Lett 1:145
85. Kawano S, Fujita N, Shinkai S (2004) J Am Chem Soc 126:8592
86. Dukh M, Saman D, Kroulik J, Cerny I, Pouzar V, Kral V, Drasar P (2003) Tetrahedron 59:4069
87. Inoue K, Ono Y, Kanekiyo Y, Ishi-I T, Yoshihara K, Shinkai S (1999) J Org Chem 64:2933
88. Nagasawa, J, Kudo M, Hayashi S, Tamaoki N (2004) Langmuir 20:7907
89. Tamaoki N, Shimada S, Okada Y, Belaissaoui A, Kruk G, Yase K, Matsuda H (2000) Langmuir 16:7545
90. Ayaghosh A, George SJ (2001) J Am Chem Soc 123:5148
91. Schoonbeek FS, van Esch JH, Wegewijs B, Rep DBA, de Haas MP, Klapwijk TM, Kellog RM, Feringa BL (1999) Angew Chem Int Ed 38:1393
92. Yasuda Y, Takebe Y, Fukumoto M, Inada H, Shirota Y (1996) Adv Mater 8:740
93. van Norstrum CF, Picken SJ, Schouten AJ, Nolte RJM (1995) J Am Chem Soc 117:9957
94. Sariciftici NS, Smilowitz L, Heeger AJ, Wudl F (1992) Science 258:1474
95. Ahmed SA, Sallenave X, Fages F, Mieden-Gundert G, Müller WM, Müller U, Vögtle F, Pozzo JL (2002) Langmuir 18:7096
96. de Jong JJD, Lucas LN, Kellog RM, van Esch JH, Feringa BL (2004) Science 304:278
97. Eastoe J, Sanchez-Dominguez M, Wyatt P, Heenan RK (2004) Chem Commun 2608
98. Sohna Sohna JE, Fages F (1997) Chem Commun 327
99. de Loos M, van Esch J, Stokroos I, Kellog RM, Feringa BL (1997) J Am Chem Soc 119:12675
100. Wang C, Hamilton AD (2002) Chem Eur J 8:1954
101. George M, Weiss RG (2003) Chem Mater 15:2879
102. van Bommel KJC, Friggeri A, Shinkai S (2003) Angew Chem Int Ed 42:980
103. Sone ED, Zubarev ER, Stupp SI (2002) Angew Chem Int Ed 41:1706

Top Curr Chem (2005) 256: 77–131
DOI 10.1007/b107172

Published online: 1 July 2005

Systematic Design of Amide- and Urea-Type Gelators with Tailored Properties

Frédéric Fages[1] · Fritz Vögtle[2] · Mladen Žinić[3] (✉)

[1]Faculté des Sciences de Luminy, Université de la Méditerranée, Case 901, 13288 Marseille Cedex 9, France

[2]Kekulé-Institute für Organische Chemie und Biochemie, Universität Bonn, Gehard-Domagk-Str. 1, 53121 Bonn, Germany

[3]Rudjer Boskovic Institute, P.O.B. 180, 10002 Zagreb, Croatia
zinic@irb.hr

Abstract The formation of gels by structurally highly diverse low molecular weight organic molecules is paradigmatically a supramolecular phenomenon. It is based on the self-assembly of certain organic molecules and involves highly specific noncovalent intermolecular interactions, in particular those inducing predominantly unidirectional aggregation. In this chapter, the design of low molecular weight gelators that incorporate single or multiple amide units as intermolecular hydrogen-bonding functionalities and methods of their preparation are given. Many efficient gelators of organic solvents and water could be prepared by the structural combination of amidic, carbamate, urea, or oxalamide groups and long aliphatic chains or aromatic groups with a large surface. The numerous potential applications in slow drug-delivery systems, the fabrication of templated materials, and in sensing devices are also discussed.

Keywords Hydrogen bonding · Self-assembly · Supramolecular chemistry · Amide gelators · Urea gelators

Abbreviations

AFM	Atomic force microscopy
BuOH	Butanol
Boc	Butoxycarbonyl
CD	Circular dichroism
DCC	*N*,*N*-Dicyclohexylcarbodiimide
DMAP	4-(Dimethylamino)pyridine
DMF	Dimethylformamide
DMSO	Dimethyl sulfoxide
EtOH	Ethanol
FTIR	Fourier-transform infrared spectroscopy
MeOH	Methanol
mgc	Minimal gelation concentration
NMR	Nuclear magnetic resonance
SEM	Scanning electron microscopy
T_{gel}	Gel melting temperature
TEM	Transmission electron microscopy
TFA	Trifluoroacetic acid
THF	Tetrahydrofuran
UV	Ultraviolet
vol	Volume
wt	Weight

1 Introduction

Amide hydrogen bonds are among the most important organizational elements that stabilize natural and synthetic supramolecular systems [1, 2]. While formation of amide–amide hydrogen bonds in water is largely entropy driven [3, 4], in lipophilic solvents it is enthalpy driven amounting to a negative free energy change ΔG of -28 ± 4 kJ mol^{-1}. Hence, the cooperative intermolecular amide–amide hydrogen bonding is favored in both hydrophilic and hydrophobic environments. Formation of gels by structurally highly diverse low molecular weight organic molecules is paradigmatically a supramolecular phenomenon. It is based on the self-assembly of certain organic molecules and involves highly specific noncovalent intermolecular interactions, in particular those inducing predominantly unidirectional aggregation. Such aggregation results in the formation of micrometer-long fibrous assemblies, which entangle into a three-dimensional gel network that entraps the solvent and results in loss of fluidity. Hence, the design of low molecular weight gelators which incorporate single or multiple amide units as intermolecular hydrogen-bonding and directional functionalities for self-assembly presents a logical and in many cases fruitful approach. In this chapter, an outline of various gelator structures and methods of their preparation is given. The review is not intended to comprehensively present the work published in the last 10 years. Its primary focus is to give an outlook of the diversity of molecular structures which were found capable of producing gels with various organic solvents and water. The common characteristic of the gelator structures covered is that such molecules incorporate amide, carbamate, urea, and oxalamide units capable of forming strong intermolecular hydrogen bonds which are either the major or auxiliary supramolecular bonds that stabilize gel aggregates. Several examples are included showing how the stepwise structural modifications could be used to tailor the properties of gels or to endow gels with advanced properties useful for various applications.

2
Amide Gelators

2.1
Monoamides and -carbamates

Perhaps structurally the most simple primary amide gelators, 3,4,5-tris-(alkoxy) benzamides **1** and **2** (Fig. 1), have been prepared by amidation of tris(alkoxy)benzoic acids and shown to gel both polar (MeOH, EtOH, DMF) and highly lipophilic organic solvents (*n*-octane, *n*-decane, toluene) at minimal gelation concentrations (mgc) lower than 2.5 wt % [5]. The gel aggregates can be embedded into cross-linked polymer matrices using monomer/cross-linker mixtures as organic solvents.

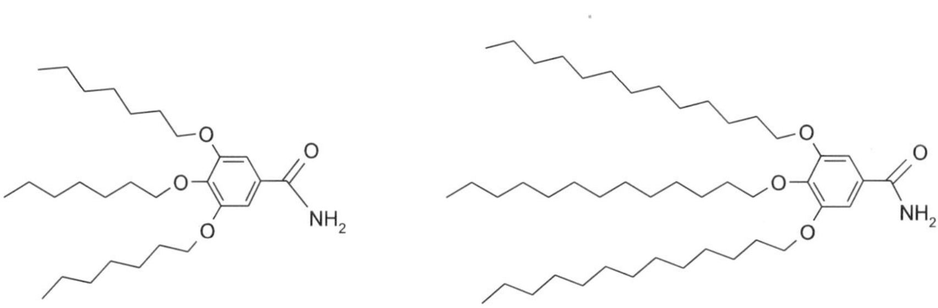

1 **2**

Fig. 1 Tris(alkoxy)benzamide gelators **1** and **2**

A series of *N*-alkyl perfluoroalkanamides were recently prepared by the Weiss group [6]. Perfluorinated alkanoic acids were transformed into acid chlorides by oxalyl chloride in benzene, and subsequent amidation with alkylamine in benzene gave *N*-alkyl perfluoroalkanamides (Scheme 1). The amide derivative **3** exhibited efficient gelation (mgc < 2.5 wt %) of various organic solvents including EtOH, 1-butanol, 1-octanol, hexane, *n*-octane, benzyl alcohol, toluene, and CCl_4.

Formation of gels by this type of compound is a consequence of the incompatibility of perfluoroalkyl and alkyl chains, which leads to the formation of lamellar aggregates additionally stabilized by intermolecular amide–amide hydrogen bonding.

12-Hydroxystearic acid (12-HSA) exhibits improved gelation properties toward various organic solvents as compared to analogous molecules lacking a hydrogen-bonding group in the hydrocarbon chain. Inspired by this observation, *N*-acyl-1,*ω*-amino acid derivatives were prepared incorporating a strongly hydrogen-bonding amide group at the terminus of the alkyl chain [7]. Within the series of prepared derivatives (Scheme 2) the carbamate **4** and amide derivatives **5**–**9** of the 11-aminoundecanoic acid (AUDA) in the form of sodium salts showed excellent gelation of polar aprotic solvents

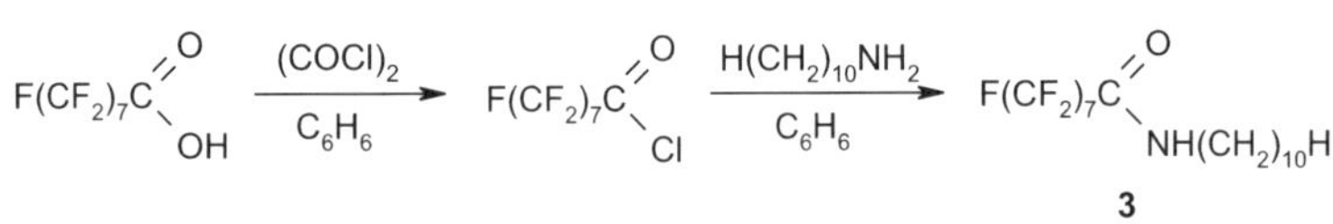

Scheme 1 Preparation of *N*-decanoyl perfluoroheptanamide **3**

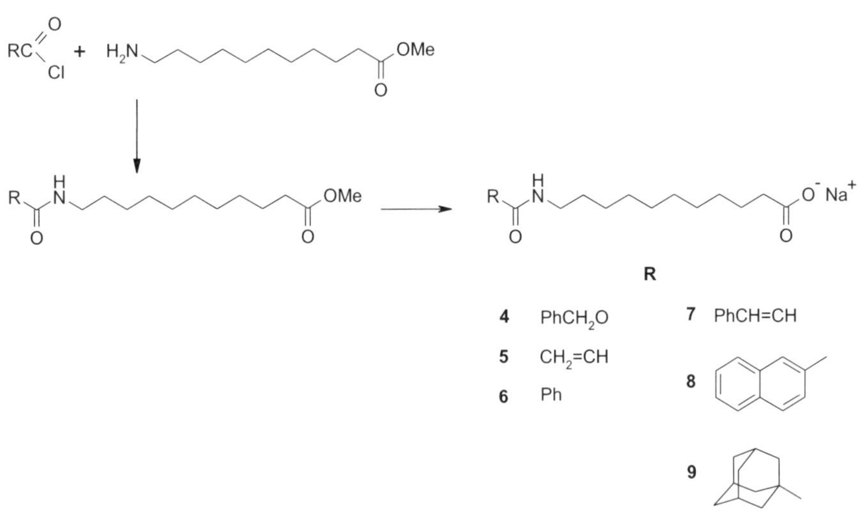

Scheme 2 Synthesis of *N*-acyl-1,*ω*-amino acid sodium salt gelators of DMF

such as DMF, DMSO, propylene carbonate, and *N*, *N*-dimethylacrylamide at concentrations of 0.4–1.0 g L^{-1}.

In the next step, AUDA-based gelators were modified by incorporation of a core peptidic unit and terminal lauric acid amide [8]. The combination of these structural fragments was expected to provide favorable properties for gelation: the presence of two core amidic functions is expected to induce self-assembly by unidirectional cooperative hydrogen bonding, and the presence of the chiral center and long aliphatic chain of the lauroyl amide should decrease the crystallization tendency of such derivatives. A series of chiral and racemic derivatives were prepared (Scheme 3) and tested for gelation as free acids and sodium salts. The synthesis starts by *N*-protection of selected amino acids using *tert*-butyloxycarbonyl anhydride and *N*,*N*-dicyclohexylcarbodiimide (DCC) condensation with the methyl ester of AUDA.

Deprotection of terminal amino acid by TFA, condensation with lauroyl chloride, and methyl ester hydrolysis under acidic or basic conditions followed by reacidification gives the free acid derivatives **10**. The sodium salts **11** are obtained from **10** using sodium hydroxide in CH_2Cl_2/MeOH mixture.

The optically active free acid derivatives and their sodium salts showed ambidextrous gelation properties, being capable of gelling not only water and

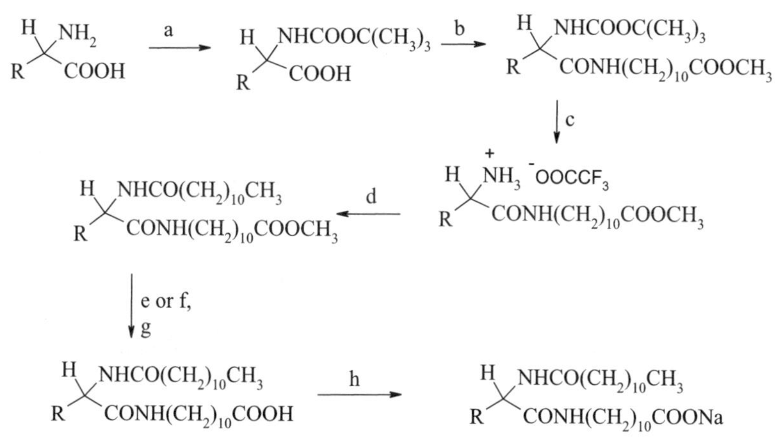

10 **11**

Leu (*S*)- and *rac*-; R= $CH_2CH(CH_3)_2$
Val (*S*)- and *rac*-; R = $CH(CH_3)_2$
PhG (*R*)- and *rac*-; R = Ph
Phe (*S*)- and *rac*-; R = CH_2Ph

Scheme 3 Preparation of free acid (**10**) and sodium salt (**11**) gelators. Reagents and conditions: (a) $(BOC)_2O$, NaOH/dioxane-water; (b) $H_2N(CH_2)_{10}COOCH_3$, DCC, Et_3N, DMAP/CH_2Cl_2; (c) TFA/CH_2Cl_2; (d) $ClCO(CH_2)_{10}CH_3$, Et_3N/CH_2Cl_2; (e) LiOH/CH_2Cl_2-MeOH; (f) 10% KOH/MeOH; (g) HCl; (h) NaOH/CH_2Cl_2-MeOH highly polar DMSO and DMF but also lipophilic aromatic solvents (toluene, *o*-xylene, tetralin) and decalin at 10^{-2}–10^{-3} mol dm^{-3} concentrations. The racemates in the form of free acid or sodium salts are generally less versatile gelators compared to the corresponding optically active forms. Striking differences, however, were observed for Na salts of *rac*-Leu-**11**, *rac*-PhG-**11** (PhG denotes phenylglycine), and *rac*-Phe-**11**, which showed excellent gelation of tetralin and decalin while their optically active isomers lack any gelation [(*S*)-Leu-**10**] or exhibit much weaker gelation [(*R*)-PhG-**10** and (*S*)-Phe-**10**]. Surprisingly, the racemic *rac*-PhG-**11** is capable of gelling a 16 times larger volume of *p*-xylene than (*R*)-PhG-**10**, and the racemic *rac*-PhG-**11**, and *rac*-Phe-**11** are able to gel much larger volumes of decalin than pure enantiomers (*R*)-PhG-**11** and (*S*)-Phe-**11**. This observation contrasts with several earlier reports which showed that pure enantiomers are generally more efficient gelators than the corresponding racemates, which tend to crystallize [9–14].

Further designs based on AUDA lead to the preparation of photochromic organogelators [15]. The covalent attachment of a photostimulable ^{2}H-chromene unit at the amino terminus of AUDA giving a urethane (**13**) or amide derivative (**15**) was accomplished by using an appropriate isocyanate or by DCC condensation of AUDA methyl ester and 3,3-diphenylnaphthopyran-8-carboxylic acid **14** (Scheme 4).

Both **13** and **15** as their sodium salts exhibited efficient gelation of DMF and DMSO. Upon UV irradiation using a 366-nm lamp, the gels of **13** and **15** became colored and began to flow due to photoinduced ring opening of the photochromic unit. After irradiation ceased, the coloration disappeared to give a colorless viscous solution which upon heating reverted to gel as the consequence of thermal ring closure into ^{2}H-chromene (Fig. 2).

A new class of pyrene-based urethane, urea, and amide gelators has been prepared (Scheme 5) [16, 17]. Tests revealed gelation of various organic solvents of medium and low polarity (*n*-BuOH, *t*-BuOH, *n*-octanol, cyclohexanol, cyclohexane, *n*-hexane, *n*-decane, *n*-dodecane, CCl_4, toluene). Variable-temperature FTIR and electronic absorption spectra show that the molecules aggregate by intermolecular hydrogen bonding and π-stacking interactions. Introduction of a chiral center (**18**) results in formation of helical aggregates

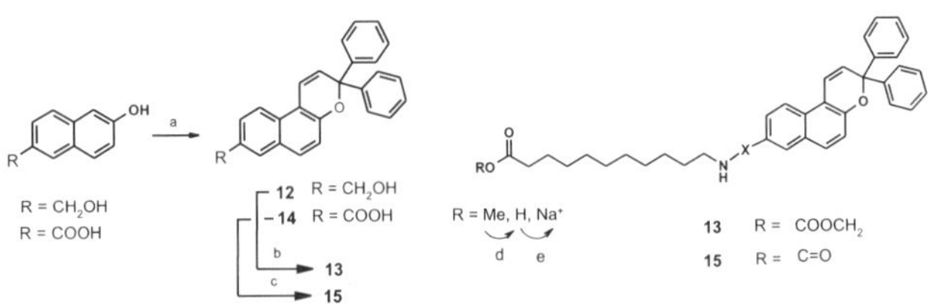

Scheme 4 Synthesis of photochromic gelators based on AUDA. Reagents and conditions: (a) 1,1-diphenylpropyn-1-ol; (b) $O=C=N-(CH_2)_{10}COOMe$, TEA; (c) HCl, $NH_2(CH_2)_{10}$-COOMe, DCC, DMAP; (d) LiOH then HCl; (e) 1 N NaOH

in cyclohexane gel, and in the CD spectra opposite Cotton effects are observed for enantiomeric **18**.

In the majority of cases of simple amide-derived gelators, an amide group is introduced as organizational element providing strong and directional intermolecular hydrogen bonding. However, the next example describes the case where the intramolecular amide hydrogen bonding induces the aggregation responsible for gelation [18]. The 5-cyano-2-(3,4,5-trioctyloxybenzoylamino)tropone (**19**, Fig. 3) prepared by acylation of 2-amino-5-cyanotropone with 3,4,5-trioctylalkoxybenzoyl chloride exhibited gelation of highly polar (DMSO, DMF, MeOH) and apolar solvents (hexane, cyclohexane, decane) at mgc in the range of 4–25 g L^{-1}. The results of ^{1}H NMR, X-ray diffraction, and X-ray crystallographic analysis consistently points toward hexagonal packing in gel aggregates. It is shown that the intramolecular NH-2-tropone carbonyl oxygen flattens the molecule, inducing efficient π–π stacking in columnar organization and hence formation of a stable hexagonal columnar phase which gives fibrous aggregates and leads to gelation (Fig. 3).

2.2 Di-, Tri- and Polyamides

2.2.1 Diamide Gelators

trans-1,2-Diaminocyclohexane served as chiral unit for construction of various bis-alkylamide gelators. Hanabusa and coworkers have prepared both

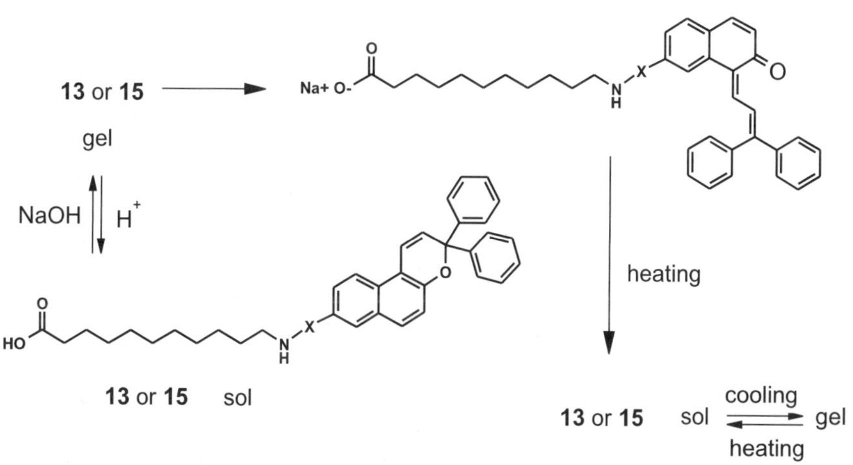

Fig. 2 Photo- and pH-controlled gel-to-sol transition of photochromic AUDA gelators **13** and **15**

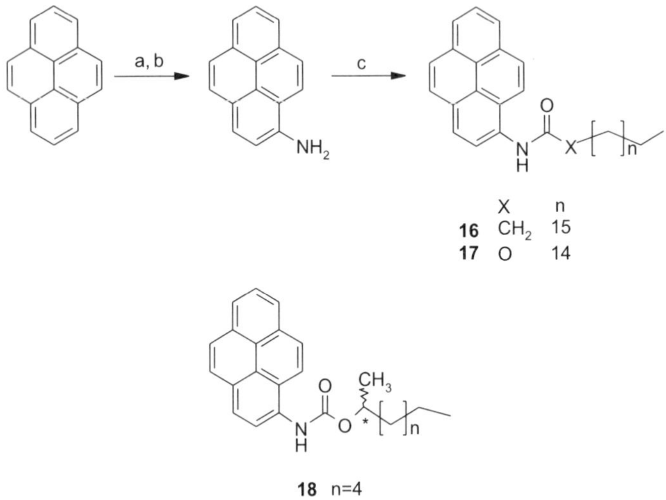

Scheme 5 Pyrene-derived urethane and amide gelators: (a) $Cu(NO_3)_2$, $CHCl_3$; (b) H_2, Pd/C; (c) **16**: 1-octadecanoyl chloride/pyridine in toluene and **17**: hexadecyl chloroformate/pyridine in $CHCl_3$

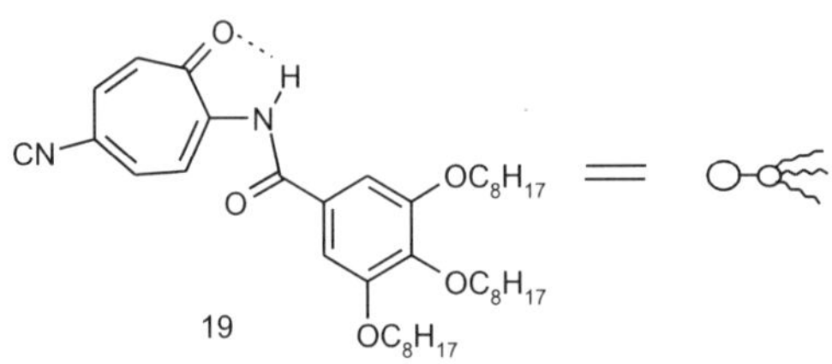

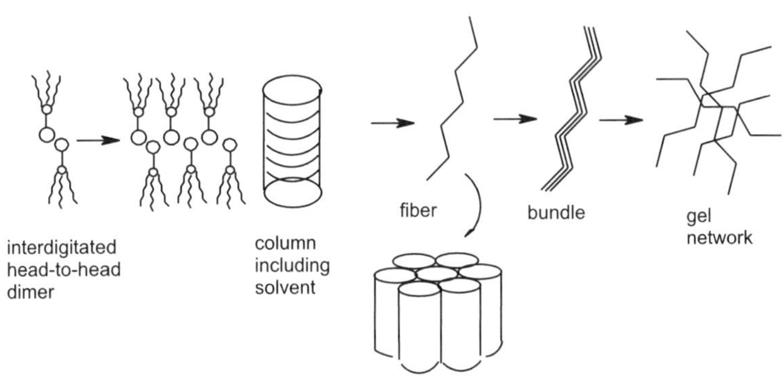

Fig. 3 Trioctyloxybenzoylamino-2-tropone derivative **19** as gelator with intramolecular hydrogen bond and the scenario of organization leading to gelation, which starts with the head-to-tail dimer followed by formation of a stacked column, hexagonal column in gel fiber, fiber bundles, and finally a gel network

enantiomers of bislauroyl amides **20** by condensation of lauroyl chloride and *trans*-1,2-diaminocyclohexane in THF in the presence of triethylamine (Fig. 4) [19]. The compounds exhibited efficient gelation of various solvents of different polarity (alkanes, chlorinated alkanes, alcohols, ketones, nitriles) with mgc in the range of 2–44 g L^{-1}. The FTIR results and molecular modeling studies suggest the formation of hydrogen-bonded molecular tapes (Fig. 4, right). CD spectroscopy of gel samples showed a strong peak which decreased upon heating and disappeared beyond the gel melting temperature; this behavior showed that chiral aggregates were present in the gel. The TEM investigation also revealed the existence of helical fibers; the helicity of the fibers was always right-handed for (1*R*,2*R*)-**20** and left-handed for (1*S*,2*S*)-**20**. The single enantiomer of **20** and the mesogenic unit containing gelator **21** were found to form two types of gels with mesomorphic compounds **22** and **23**. Both **20** and **21** are capable of gelling anisotropic and isotropic states of the mesogens [20, 21].

The same type of gelator was functionalized with *β*-diketonato ligands (**24**, Fig. 5). Although **24** is unable to gel methanol, its mixture with **20** in the molar ratio 1 : 4 is capable of gelling at a mgc of 3 wt %. The FTIR measure-

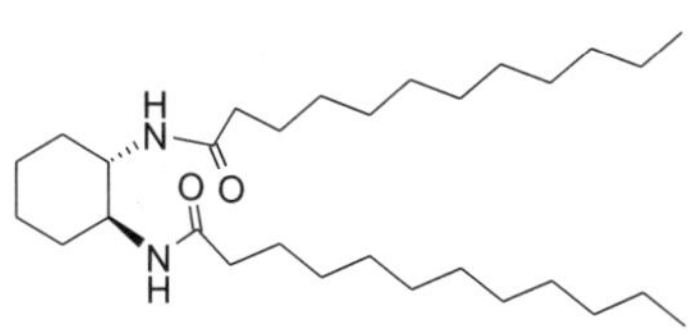

20

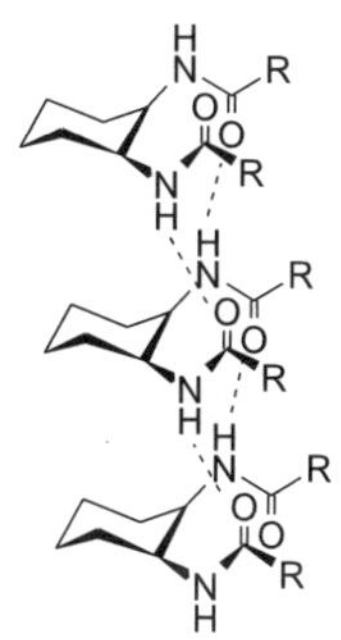

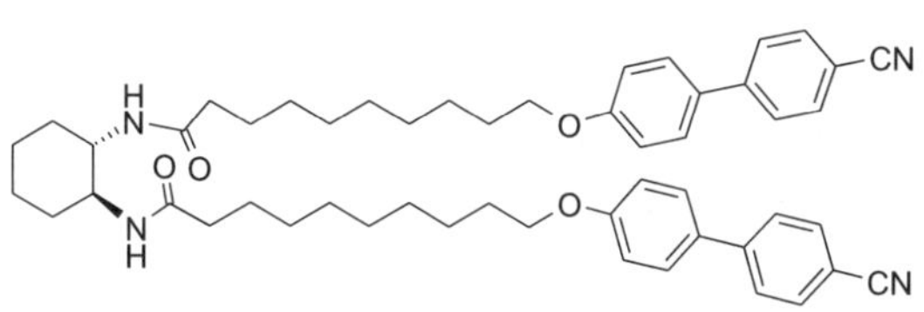

21

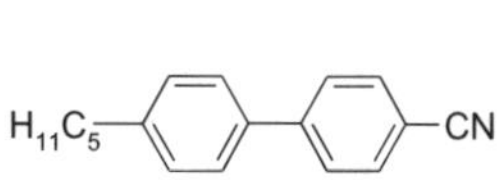

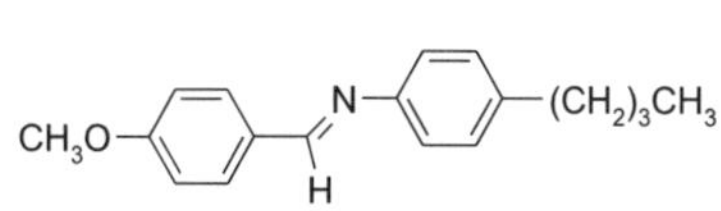

22 **23**

Fig. 4 *N*,*N*-bis(lauroyl)-*trans*-1,2-cyclohexylamine **20** and intermolecular hydrogen bonding motif between self-complementary cyclohexylbisamide units in gel aggregates; gelator **21** with mesogenic cyanobiphenyloxy units and the mesomorphic compounds **22** and **23** giving room-temperature liquid crystals ments indicated that molecules of **24** were fully integrated in the assembly of **20**. Addition of triethylamine over the methanol gels containing divalent metal cations (Cu^{2+}, Ni^{2+}, Co^{2+}) resulted in their complexation with β-diketonato units of incorporated **24**, inducing increased mechanical strength and thermal stability of the gels [22].

The mechanical strength and thermal stability of gels can also be improved by the photoinduced polymerization of gelator molecules bearing polymerizable functions. Gelator **25** with two diacetylenic groups was prepared in 50% yield by condensation of (1*R*, 2*R*)-(-)-diaminocyclohexane and 12-tricosadiynoic acid in the presence of *N*-ethyl-*N*′-dimethylaminopropylcarbodiimide (Scheme 6) [23]. The gelator showed excellent gelation of cyclohexane, hexane, ether, methanol, 1-butanol, and acetonitrile at a concentration of 5 wt %. Irradiation of the hexane gel by UV light (a 500-W high-pressure Hg lamp) at room temperature resulted in a color change giving blue gel with an increased gel-to-sol transition temperature compared to nonirradiated gel. This observation was explained by the intermolecular polymerization within the gel fiber structure.

Chiral gelators based on the *trans*-1,2-diaminocyclohexane unit were successfully used as templates for the preparation of silica or transition metal oxide nanostructures: helical fibers or tubes [24–26].

Simple long-chain alkanoic acid bisamides of 3,4-diaminobenzophenone (**26–28**; Scheme 7) showed moderate gelation of various polar and apolar organic solvents. It is interesting that irradiation of the 2-propanol gel of **26** results in photoreaction to give pinacol. Since the photoproduct is soluble in the solvent, the irradiation leads to gradual transformation of the gel into a solution [27].

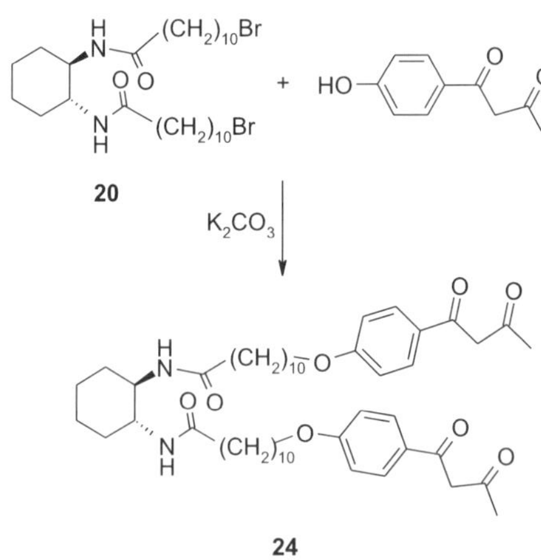

Fig. 5 Bis-β-diketonato gelator **24**

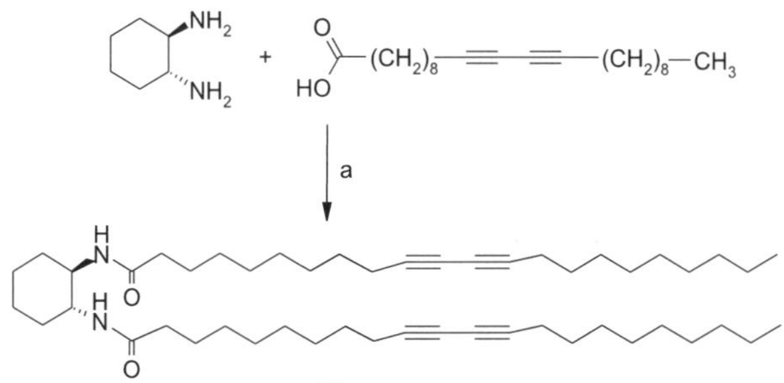

25

Scheme 6 Preparation of bis(diacetylenic) gelator **25**: (a) *N*-ethyl-*N*′-dimethylamino-propylcarbodiimide (EDCI), dry dichloromethane

A series of the α,ω-polymethylene diamines having even or odd numbers of methylene groups were converted into dodecanoic acid diamides **29** and **30** (Fig. 6), which showed gelation of aliphatic and aromatic solvents, methyl oleate, soybean oil, ethyl acetate, and acetonitrile [28]. It was found that the even and odd numbers of bridging methylenes are decisive for the gelation: for gelation of mesitylene by **29** (n = even number) the longer the bridging aliphatic chain, the poorer the efficiency. In contrast, **30** (n = odd number) exhibits efficient gelation of mesitylene at a mgc of less than $10\ \mathrm{g\,L^{-1}}$. Also, for other liquids **30** was superior to **29**. The SEM images of **29** and **30** xerogels show flat, ribbonlike structures and fine woven fibers, respectively. The observed gelation and morphological differences were explained

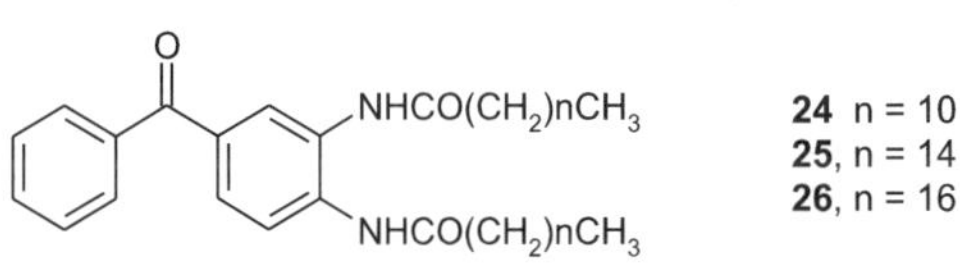

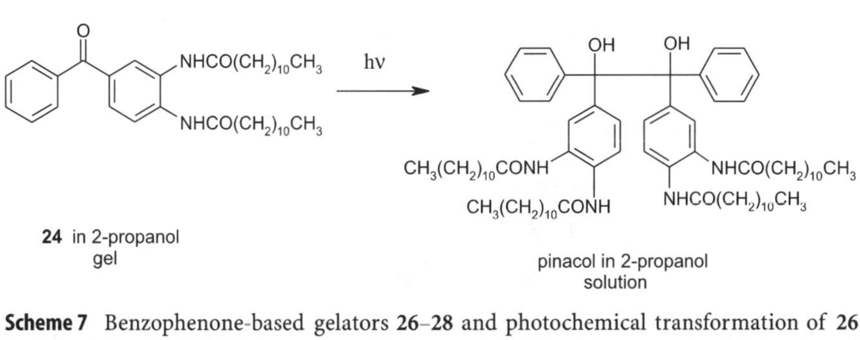

Scheme 7 Benzophenone-based gelators **26**–**28** and photochemical transformation of **26** by irradiation (400-W high-pressure Hg lamp at 15 °C) of 2-propanol gel into pinacol resulting in gradual decomposition of the gel into solution

by the self-complementarity of the diamides with an even number of bridging methylenes, which induces the self-assembly by hydrogen bonding into ribbonlike assemblies, while in the case of the non-self-complementary odd amides the self-assembly involves at least four hydrogen-bonded gelator molecules leading to woven tiny fibers (Fig. 6).

The introduction of stereogenic centers in this type of gelator results in new morphologies of gel fibers such as twisted ribbons and coiled coils. Chiral diamides **29-Me** and **30-Me** were prepared by acylation of the α,ω-polymethylene diamines with (2*S*)-2-methyldodecanoyl chloride in a two-phase ether–water system in the presence of sodium bicarbonate (Scheme 8). Formation of twisted ribbons and coiled coils was explained by chiral van der Waals interactions between terminal alkanoic chains induced by introduction of chiral centers.

Symmetrical dialkoxybenzoic acid diamides (**31**,**32**) as gelators were prepared by alkylation of 3,5-dihydroxybenzoic acid methyl ester with 11-bromo-*N*-decylundecanamide under phase-transfer catalysis conditions with excess K_2CO_3 [29]. The nonsymmetrical derivative **33** differing in directionality of amide groups in the side chains was prepared from the acid derivative **34** in three steps (Scheme 9). The methyl ester derivative **31**, the free acid **32**, and the nonsymmetrical methyl ester **33** exhibited gelation of benzene, toluene, and *p*-xylene at a mgc of 3 wt %. Freeze-fracture electronic microscopy revealed platelet-like aggregates with 150–180 –nm widths in the toluene gels of **31** and **33**. In contrast, in the toluene gel of the free acid derivative **32** tiny fibers of 17-nm width were observed. This sharp difference in the morphology indicates that the acid group prevents the lateral packing of fib-

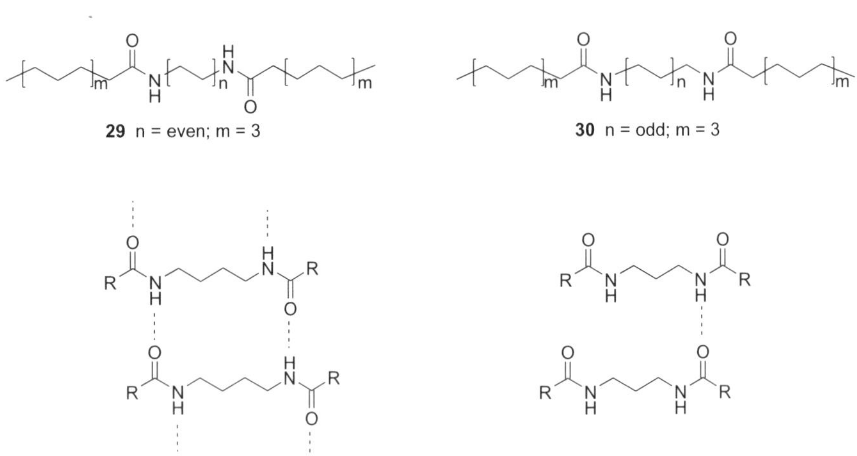

Fig. 6 Diamide gelators with even and odd numbers of bridging methylene groups and indicated self-assembly by amide–amide hydrogen bonding

rils into larger objects, as in the case of **31** and **33**, and limits aggregation to tiny fibers.

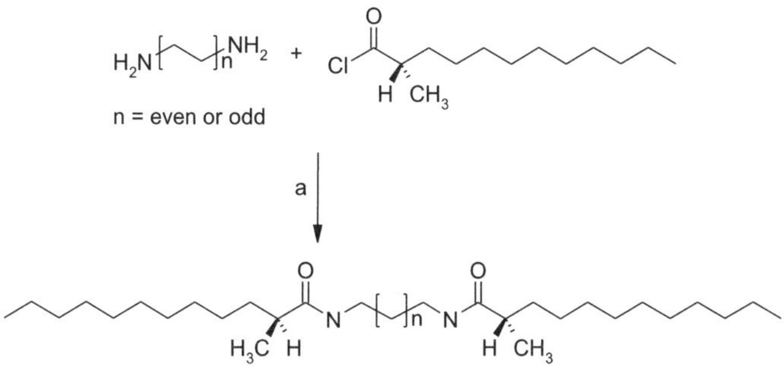

29-Me n = even
30-Me n = odd

Scheme 8 Preparation of chiral even (**29-Me**) and odd (**30-Me**) diamide gelators: (a) Na_2CO_3, ether/water

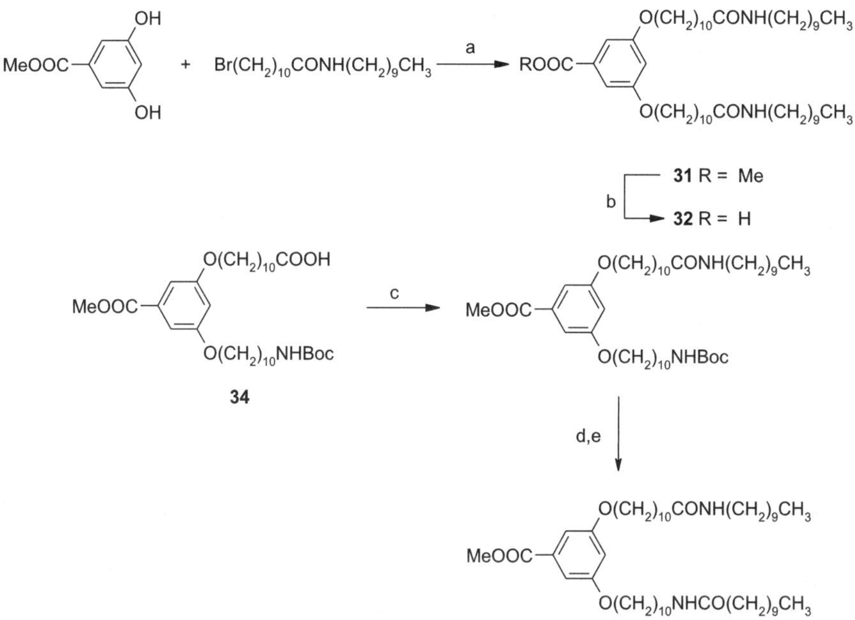

Scheme 9 Preparation of symmetrical gelators **31** and **32**: (a) K_2CO_3; (b) LiOH/MeOH. Three-step synthesis of the nonsymmetrical gelator **33**: (c) EDCI (3 equiv.), $CH_3(CH_2)_9$-NH_2 (1 equiv.), HOBu (0.16 equiv.), NMP; (d) 4 M HCl in EtOAc; (e) $CH_3(CH_2)_9COOH$ (1 equiv.), EDCI (2 equiv.)

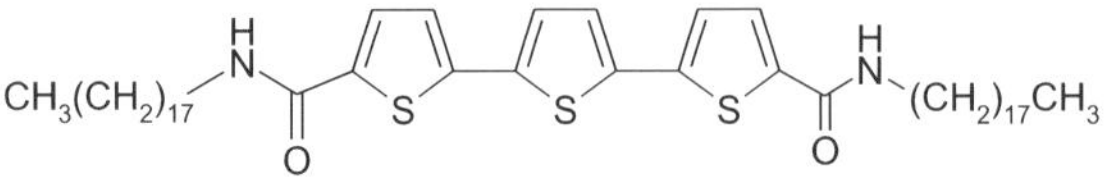

35

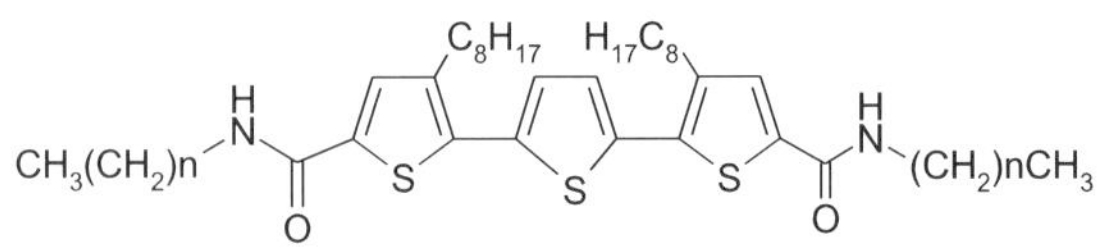

36 n = 4, 7, 15, 17

Fig. 7 Nongelling (**35**) and gelling (**36**) terthiophene diamide derivatives

The terthiophene-based diamide derivatives **35** and **36** (Fig. 7) bearing long alkyl chains were prepared [30]. Only derivatives **36** having *n*-octyl substituents at the 3,3″-terthiophene positions were found to gel *n*-alkanes (C_6–C_8; *i*-C_8; C_{11}; C_{12}; C_{14}) at mgc values in the range 3.6–5.3 g L^{-1}.

2.2.2 Triamide Gelators

N,N',N''-tristearyltrimesamide (**37**) was prepared by the condensation of trimesoyl chloride and stearylamine and shown to gel 1,2-dichloroethane, DMF, nitrobenzene, *N*-methylformamide, benzonitrile, and DMSO at mgc values of 5–41 g L^{-1}. The secondary amide derivative **38** and the derivatives with C_8 and C_3 alkyl chains were found to be incapable of gelling [31]. The same type of compounds (including **37** and **39–41**, Fig. 8) were also studied by Hanabusa [32] and found to produce highly viscous solutions or crystallize, i.e., **37** from DMSO, DMF, and nitrobenzene.

Much more efficient triamide gelators were prepared from *cis*-1,3,5-cyclohexanetricarboxylic acid and long-chain alkylamines (Fig. 8) [33]. Gelators **42** and **43** exhibited excellent gelation of various lipophilic solvents (hexane, cyclohexane, benzene, toluene, chlorobenzene, and pyridine) and also of kerosene.

Novel triamide gelator **44** based on 4,4′,4″-triaminotriphenylamine and stearic acid was prepared by the group of Shirota [34]. Triaminotriphenylamine was reacted with high-purity stearic acid in mesitylene at 164 °C for 8 h with constant removal of water (Scheme 10).

Compound **44** showed efficient gelation of aromatic solvents, DMSO, and 1,2-dichloroethane with mgc values of 1.3–5.8 g L^{-1}. Gels of **44** in the presence of *n*-Bu_4NClO_4 exhibited high ionic conductivities (σ) comparable to those of solutions of tetrabutylammonium perchlorate. Gels were shown to function as electrochromic materials; upon electrochemical oxidation a green-

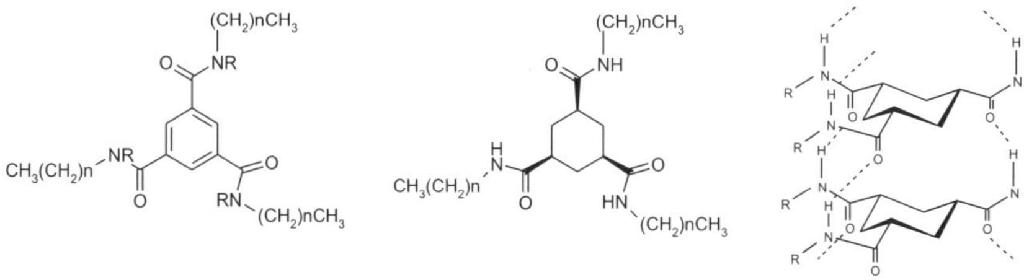

37 n = 17 R = H
38 n = 17 R = Me
39 n = 2 R = H
40 n = 7 R = H
41 N = 11 R = H

42 n = 5
43 n = 11
44 n = 17

Fig. 8 Triamide gelators based on 1,3,5-benzenetricarboxylic acid and *cis*-1,3,5-cyclohexanetricarboxylic acid and proposed unidirectional hydrogen-bonded organization in the gels of **42** and **43**

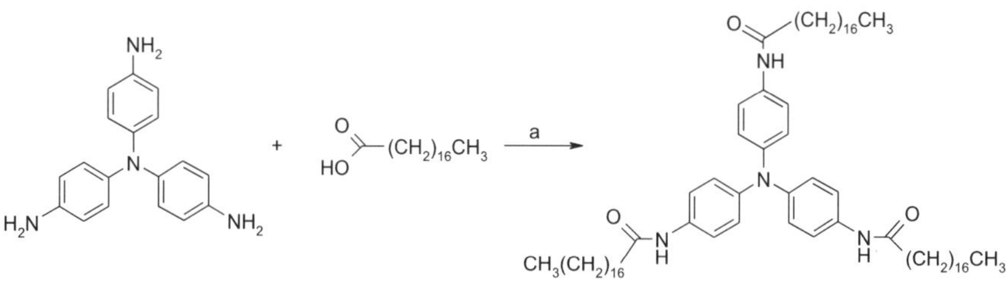

44

Scheme 10 Preparation of triamide gelator **44**: (a) mesitylene, 165 °C, 8 h, under nitrogen

colored triphenylamine radical cation is formed, which upon reduction becomes colorless. The same group prepared more complex gelator molecules **45** and **46**, which also possessed electroactive triphenylamine units, by repeated Ullman condensations and studied their electrochemical properties (Scheme 11) [35, 36].

Triamide derivatives **47–50** (Fig. 9) intramolecularly organized in C_3-symmetrical disks have been prepared and shown to undergo helical supramolecular organization by cooperative stacks, which results in gelation of some lipohilic solvents [37]. The "sergeants-and-soldiers" principle discovered for covalent polymers, which leads to strong amplification of chirality when a small amount of a chiral monomer is added to achiral monomers, is also observed for the mixtures of achiral and chiral **47**/**48** and **49**/**50** [38, 39]. Addition of only 2.5% of chiral **50** to achiral **49** generates a strong Cotton effect in hexane. This shows that chiral **50** induces organization of many achiral molecules of **49** into a helical columnar aggregate.

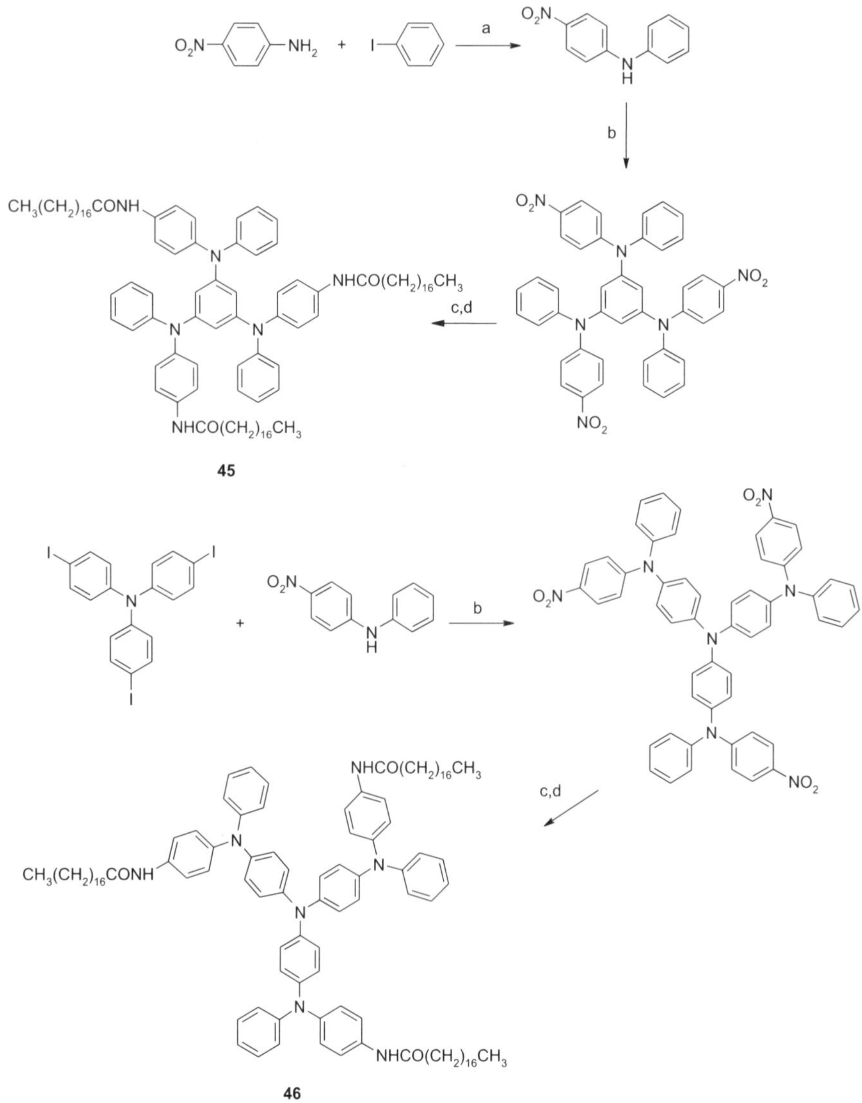

Scheme 11 Synthesis of electrochromic gelators **45** and **46**: (a) 188 °C, Cu powder, K_2CO_3, 18-crown-6; (b) 160 °C, mesitylene, Cu powder; (c) H_2/Raney nickel; (d) THF, stearoyl chloride, Et_3N

2.2.3
Polyamide Gelators

The hexaamide gelators **51** and **52** based on the triphenylene unit were prepared by the reaction of 2,3,6,7,10,11-hexahydroxytriphenylene and 2-

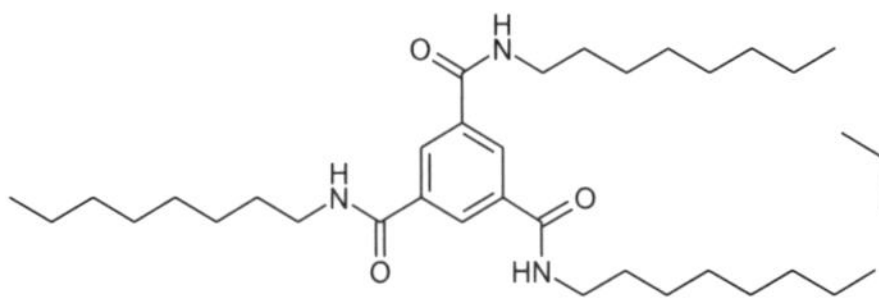

47

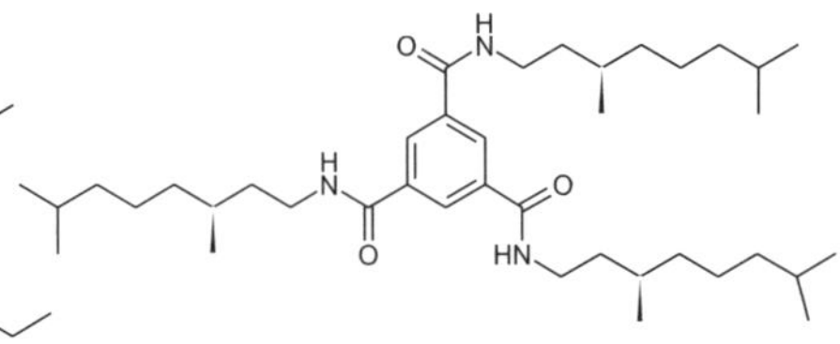

48

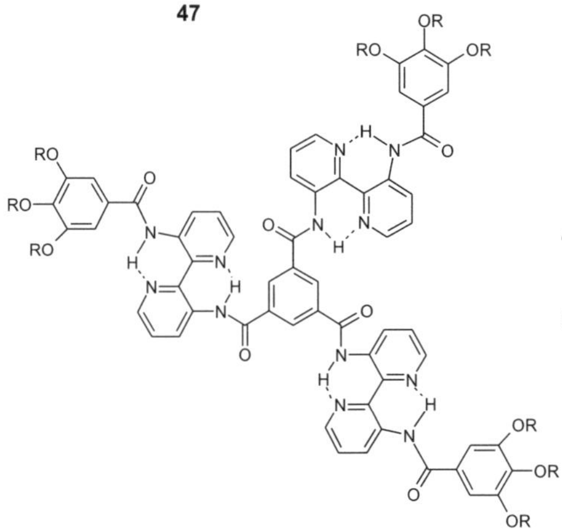

49 R *= $C_{12}H_{25}$

50 R =

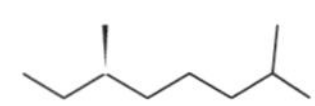

Fig. 9 C_3-symmetrical triamide gelators: achiral **47** and **49**; chiral **48** and **50**

bromo-*N*-alkylacetoamide (Scheme 12) [40]. Both compounds efficiently gelled *n*-hexane, *n*-octane, cyclohexane, and *p*-xylene at a concentration of 13.4 wt/vol %. However, **51** exhibited gelation of *n*-hexane and cyclohexane at mgc values of 4.0 and 2.7 wt/vol %, respectively. It was observed that the cyclohexane gels of **51** and **52** exhibited a yellow and greenish-blue color, respectively, while their sols were blue. Fluorescence spectroscopy of the gels showed strong excimer emission around 525.0 nm in the **51** cyclohexane gel but not in the **52** gel.

The results suggest self-organization of **51** with eclipsed overlap of triphenylene units in the hydrogen-bonded aggregates, resulting in excimer formation and staggered overlap of the same units in the aggregate of **52**.

The tetraphenylporphyrin derivatives **53** and **54**, which bear eight amide functions on phenyl substituents, were prepared by the Shinkai group as novel gelators capable of self-assembly via π–π stacking of the porphyrins supported by multiple hydrogen bonding between the lateral amide groups [41]. Such derivatives were brilliantly designed to form capsule molecules stabilized by hydrogen bonding between four amide groups of one porphyrin unit and four amide groups of another porphyrin molecule (Fig. 10). The distance between the porphyrins in the capsule is comparable to the size of C_{60} fullerene. It was found that a mixture of C_{60} and **53**

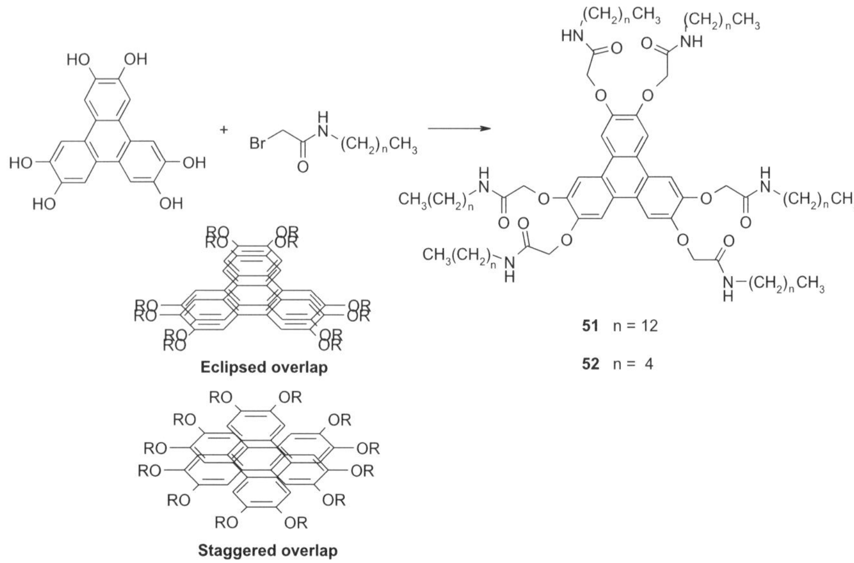

Scheme 12 Hexaamide gelators **51** and **52** prepared by condensation of 2,3,6,7,10-hexahydroxytriphenylene and 2-bromo-*N*-alkylacetoamide, and indicated eclipsed and staggered overlap of triphenylene units in the gel aggregates

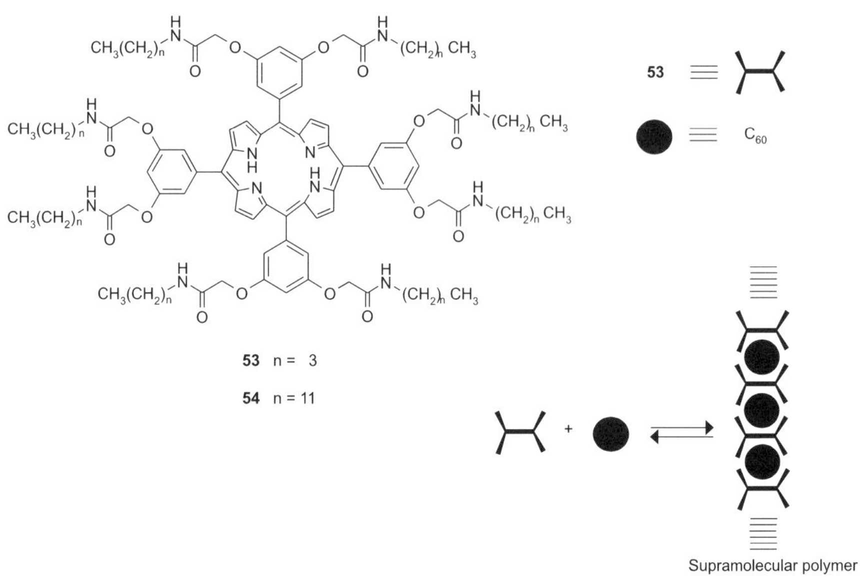

Fig. 10 Octaamide–porphyrin gelators **53** and **54** and supramolecular polymer formed by **53** and C_{60} in benzene gel

in a 1 : 2 molar ratio is capable of gelling benzene, toluene, *p*-xylene, and anisole, and that the thermal stability (T_{gel}; °C) of C_{60} + **53** benzene gel is much higher (120 °C) than that of **53** benzene gel. Experimental evidence is provided that the fibrous supramolecular polymer with encapsulated C_{60} is formed, and that the addition of fullerene changes the two-dimensional aggregation mode of **53** in the gel into the one-dimensional mode [42]. Such composed porphyrin–fullerene gels are of high interest for development of photofunctional gels with electron-transfer properties.

3 Amino Acid Derivatives and Peptides

3.1 Long Alkyl Chain Amino Acid Derivatives

Derivatization of amino acids with alkyl chains of variable length or various aromatic groups gives amphiphilic molecules capable of forming elongated aggregates which cause gelation of certain solvents [43].

A series of the 4-hexadecanoyl-2-nitrophenyl esters of *N*-benzoyloxycarbonyl-protected amino acids have been prepared by Hanabusa according to Scheme 13 [44]. The L-Ala derivatives **55**, **57**, and **58** as well as the L-Glu derivative **56** showed very efficient gelation of methanol and cyclohexane at the concentrations of 0.5 and 0.25 wt %, respectively.

Modifications of the *N*-protecting groups and the introduction of ester alkyl chains with less than ten C atoms resulted in loss of any gelation ability. The spectroscopic evidence (FTIR) for intermolecular hydrogen bonding between urethane groups in the gel aggregates is provided. It is shown that in the spectra of gels strong CD peaks appear which, however, disappear in the sol phase. This observation clearly shows that enhancement of chirality occurs upon aggregation.

Within the series of *N*-benzyloxacarbonyl (Z)-protected alkylamides of phenylalanine (**59**, **60**) and dodecamethylene (**61**), and alkyldiacetylene (**62**)-bridged bis(*N*-Z-phenylalanine) amides versatile gelators of various organic solvents were found [45]. The synthesis of gelators is outlined in Scheme 14.

Simple long-chain alkanoic acid amides of amino acids represent typical amphiphilic structures with carboxylic acid polar head groups. Medium and long alkyl chain amides of L-Ala (**63**–**65**) showed excellent gelation of many lipophilic solvents due to organization into inversed bilayer aggregates [46]; *N*-lauroyl-L-Ala (**64**) was found to efficiently gel various higher alkane and aromatic solvents, including commercial fuels, at mgc values between 2 and 12 g L^{-1} (Fig. 11). If two phasic fuel–water systems are tested the gelator selectively gels the oil phase, revealing a potential use of gelators in sea oil

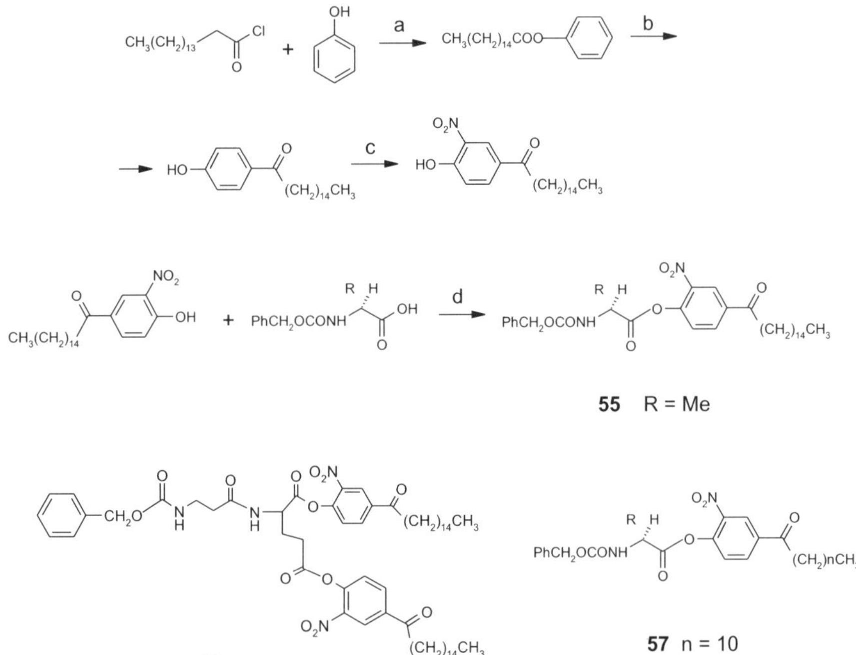

55 R = Me

56

57 n = 10
58 n = 16

Scheme 13 Synthesis of 4-alkyl-2-nitrophenyl esters of *N*-benzoyloxycarbonyl-protected L-Ala **55**,**57**, and **58** and *β*-alanyl-L-Glu derivative **56**: (a) Δ; (b) $AlCl_3$, HCl; (c) HNO_3/H_2SO_4; (d) DCC

transport accidents [47]. Recently, the methyl ester of **64** was used in combination with soybean oil, ethanol, and a model fluorescent drug for the preparation of solid hydrophobic implants possessing sustained drug delivery [48]. Analogous alkanoic acid amides of L-Ser are able to gel water and lipophilic solvents, giving tubules or helical ribbons as major supramolecular aggregates in both media [49, 50].

Another amino acid frequently used in the preparation of gelators in combination with long aliphatic chains is L-Lys, thus allowing additional functionalization at the butylamino group [51, 52]. The bolaamphiphile constructed from L-Lys and *ω*-aminododecanoic acid (**66**) forms hydrogels by organizing into stable micellar fibers [53]. The L-Lys-based gelators **67** and **68** containing viologen units showed gelation of aromatic solvents and exhibited photosensitized charge separation in the presence of tris(2,2′-bipyridine)ruthenium(II) and triethanolamine [54]. Following this observation, new tris(2,2′-bipyridine)-type ruthenium(II) complexes **69** and **70** bearing L-Lys-containing side chains were prepared and shown to gel aromatic solvents at mgc values of 2–4 g L^{-1} [55]. The gelling complexes were prepared as shown in Scheme 15.

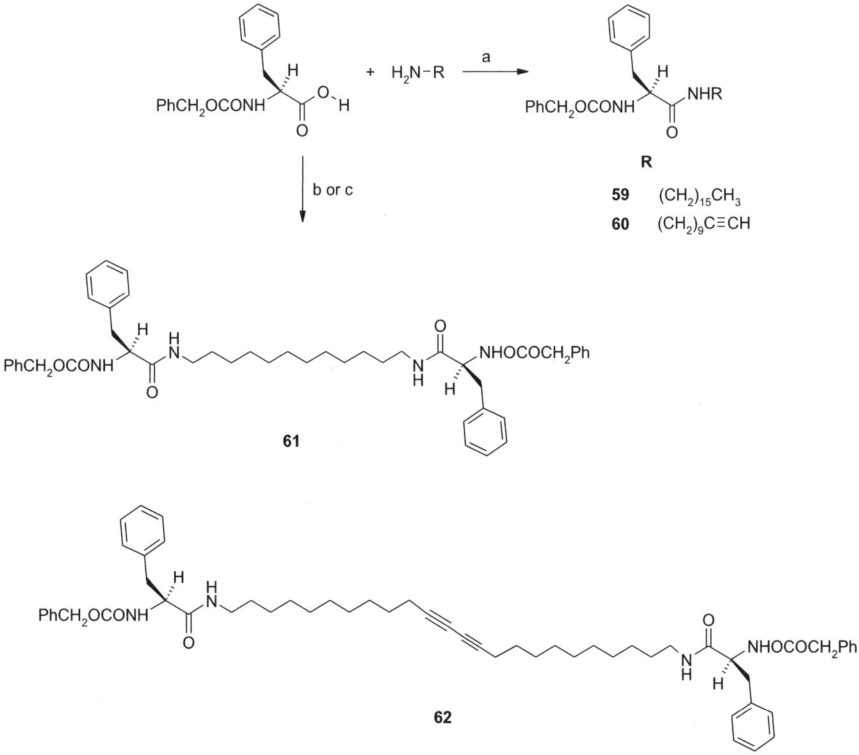

Scheme 14 Synthesis of gelators **59–62**: (a) DCC, DMAP, $CHCl_3$; (b) 1,12-diaminododecane, DCC, DMAP, $CHCl_3$; (c) oxidative dimerization of **60**, $CuCl_2$, dry MeOH, dry pyridine

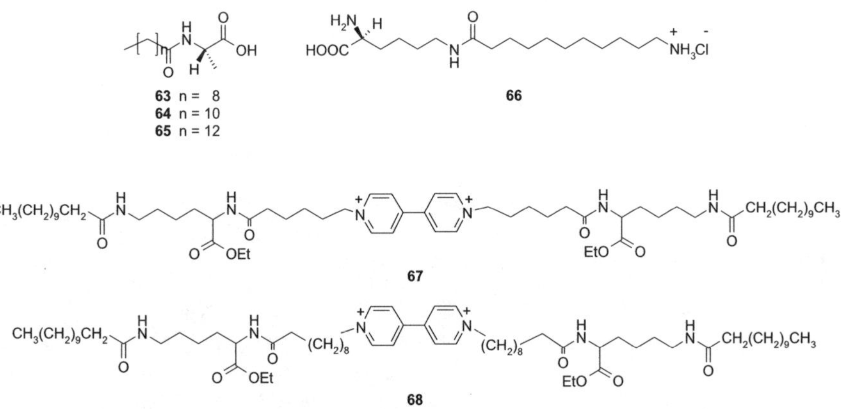

Fig. 11 Structures of selected L-Ala- and L-Lys-based gelators

It was found that **70** forms a gel with toluene and other aromatic solvents while **69** gives isotropic solutions; however, the TEM images showed that both compounds give fibrous aggregates in toluene. Upon addition of methyl viologen (MV^{2+}) to the methanol/toluene system containing either **69** or **70** below its mgc and irradiation by visible light, the methyl viologen radical ($MV^{+\cdot}$) was formed at two and five times higher rates as compared to that with $Ru(bpy)_3{}^{2+}$ as the sensitizer. It was confirmed that in the system with the self-assembled sensitizers **69** or **70**, the lifetime of $MV^{+\cdot}$ is long and the back reduction of Ru(III) to Ru(II) by triethanolamine as sacrificial electron donor is fast as compared to the simple $Ru(bpy)_3{}^{2+}$-containing system.

A series of efficient hydrogelators based on the core L-Lys unit and two long aliphatic chains, one carrying a positively charged heteroaromatic group, have been prepared (Scheme 16). Synthesis starts with esterification of N^{ε}-lauroyl-L-lysine by ethanol, methanol, or 11-bromoundecanol followed by an acylation step with alkanoyl chloride or *n*-bromoalkanoyl chloride and subsequent quaternization of pyridine or methylimidazole [56, 57]. Such gelators are able to gel water at a mgc of 1–3 $g\,L^{-1}$.

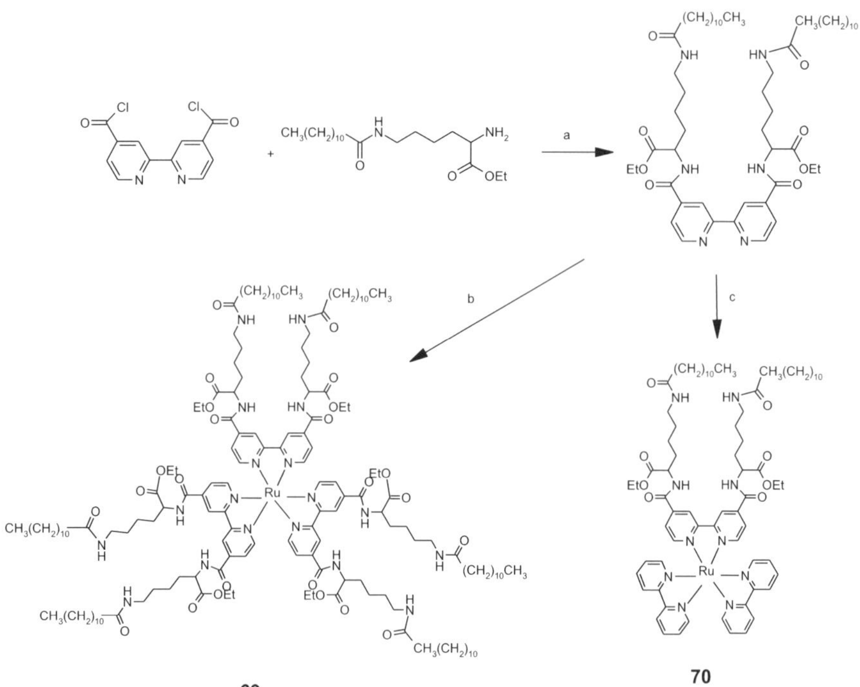

Scheme 15 Synthesis of tris(2,2′-bipyridine)-type ruthenium(II) gelators **69** and **70**: (a) dry THF and Et_3N; (b) $RuCl_3 \cdot 3H_2O/EtOH$; (c) $Ru(BPy)_2Cl_2 \cdot 2H_2O/80\%$ EtOH

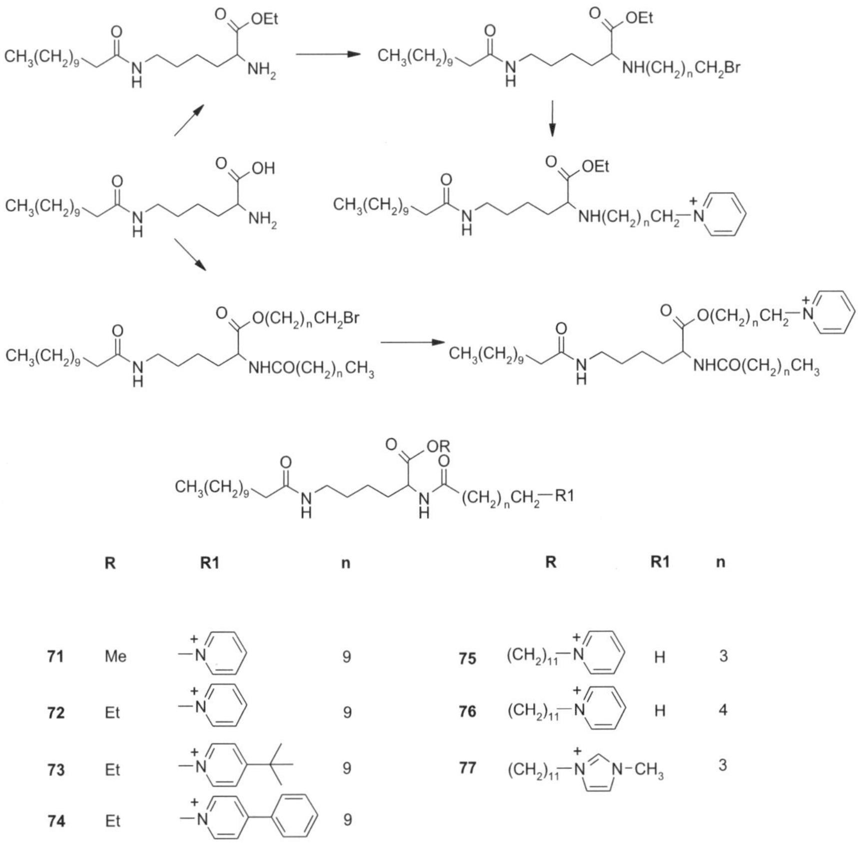

Scheme 16 Synthesis of N^{ε}-lauroyl-L-Lys-based hydrogelators with quaternized heteroaromatic groups in α-amino (**71–74**) and ester (**75–77**) alkyl branches

A very interesting example of dendritic two-component gels was recently published by Smith et al. [58, 59]. Although neither **78** (Fig. 12) nor α,ω-diaminoalkanes are capable of gelling, their 2 : 1 mixture results in large changes of gel melting temperatures (T_{gel}) from 105 °C for C_{12} to 41 °C for C_8, while variation of the dendrimer : diamine ratio results in a change of gel morphology. This example reveals the approach to gel systems with tunable properties.

The octadecylamide of *N*-benzyloxycarbonyl-L-isoleucine **79** and *N*-benzyloxycarbonyl-L-Val-L-Val alkylamide **80**, as well as some other *N*-Z-dipeptide alkylamides [60], exhibited excellent gelation properties toward many organic solvents of low, medium, and high polarity at mgc values in the range of 5–40 g L^{-1}. The compound was used for the preparation of organogel electrolytes and porous titania with fibrous structure by sol–gel transcription of gel fibers using polymerization of titanium tetraisopropoxide [61, 62].

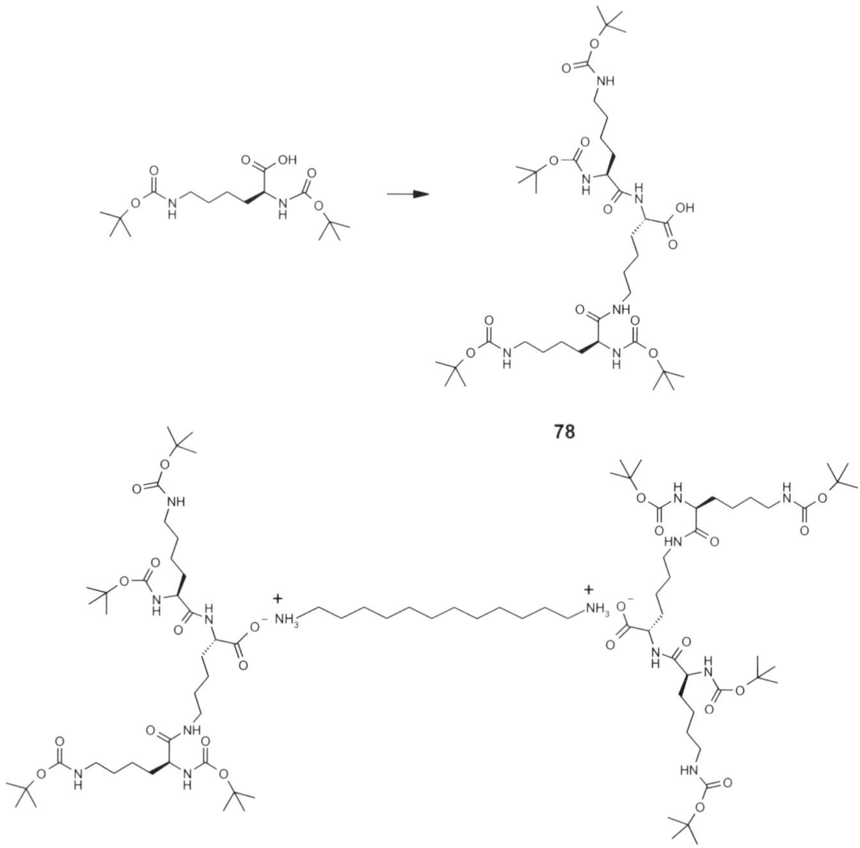

Fig. 12 Structure of two-component gelation system based on the interaction between **78** and diaminoundecane

The L-Val n-C_{12}- and n-C_{18}-alkylamide derivatives of *o*-, *m*-, and especially *p*-benzenedicarboxylic acid (Fig. 13) are versatile gelators of many highly polar (DMSO, DMF, EtOH) and apolar (carbon tetrachloride, chloroform, benzene) organic solvents and even kerosene, silicone oil, and soybean oil [63]. The mgc values of **84** are in the range of 1–24 $\mathrm{g\,L^{-1}}$.

The cationic surfactants based on Leu (**85–91**, Fig. 14) with a *p*-toluenesulfonate counterion exhibited gelation of hexane, octane, decane, dodecane, kerosene, light oil, and silicone oil at mgc values in the range of 0.2–3 wt %. Fluorescence studies and the observation of a strong CD peak suggest the formation of helical aggregates by stacking of *p*-toluenesulfonate groups [64].

In his seminal paper published in 1985, Kunitake showed that some L-Glu-derived double-chain ammonium amphiphiles **92** initially formed vesicles in water which upon ageing transferred into fibrous aggregates such as helices, helical ribbons, and tubules of micrometer lengths [65]. Ihara and cowork-

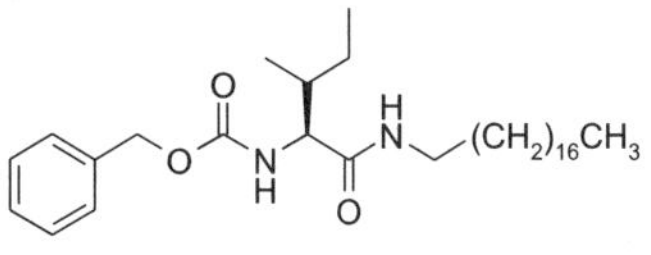

79

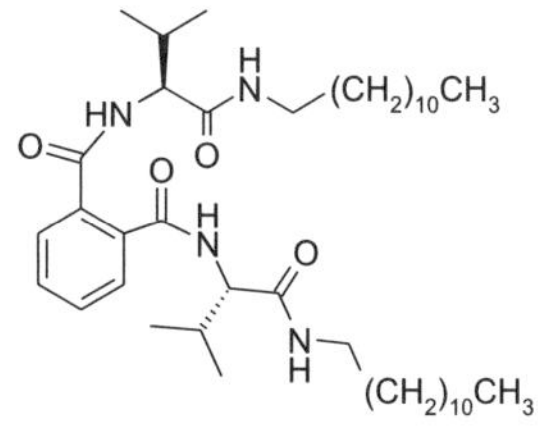

81

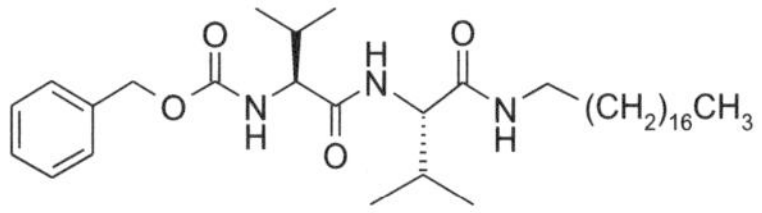

80

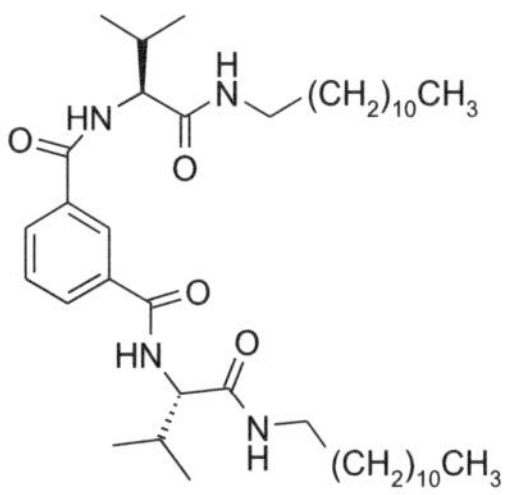

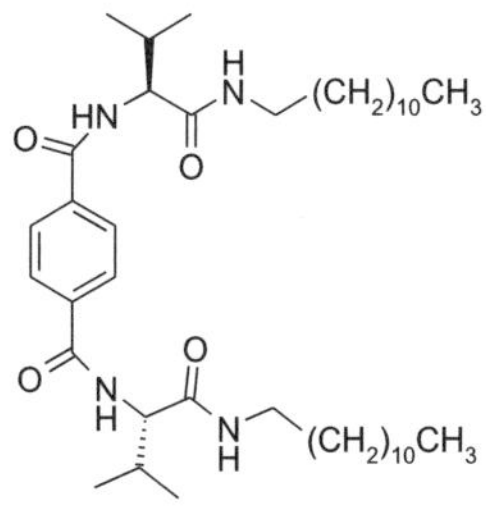

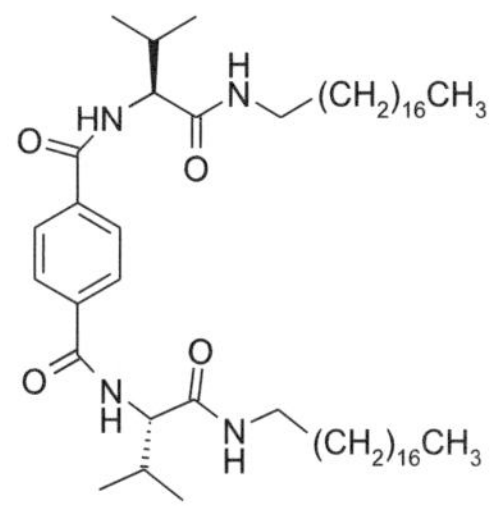

82 83 84

Fig. 13 *N*-Z-Ile (**79**), *N*-Z-Val-Val (**80**), and L-Val benzenedicarboxylic acid-based gelators **81–84**

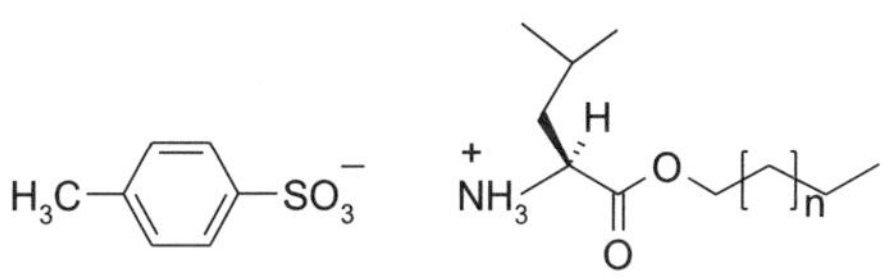

85 - 91 n = 10 -16

Fig. 14 Leu-based cationic surfactant gelators

ers have shown that the uncharged double alkyl chain L-Glu derivatives **93** and **94** (Fig. 15) form gels with benzene. The 1 : 1 mixture of **93** and **94** in benzene showed a clear phase separation phenomenon that is characteristic for biomembranes [66]. Introduction of an isoquinoline unit (**95**) capable of complexing Cu(II), Co(II), and Zn(II) results in enhanced chirality of the gel aggregates in the presence of Cu(II), while the latter two cations destroy aggregation [67].

A very interesting example of a hydrogel showing a pH-responsive volume phase transition has been recently reported by Hamachi [68]. Addition of 10 mol % of **96** into the hydrogel formed by glycosylated L-Glu gelator **97** conserves the thermally induced swelling–shrinkage volume transition at pH 4, while at pH 7 only gel-to-sol transition could be observed (Fig. 16). Such

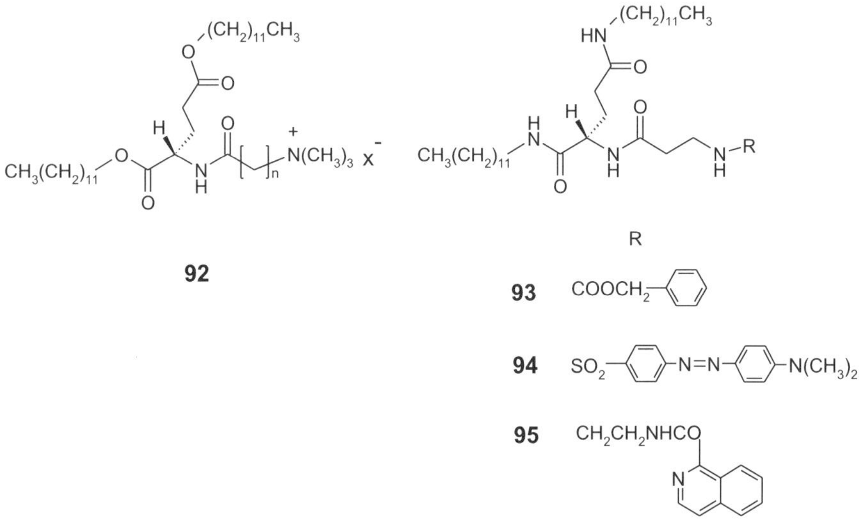

Fig. 15 Double alkyl chain L-Glu-based gelators

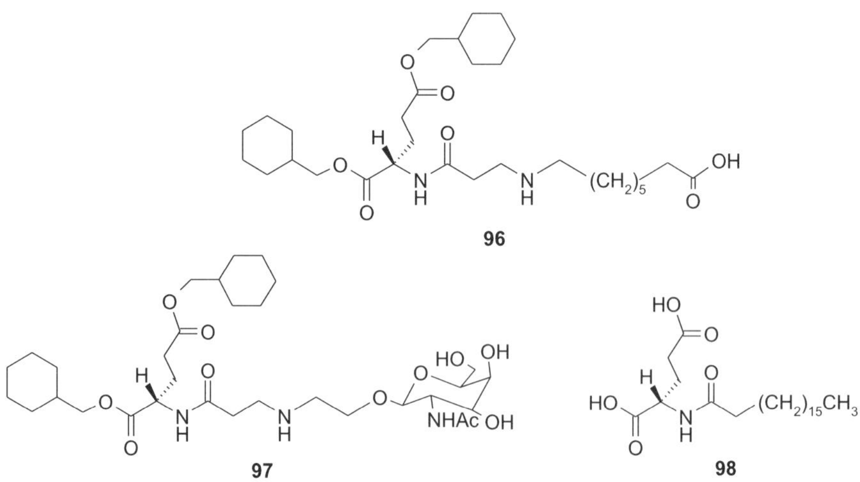

Fig. 16 L-Glu-based gelators **96** and **97** giving hydrogels with pH-responsive thermally induced swelling–shrinkage volume phase transition, and acidic multi-H-bonding gelator **98** capable of different self-assembly in chloroform (nanofibers) and water/ethanol 1 : 1 mixture (nanodisks)

a pH-controlled volume phase transition could be of high importance for the development of new hydrogel-based drug delivery systems.

N-stearoyl-L-glutamic acid **98** (Fig. 16) is able to gel chloroform and also a highly polar water/ethanol 1 : 1 mixture [69]. AFM, SEM, and spectroscopic investigations revealed that the same gelator self-assembles into fiber bundles in chloroform and into spiral bilayered nanodisks in water/ethanol. These results demonstrate that in the case of multi-H-bonding amphiphilic gelators, the properties of a solvent control intramolecular/intermolecular hydrogen bonding styles, which in turn results in different morphologies of gel fibers and consequently different properties of the gels.

A highly interesting group of biocompatible hydrogelators was designed by Feringa and van Esch [70, 71]. The approach is based on the 1,3,5-cyclohexanetricarboxylic acid scaffold onto which various lipophilic amino acid units were attached to give 1,3,5-triamide *cis,cis*-cyclohexane derivatives (Scheme 17).

Gelators **99–104** exhibited impressively efficient gelation of water at mgc values of 0.06, 0.08, 0.03, and 0.07 wt %, respectively, with **101** being the most efficient low molecular weight hydrogelator to date. In water, fibrous gel aggregates are formed by favorable hydrophobic interactions between lipophilic amino acid substituents and by formation of three amide-type hydrogen bonds per gelator molecule occurring within the lipophilic core of the aggregate. The acidic gelators **99** and **100** as well as the basic ones, **103** and **104**, show pH-reversible gel-to-sol transitions, which makes them of interest for various medicinal applications. In the pH range of 3.2–4.0 the hydrogel of the acidic gelator **99** turns into a sol due to formation of carboxylate anions, leading to strong repulsive interactions in the aggregates (Fig. 17). However, a considerably higher pH range (4.3–5.8) is needed to form a sol

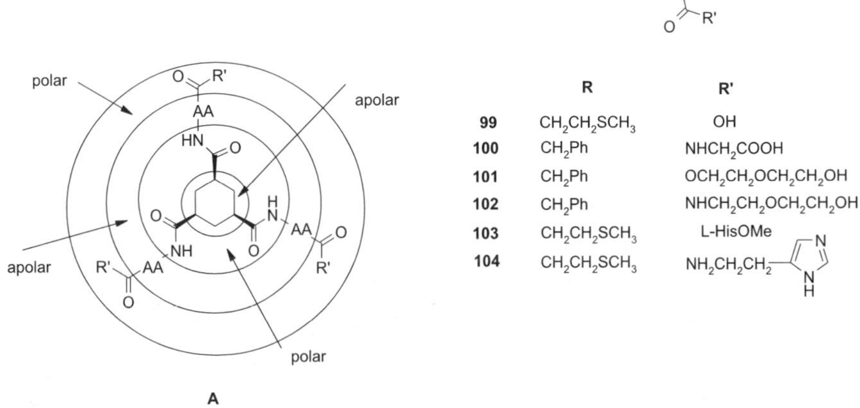

	R	R'
99	$CH_2CH_2SCH_3$	OH
100	CH_2Ph	$NHCH_2COOH$
101	CH_2Ph	$OCH_2CH_2OCH_2CH_2OH$
102	CH_2Ph	$NHCH_2CH_2OCH_2CH_2OH$
103	$CH_2CH_2SCH_3$	L-HisOMe
104	$CH_2CH_2SCH_3$	$NH_2CH_2CH_2$–(imidazolyl)

Scheme 17 Synthesis of 1,3,5-triamide cyclohexane hydrogelators and (**A**) schematic representation of the design principle showing polar and apolar domains

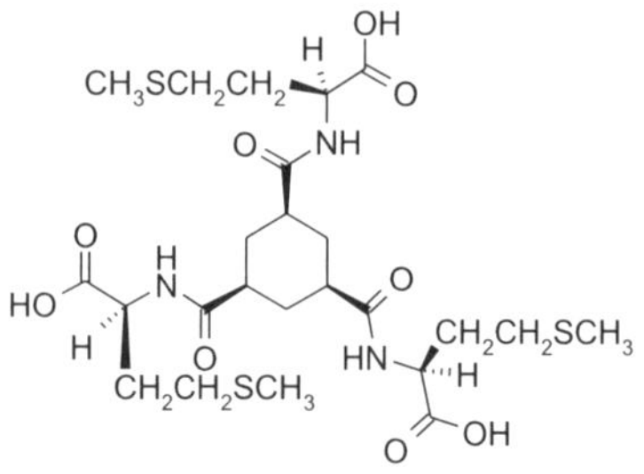

99

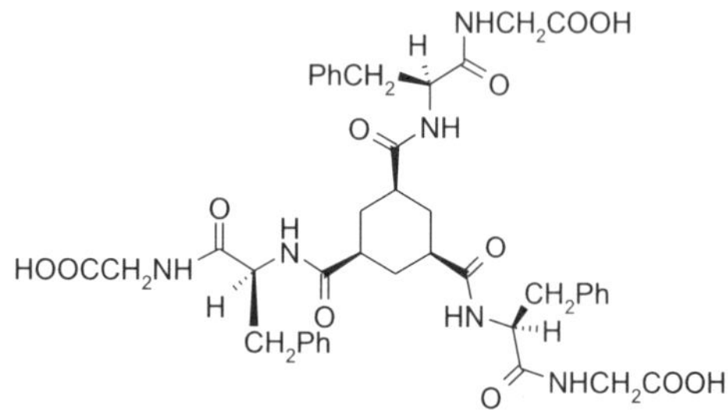

100

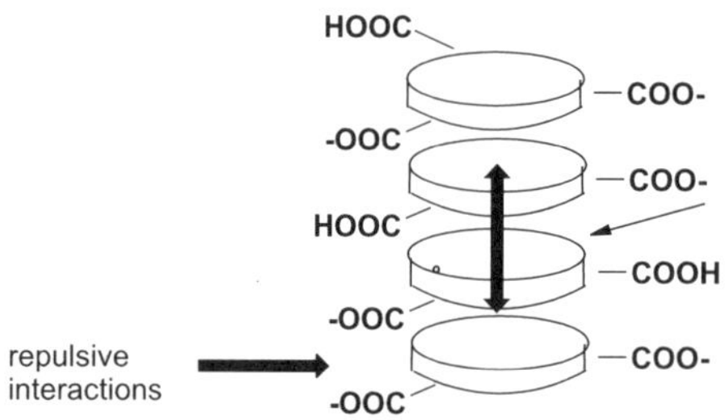

Fig. 17 Acidic hydrogelators **99** and **100** and schematic representation of the attractive and repulsive interactions in the gel aggregates

for the similar acidic gelator **100**. The higher pH range in the latter case is a consequence of stronger attractive interactions in the gel aggregates due to an additional amide-type hydrogen bond that can form between Gly units **100** (R′ = $NHCH_2COOH$). This important result shows that the pH-dependent behavior of gelators can be tuned by selection of groups with different pK_a values and also by changing the strength of the intermolecular attractive interactions (Fig. 17).

3.2 Cyclic Amino Acid Derivatives

The gelation properties of small depsipeptide **105** containing two L-Leu units was observed by Kellogg in 1993 [72]. The series of unsymmetrical diketopiperazine derivatives prepared by cyclization of various dipeptide esters of two different amino acids showed versatile gelation properties (Scheme 18) [73]. Gelators were prepared from ethyl esters of *N*-benzyloxycarbonyl-protected dipeptides by removal of the protecting group and refluxing of the amino ester in 1,3,5-trimethylbenzene. Interestingly, the diketopiperazines containing identical amino acids lack any gelation property. Within the series, the Phe, Leu (**106**) and Gly, and γ-ethylGlu (**107**) diketopiperazines showed the most efficient gelation of various organic solvents of different polarity at mgc values of 1–11 $g\,L^{-1}$. Incorporation of branched

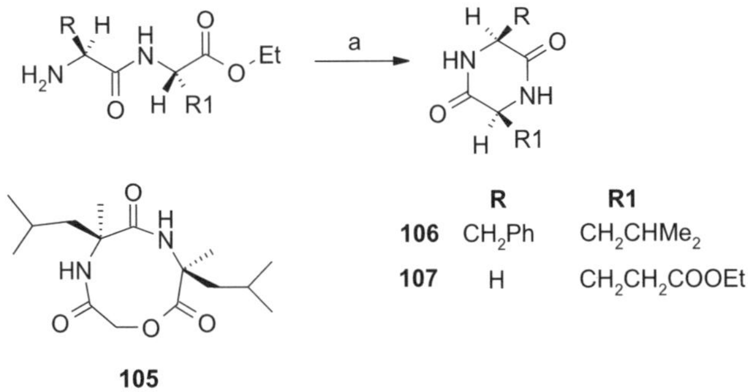

Scheme 18 Preparation of diketopiperazine gelators: (a) reflux in 1,3,5-trimethylbenzene

amino acids such as L-Glu or L-Asp in the diketopiperazine improves its gelation ability [74].

Escuder and Luis prepared a series of peptidomimetic cyclophanes constructed from *m*-phenylene and two identical amino acid units bridged by diaminopolymethylenes of variable length [75, 76]. The macrocycles were prepared by condensation of activated *N*-carbobenzoxy-protected amino acids with α,ω-diamino alkanes and macrocyclization with 1,3-bis(bromomethyl)-benzene (Scheme 19).

The C_2-symmetric Val and Ile derivatives **108–110** showed excellent gelation of aromatic solvents at mgc values in the range of 0.3–0.5 wt %. Transcription of molecular chirality into the supramolecular level has been observed by SEM and temperature-dependent CD spectroscopy. Such gelator molecules organize into columnar aggregates by formation of six hydrogen bonds per gelator molecule, and the cooling rate on going from sol to gel was

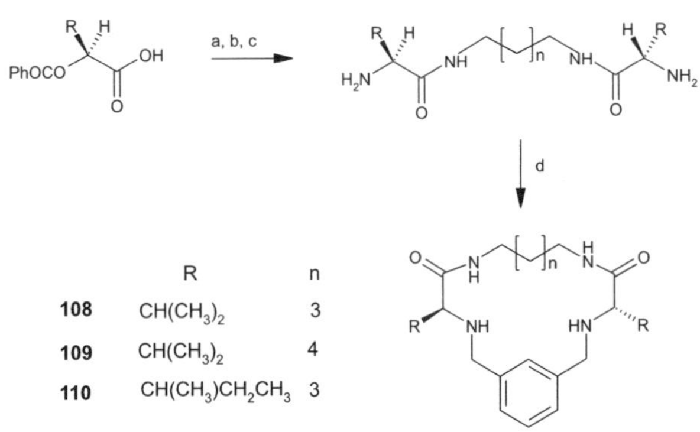

Scheme 19 Preparation of biomimetic cyclophane-type gelators: (a) *N*-hydroxysuccinimide, DCC, THF, 0 °C; (b) $H_2N(CH_2)_nNH_2$, DME, room temp.; HBr in AcOH (33%), basic workup; (c) 1,3-bis(bromomethyl)benzene, tetrabutylammonium bromide, CH_3CN, reflux

shown to have a strong influence on gel morphology. Fast cooling of **108** sol in benzene gave more helices of smaller dimensions while a slow cooling rate induced coiling of helices into larger fibers.

3.3 Bola-Type Amino Acid Gelators

A series of *N*,*N*′-*p*-toluoyl L-cystine derivatives have been prepared by Menger and it was shown that some members of the series are highly efficient hydrogelators [77,78]. The structure of the gelators resembles that of bolaamphiphiles [79] having two polar terminal groups (COOH or $CONH_2$) and a lipophilic interior. The best gelators in the series (**111**–**113**) were prepared as outlined in Scheme 20.

These highly efficient hydrogelators were designed by increasing hydrogen-bonding interactions between polar end groups ($CONH_2$) and most probably the aromatic stacking interactions between *N*-aroyl units. Compound **113** is able to gel the 75% water/25% DMSO mixture at 0.25 mmol dm^{-3} concentrations while **111** and **112** gelled the 95% water/5% DMSO mixture at 0.5 mmol dm^{-3} concentrations.

Another type of "bola gelator" incorporating two amino acid units as polar parts has been prepared by DCC condensation of *N*-protected amino acids

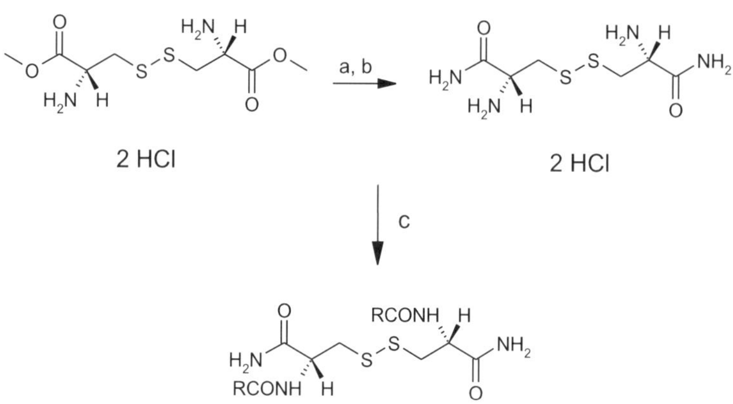

	R
111	phenyl
112	*p*-toluyl
113	2-naphtyl

Scheme 20 Synthesis of L-cystine-based hydrogelators **111**, **112** and **113**: (a) liq. NH_3, −33 °C; (b) HCl/MeOH; (c) **111**: water, NaOAc, benzoyl chloride; **112**: Et_3N, DMSO/$CHCl_3$, *p*-toluoyl chloride, **113**: Et_3N, DMSO/$CHCl_3$, 2-naphthoyl chloride

with selected α,ω-oligomethylene diamines (Fig. 18) [80]. The L-Val (**114–116**) and L-Ile (**117, 118**) based gelators exhibited versatile gelation properties toward various alcohols, aromatic, ether, and chlorinated solvents, DMF and DMSO, silicone oil, and hydrocarbon fuels.

Further synthetic modification of this type of gelator by shortening of the polymethylene bridge and introduction of charged pyridinium groups at both termini (**119–121**) gave excellent hydrogelators [81]. Gelation of water is clearly dependent on the counterion hydrophilicity so that the dichloride **119** is too soluble; however, the dibromide **120** (mgc $25\,g\,L^{-1}$) is a moderately efficient gelator and the hexafluorophosphate **121** (mgc $2\,g\,L^{-1}$) is a 15 times more efficient hydrogelator than **120**. The gelation ability of **119** and **120** increases for aqueous solutions of inorganic acids (HCl) and salts (NaCl, $MgCl_2$, $CaCl_2$) presumably due to the reduction of electrostatic repulsions in the gel aggregates; such an effect, however, was not observed for the hexafluorophosphate **121**.

A series of bis(amino acid) and bis(amino alcohol) oxalic acid amide derivatives (Fig. 19) have been prepared and shown to exhibit efficient gelation of lipophilic solvents and also of highly polar water or water/DMSO and water/DMF mixtures (ambidextrous gelators) [82–84].

This type of bola gelator lacks any long alkyl chains and its design is based on the strong hydrogen bonding and hence directing ability of the self-complementary oxalamide units also used in the design of molecular solids [85]. The optically active and racemic gelators as well as the *meso*-

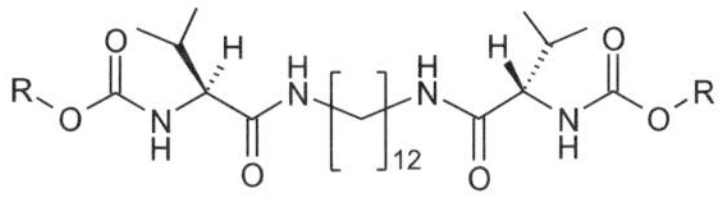

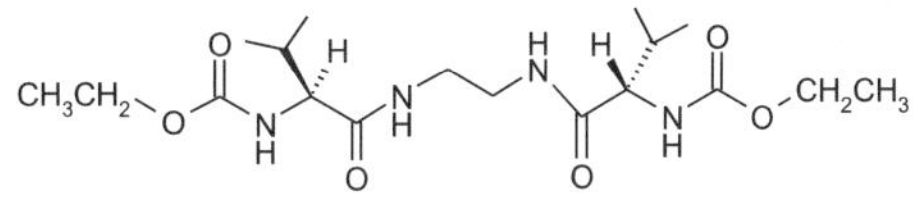

116

	R
114	CH_2Ph
115	Et

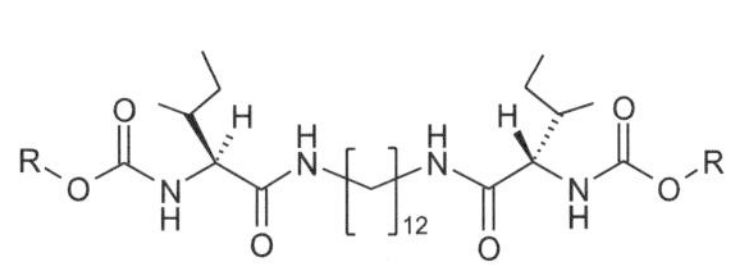

	R
117	CH_2Ph
118	Et

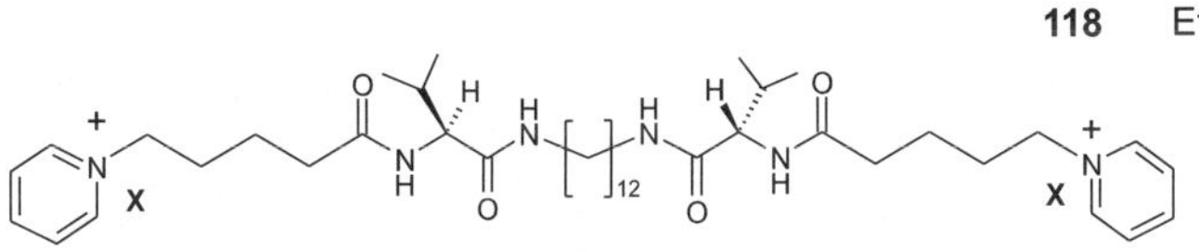

119, X = Cl^- **120**, X = Br^- **121**, X = PF_6^-

Fig. 18 Bola-type L-Val (**114–116**) and L-Ile (**117, 118**) organic solvent gelators and doubly charged hydrogelators **119–121**

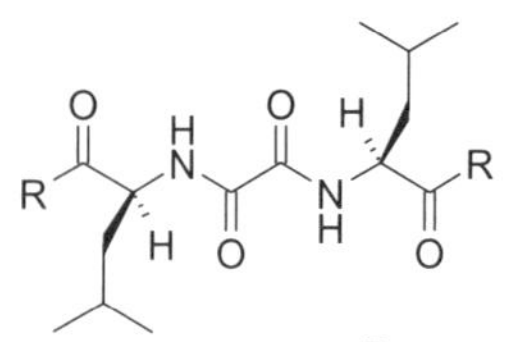

	R
(*S*,*S*)-122	OH
(*S*,*S*)-123	NH_2

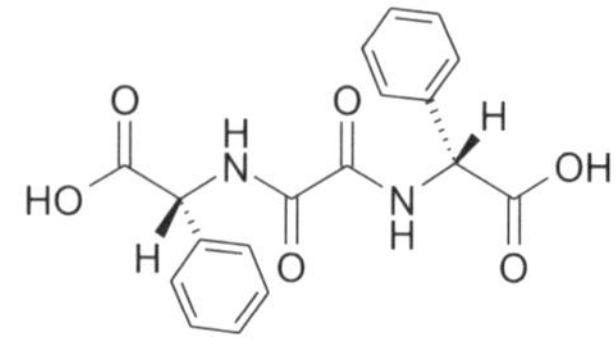

(*R*,*R*)-124

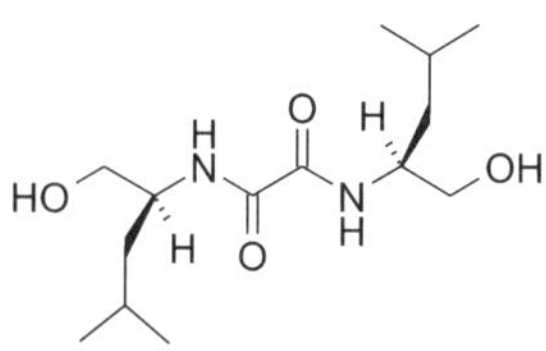

(*S*,*S*)-125

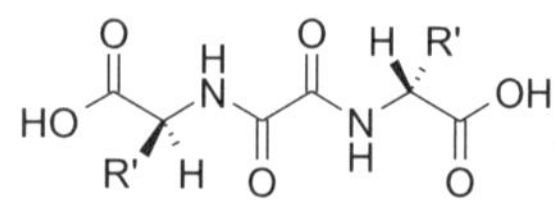

meso-126

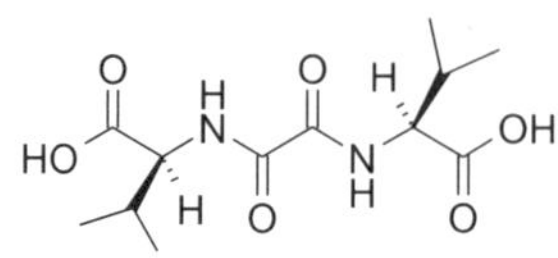

(*S*,*S*)-127

Fig. 19 Bola-type bis(amino acid and amide) **122–124** and **127**, bis(amino alcohol) **125**, and *meso*-diastereomer **126** oxalic acid amide gelators

diastereomers were prepared by reaction of oxalyl chloride with single-enantiomer or racemic amino acid in a biphasic dichloromethane/aqueous NaOH system. The bis(L-LeuOH) **122**, bis(L-LeuNH$_2$) **123**, bis(D-PhgOH) **124**, and bis(leucinol) **125** showed the best gelation properties within each series. Investigation of stereochemical influence on gelation properties revealed that single enantiomers are generally better gelators than the racemates, which tend to crystallize, while as a rule the achiral *meso*-diastereomers **126** lack any gelation ability. However, a notable exception was found for *rac*-**125**, which is capable of gelling CCl_4, *m*- and *p*-xylene, and also up to a five times larger volume of toluene than the (*S*, *S*)-**125** giving in each case stable gels not prone to crystallization. Spectroscopic investigations and single crystal X-ray structures of both the bis(amino acid) and bis(amino alcohol) gelators revealed that these types of gelators tend to form inversed bilayers in organic solvents [83, 84]. For *rac*-**125**, organization into *meso* bilayers consisting of (*R*)- and (*S*)-enantiomers is found in unstable gels which tend to crystallize upon ageing. The observation that *rac*-**125** efficiently gels toluene, some other aromatic solvents, and CCl_4 giving stable gels was explained by the formation of enantiomeric bilayers in the latter solvents each consisting of the single enantiomer [84]. Interestingly, (*S*, *S*)-**122** and the bis(ValOH) derivative (*S*, *S*)-**127** were found to form gels of exceptional thermal stability with some organic solvents of medium and low polarity. Their CCl_4 gels can be heated to temperatures 50 °C higher than the boiling point of the solvent without apparent gel-to-sol transition [86]. Based on the constitutional and conformational similarity to the oxalamide-type gelators, the gelation property of bis(PheOH) fumaric acid amide **128** and lack of any gelation for the bis(PheOH) maleic acid amide **129** was predicted (Fig. 20). Indeed, the acidic aqueous solution of **129** was found to contain only 2–10 μm-sized microspheres; upon UV irradiation, however, gel is instantaneously formed due to fast photoisomerization of **129** into **128** which self-assembled into fibrous aggregates (Fig. 20) [87]. This example shows that photoisomerization at the molecular level induces morphological transformation at the supramolecular level exemplified by the conversion of microspheres to gel fibers.

Monoalkyloxalamide derivative **130** forms gels with various chlorinated and aromatic solvents at mgc values in the range of 3–6 g L^{-1} [88]. The gelator organizes into ordered inversed bilayer primary aggregates which are subsequently juxtaposed and interlocked by van der Waals interactions giving gel fibers. Also, bis(*ε**N*-lauroyl-L-Lys) oxalyl acid amide derivatives prepared recently from *ε**N*-lauroyl-L-LysOH and oxalyl chloride showed excellent gelation of various alcohols, ketones, ethers, aromatic and chlorinated solvents, and polar solvents DMSO and DMF [89]. The dicarboxylic acid derivative **131** and linear (**132**) and branched (**133**) alkyl ester derivatives showed gelation of a similar set of solvents, but branched alkyl esters **133** and **134** showed in-

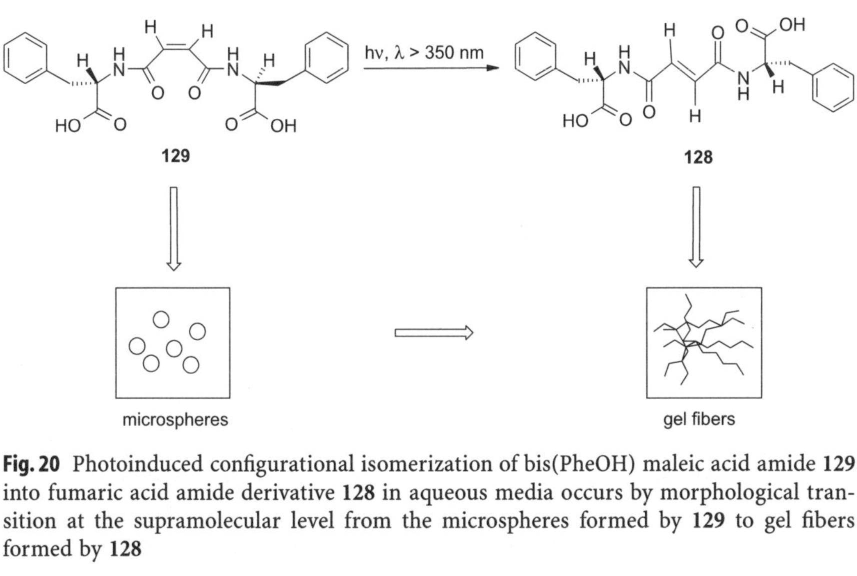

Fig. 20 Photoinduced configurational isomerization of bis(PheOH) maleic acid amide **129** into fumaric acid amide derivative **128** in aqueous media occurs by morphological transition at the supramolecular level from the microspheres formed by **129** to gel fibers formed by **128**

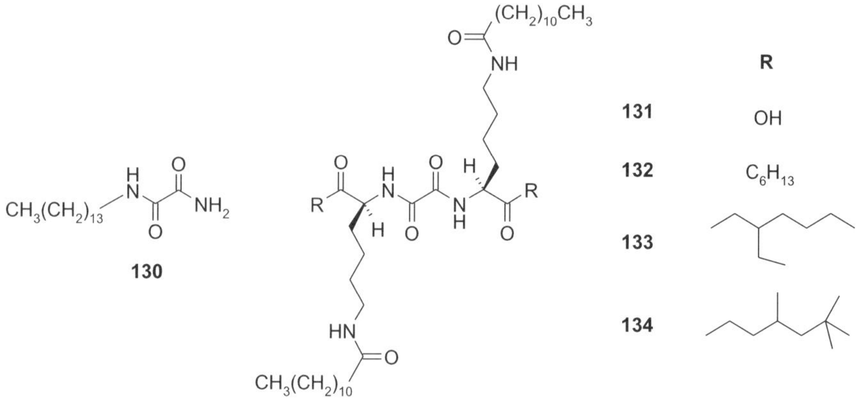

Fig. 21 Monoalkyloxalamide gelator **130** and bis(*N*-lauroyl-L-Lys) oxalyl acid amide gelators **131–134**

creased thermal stability of cyclohexane gels compared to **132**; **134** can gel cyclohexane even at 70 °C at a mgc of 0.7 wt % (Fig. 21).

3.4 Small Peptidic Gelators: Toward Biomaterials

At pH 3–5, the *N*-(fluorenylmethoxycarbonyl) dipeptides (Fmoc-dipeptides, **135** and **136**, Fig. 22) form stable hydrogels at 4 mM concentrations [90]. The CD spectra of **135** and **136** hydrogels are nearly mirror images and show Cotton effects at 219 (n–π^* transition) and 246–304 nm (π–π^* transition) indicating superhelical organization of gel aggregates. The hydrogel turns immediately into the sol at pH 6 showing reversible gel-to-sol transition. In addition to the pH-reversible transformation, the **135** hydrogel respond to

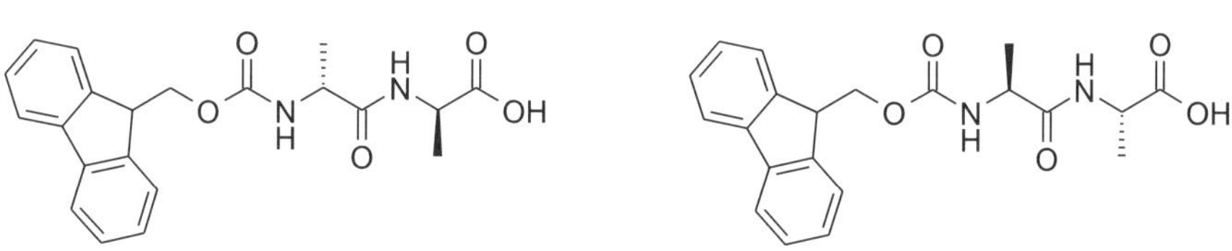

135 **136**

135 hydrogel + vancomycin ⟹ **135**-vancomycin complex solution

Fig. 22 Enantiomeric Fmoc-D-Ala-D-Ala (**135**) and Fmoc-L-Ala-L-Ala (**136**) hydrogelators and enantioselective vancomycin-induced **135** hydrogel-to-solution transition a specific biological ligand–receptor interaction: addition of 1 equiv. of vancomycin, the antibiotic known to bind with the D-Ala-D-Ala terminus of the peptidoglycan bacterial cell wall precursor, turns the hydrogel into a clear solution. However, the hydrogel of the enantiomer **136** showed no change upon adding 1 equiv. of vancomycin, in accord with the known fact that the antibiotic cannot bind with the L-Ala-L-Ala fragment. These results are a good example of the stereoselective biological ligand–receptor interactions which can be used to induce phase transition, and offer a new design principle toward responsive supramolecular gels.

A tripeptide **137** possessing two noncoded amino acids (Aib and β-Ala) showed efficient gelation of cyclohexane, benzene, toluene, and *o*-dichlorobenzene [91]. Its gelation ability has been explained on the basis of the X-ray crystal structure showing organization of β-turn-containing molecules into double-columnar fibrous aggregate by intermolecular hydrogen bonding (Fig. 23). In contrast, the tripeptide **138** containing Gly instead of Aib forms an extended conformation, which organizes into an antiparallel β-sheet structure by intermolecular hydrogen bonding and gives crystals instead of gels in the same solvents.

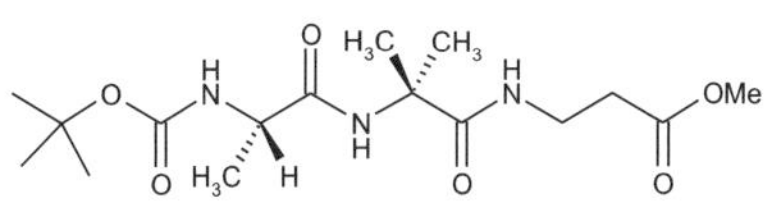

137

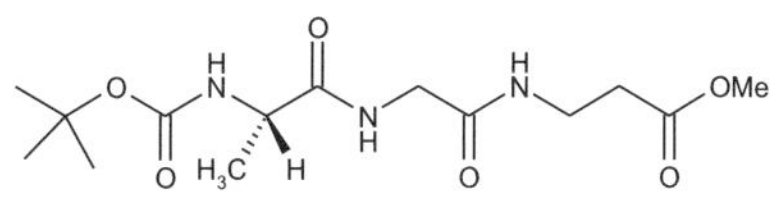

138

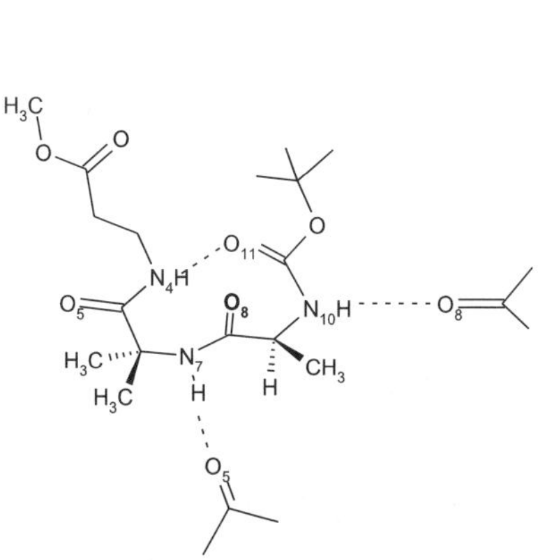

137

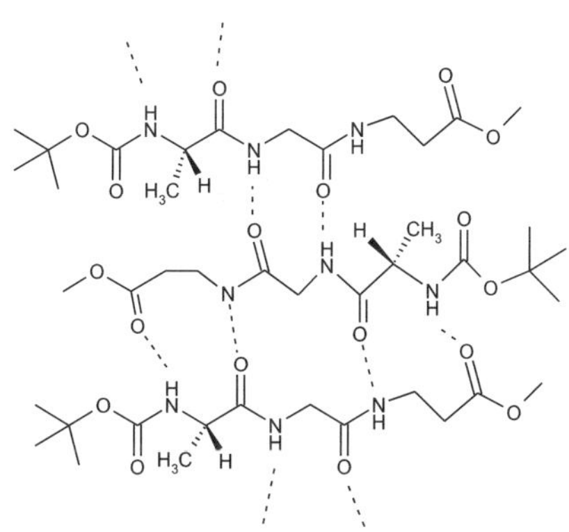

138

Fig. 23 β-Turn-forming gelator **137** which organizes into double-columnar fibrous aggregate by two intermolecular hydrogen bonds ($N_7H\cdots O_5=C$ and $N_{10}H\cdots O_8=C$), and nongelling tripeptide **138** with Gly instead Aib unit which organizes into an antiparallel β-sheet structure that leads to crystallization

In a search for an accurate model of amyloid fibrils related to various prion diseases including Alzheimer's, Huntington's, and Parkinson's diseases, the amphiphilic tripeptides **139–142** were prepared [92, 93]. They were found to form gels with solvents of low polarity (benzene, CCl_4, cyclohexane, dielectric constants $\varepsilon < 3$); however, in a solvent of medium polarity such as chloroform they do not aggregate. In water which possesses a high ε value of 78.39 they give translucent dispersions. Their self-assembly into bilayered aggregates in water is governed by hydrophobic interactions supported by intermolecular hydrogen bonding between tripeptide fragments, giving parallel β-sheets. It seems that the aggregates are of insufficient length to accomplish gelation. In CCl_4, however, inversed bilayers are formed possessing a similar β-sheet hydrogen-bonding organization which also represents the dominant intermolecular interaction in solvents of low polarity (Fig. 24).

Another approach to amyloid models is based on oxidized glutathione **143** [94]. The compound yields transparent gels in DMSO and in water/DMF and water/MeOH mixtures giving extended antiparallel β-sheetlike aggregates by intermolecular hydrogen bonding (Fig. 25). The disulfide bridge functions as a covalent preorganizational element inducing an intermolecu-

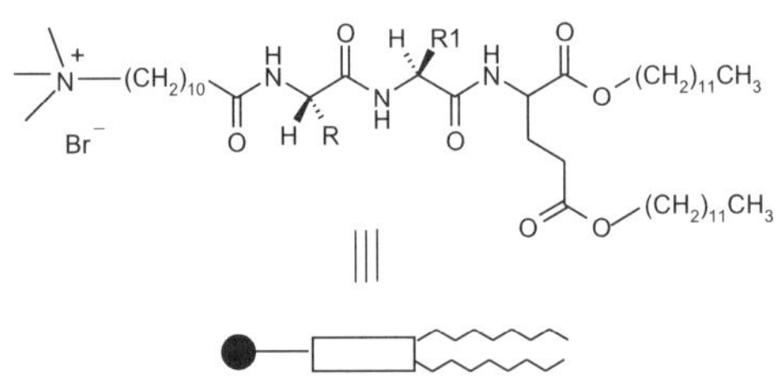

	R	R1
139	CH_3	CH_3
140	$CH(CH_3)_2$	$CH_2CH(CH_3)_2$
141	$CH(CH_3)_2$	$CH(CH_3)_2$
142	CH_2Ph	CH_2Ph

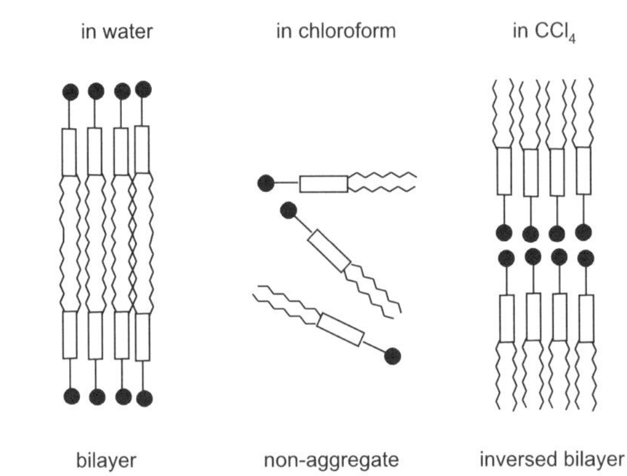

Fig. 24 Amphiphilic tripetide gelators **139–142** and schematic representation of their primary aggregation in water (bilayers), CCl_4 (inversed bilayers), and lack of aggregation in chloroform

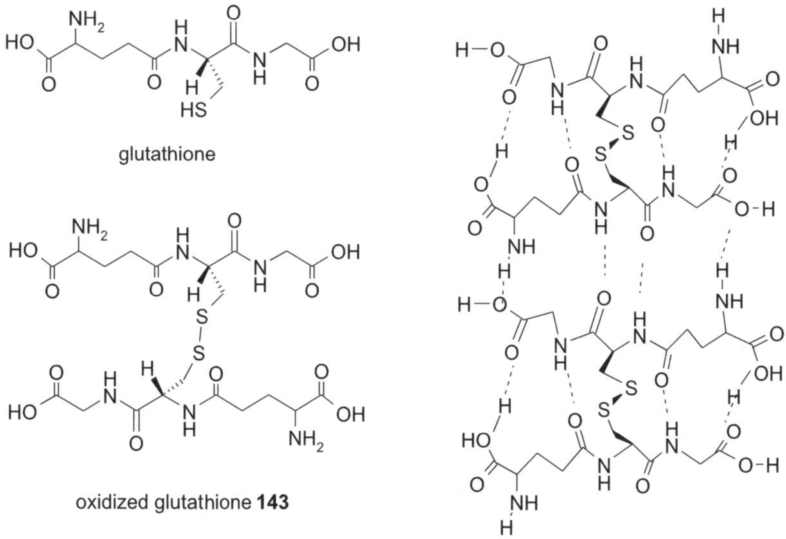

Fig. 25 Glutathione, oxidized glutathione gelator **143**, and extended β-sheet structure giving gel fibers in organic solvents

lar β-sheet type of hydrogen bonding, and the geometry of the γ-glutamyl residue enables specific intermolecular recognition through hydrogen bonding. Slices of DMSO gel were stained by alkaline Congo Red and observed under crossed polarizers. A strong green birefringence was observed as in the case of the amyloid fibrils.

A series of oligomers of amino acids **144–146** were prepared and shown to exhibit efficient gelation of various organic solvents ranging from polar lower alcohols, DMF, and DMSO to apolar aromatic solvents [95]. The oligomers were prepared by decarboxylation of *N*-carboxyanhydrides of selected amino acids in the presence of an alkylamine (Scheme 21) [96]. The degree of oligomerization was determined from the relative intensities of selected NMR signals. Oligo(glycine) and oligo(L-alanine) lack any gelation ability while oligo(L-Phe), oligo(L-Ile), and oligo(L-glutamate esters) are efficient gelators. The glutamate esters with a short terminal alkylamino chain (**146**; $m = 3$) showed better gelation of aromatic solvents than the derivatives with a longer chain, and gelation was achieved only for derivatives with a degree of oligomerization less than 10.

The 24-residue peptide K24 related to the transmembrane domain of the IsK protein was found to readily form β-sheet tapes in lipid bilayers. The same peptide formed stable transparent gels with methanol or 2-chloroethanol at a concentration of $3\,\mathrm{g\,L^{-1}}$ by the same type of organization [97]. TEM of the gels revealed entangled polymeric tapes of single molecule thickness (0.7 nm). The much shorter, de novo 11-residue peptide with primary sequence MeCO-Gln-Gln-Arg-Phe-Gln-Trp-Gln-Phe-Glu-Gln-

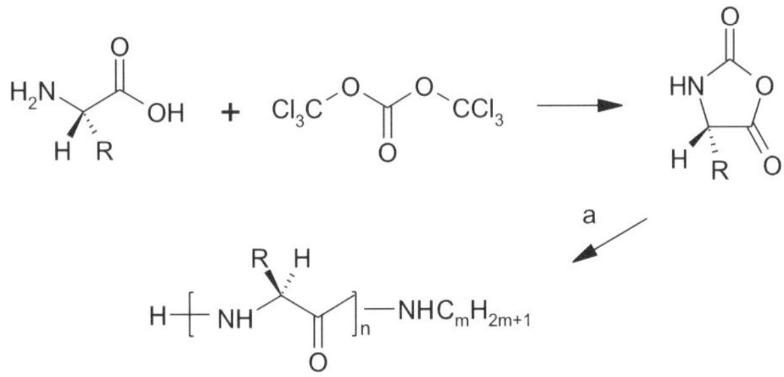

	R	n	m
144	CH(Me)Et	5.0	18
145	CH_2Ph	5.1	12
146	CH_2CH_2COOEt	3.5	3

Scheme 21 Synthesis of *N*-carboxyanhydrides and their oligomerization into **144–146**: (a) $C_mH_{2m-1}NH_2$

Gln-NH_2 (**147**) was designed to form β-sheet polymer tapes in water based on the following design principles: (1) cross-strand attractive forces (hydrophobic, electrostatic, hydrogen bonding) between side chains; (2) lateral recognition between adjacent β-strands to constrain their unidirectional self-assembly, and (3) strong solvation of the surface of the tapes to control solubility. The proposed antiparallel β-sheet arrangement in the gel is shown in Fig. 26.

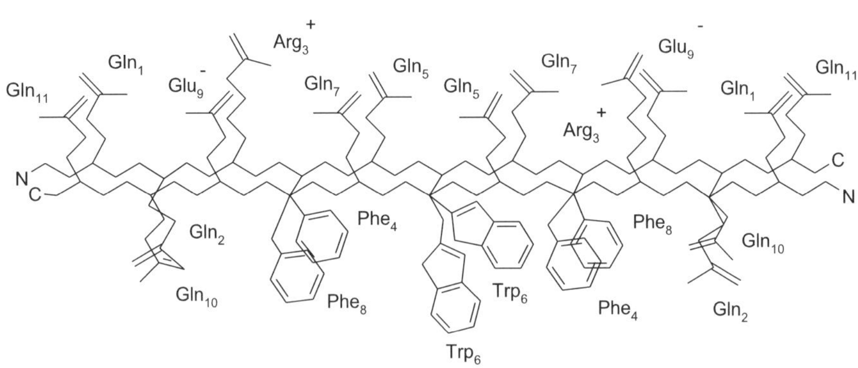

Fig. 26 Schematic representation of antiparallel β-sheet arrangement of 11-residue peptide **147**. C and N denote N and C termini. Electrostatic interactions between negatively charged Glu_9 and positively charged Arg_3 side chains and stacking interactions between Phe_4-Phe_8 and Trp_6-Trp_6 are indicated

A theoretical model that can predict the morphology (tapes, ribbons, fibrils, fibers) and properties of self-assembled peptides from the molecular parameters of the peptide monomers was recently proposed [98]. Self-assembling peptide polyelectrolyte complexes capable of forming nematic hydrogels have been prepared by the same group of researchers [99]. Mixing of aqueous solutions of 11-residue cationic and anionic peptides results in the formation of polyelectrolyte β-sheet complexes of 1 : 1 stoichiometry, which further self-assemble into a fibrillar network producing a nematic hydrogel. The network is quite robust to variations of pH or peptide concentration so the hydrogel is stable in the pH range of 1–12. The latter results open a new avenue for exploiting protein-like self-assembly in the production of biocompatible soft materials with numerous possible applications, for example, encapsulation, immobilization, and separation of cells, proteins, antibodies, or enzymes.

The first antibiotic hydrogelator, the vancomycin–pyrene conjugate **148**, has been recently prepared (Fig. 27) [100]. In water, the conjugate organizes into helical fibers by multiple hydrogen bonding between vancomycin units and π–π stacking between pyrenes. It was observed that **148** is 8–11 times more active against VRE than vancomycin; it is hypothesized that the conjugate **148** aggregates into polymerlike structures at the cell surface. The approach opens new exciting possibilities for the preparation of various hydrogelator–biologically active compound conjugates, which may result in improved properties of drugs [101].

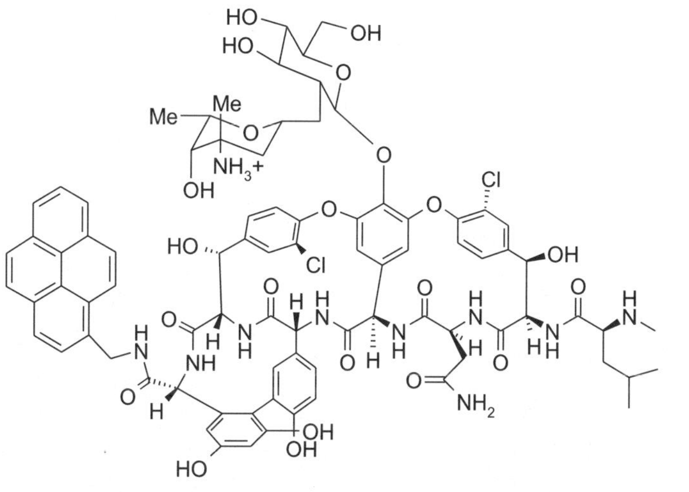

148

Fig. 27 A hydrogelator, vancomycin–pyrene conjugate **148**, exhibiting enhanced antibacterial activity compared to that of vancomycin

4 Urea-Derived Gelators

4.1 Bisurea Derivatives

The tendency of the self-complementary substituted urea fragments to self-assemble into one-dimensional arrays of hydrogen bonds was used in the design of molecular solids [102, 103]. Synthesis of *trans*-(1*R*,2*R*)-1,2-bis(dodecylureido)-cyclohexanes **149** and **150** having two ureido units and long alkyl chains gave excellent gelators of various polar (MeOH, EtOH, DMF, DMSO) and apolar (chloroform, tetrachloromethane, benzene, toluene) organic solvents (Fig. 28) [104]. The compounds prepared from *trans*-(1*R*,2*R*)-1,2-diamino cyclohexane and the respective alkylisocyanate showed a gelation efficiency in the range of 2–15 $g\,L^{-1}$. In the *trans*-bis(ureido) derivatives the urea units are in the antiparallel arrangement and hence self-complementary so that the one-dimensional self-assembly results in the formation of linear aggregates. In contrast, the *cis* derivative **151** is not self-complementary and hence is unable to self-assemble and lacks any gelation ability. The rheology and thermotropic properties of gels formed by **149** and various primary alcohols have been studied in detail [105].

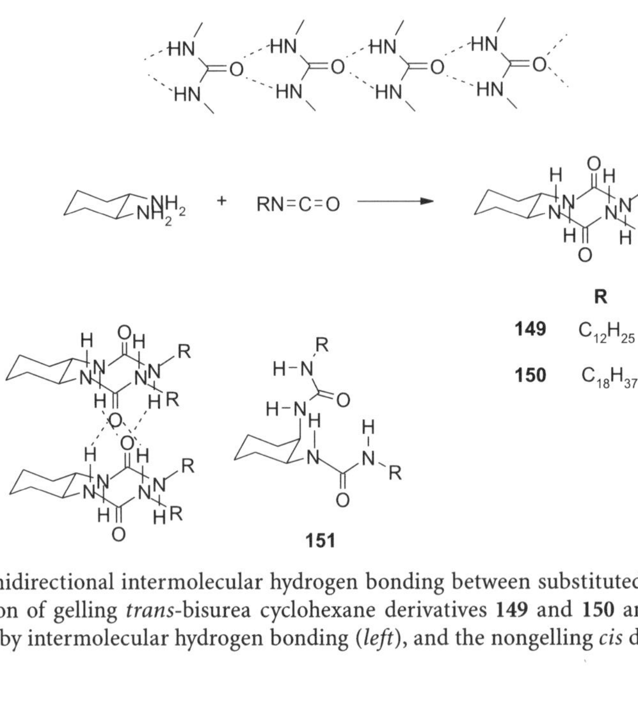

151

Fig. 28 Unidirectional intermolecular hydrogen bonding between substituted ureas (*top*), preparation of gelling *trans*-bisurea cyclohexane derivatives **149** and **150** and their self-assembly by intermolecular hydrogen bonding (*left*), and the nongelling *cis* derivative **151**

The achiral bisurea derivatives **152–154** possessing terminal benzyl groups and **155–157** with terminal alkyl chains of variable lengths were studied by van Esch and Feringa (Scheme 22) [106, 107]. The compounds exhibited efficient gelation of 1- and 2-octanol, toluene, tetralin, *n*-butyl acetate, cyclohexanone, and hexadecane. Gelators were prepared in high yields by reaction of alkyl or benzyl isocyanates with the respective amines.

Electronic microscopy investigation of gels revealed that the gel fibers have a multilayered structure. The STM study of monolayers of **156** physisorbed at the graphite/1-octanol interface showed at submolecular resolution the presence of slightly bent ribbons of 5-nm thickness formed by hydrogen bonded all-*trans* molecules. Subsequently, such ribbons stack to form mul-

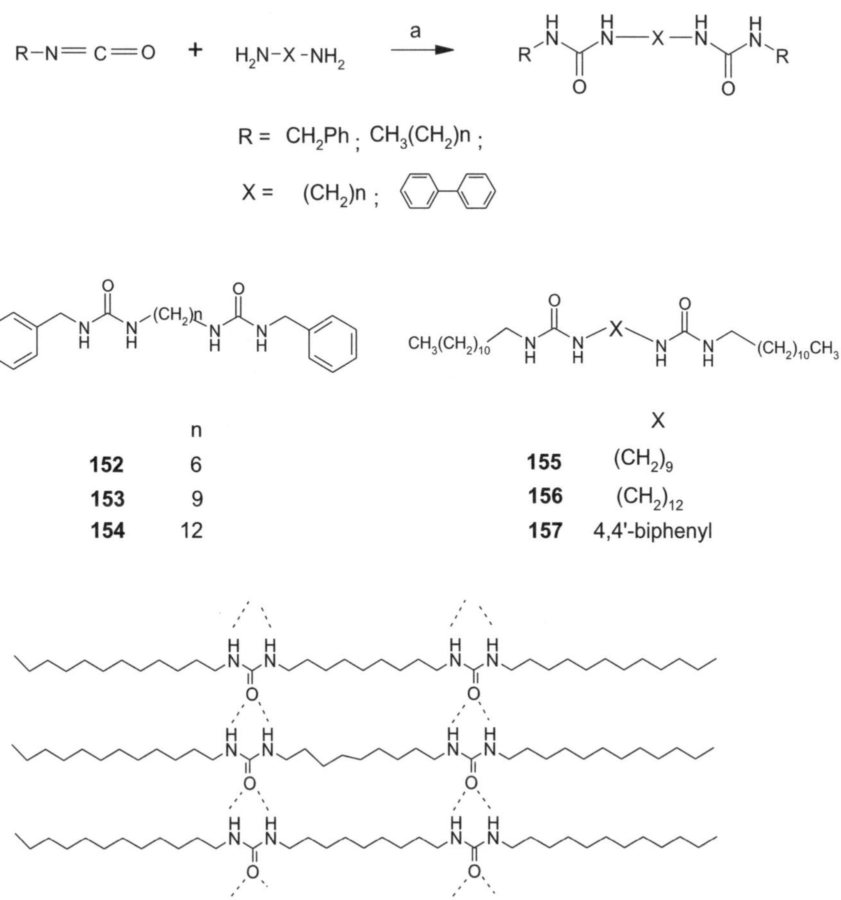

Scheme 22 Synthesis of bisurea gelators **152–157**: (a) dichloromethane, room temp. (*top*). Ribbonlike organization of all-*trans*–**155** by intermolecular hydrogen bonding involving both urea units (*bottom*)

tilayered sheets which finally stack to give gel fibers. Further STM studies of self-assembled monolayers of **155** and **156** revealed different contrasts of urea groups for odd and even numbers of C atoms in the spacer, and this contrast variation was correlated with the orientation of the molecules within the monolayer [108]. The two-component self-assembly of selected bisurea derivatives aimed at patterning organic monolayers was also studied by STM by the same group. It was shown that the variation in position of urea groups, molecule length, odd–even number of spacer atoms, and shape complementarities of one component controls the two-dimensional patterning from randomly intermixed systems to phase separation [109]. The same group also prepared chiral (1*R*,2*R*)-bis(amido)- (**158**) and bis(ureido)- (**159**) cyclohexane gelators bearing terminal polymerizable methacrylate functions (Fig. 29) [110, 111].

While **158** gelled only tetralin at high concentration (mgc 25 g L^{-1}) the bis(ureido) derivative **159** readily formed gels with a variety of organic solvents including cyclohexane, butyl acetate, benzene, tetralin, and 1,2-dichloroethane at mgc values in the range of 2–20 g L^{-1}. Butyl acetate gel of **159** was irradiated (200-W high-pressure Hg lamp) in the presence of 2-phenylacetophenone resulting in complete photopolymerization of the gel. In this way the supramolecular gel containing fibers formed by self-assembly of **159** is transformed into a polymer gel showing a more than 100 °C higher gel melting temperature than that of the unpolymerized gel. The freeze-dried polymerized gel gave a highly porous material (density of only 0.005 kg dm^{-3}) retaining the shape of the vessel and having the characteristics of an organic

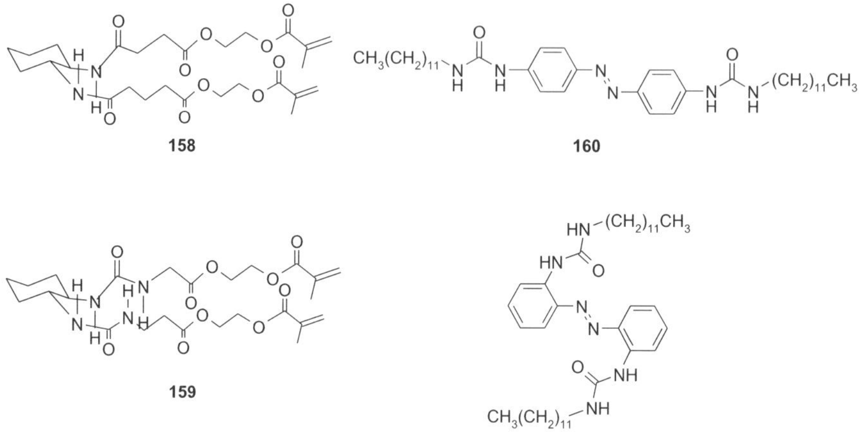

Fig. 29 Polymerizable bis(amido)- **158** and bis(ureido)- **159** cyclohexane gelators containing methacrylate end groups, and azobenzene gelators **160** and **161**

aerogel; resolvation by addition of methanol or chloroform gives a transparent gel again without any change of shape.

Introduction of a 4,4′-azobenzene spacer between two urea units gives the highly efficient gelators **160** and **161** (Fig. 29) that are capable of hardening highly polar solvents (DMSO, EtOH, 2-propanol), aromatic solvents, and highly lipophilic hexane. Compound **161** is especially efficient as it is able to gel hexane at a 0.2 g dm^{-3} concentration by organizing into two different morphs depending on kinetic factors and the polarity of the solvent [112].

Chiral recognition was found to occur through cooperative interactions in the aggregates and gels of **162** and **163** [113]. Coaggregation of (*R*, *R*)-**162** and (*R*, *R*)-**163** was less favorable than the coaggregation of (*S*, *S*)-**162** and (*R*, *R*)-**163**. Since the dimerization constants are not affected by the chirality of the components and the association constant for the (*S*, *S*)-**162** and (*R*, *R*)-**163** coaggregates, $K_{R,S}$, is almost twice as large as $K_{R,R}$, it is evident that cooperativity is the most important factor for chiral recognition in these systems. The

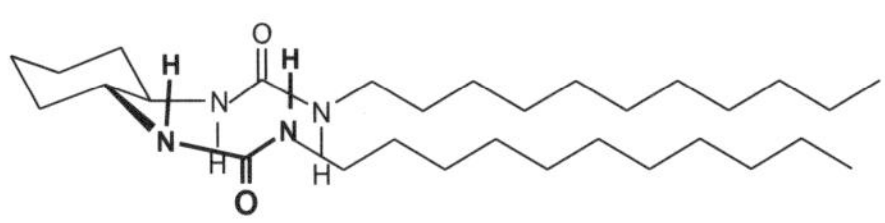

(*R*,*R*)-**162**

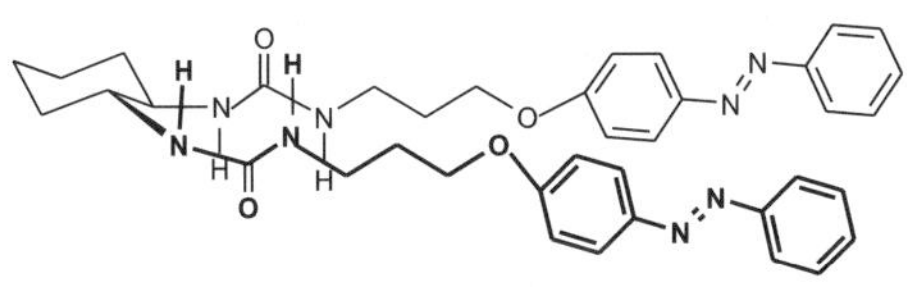

(*R*,*R*)-**163**

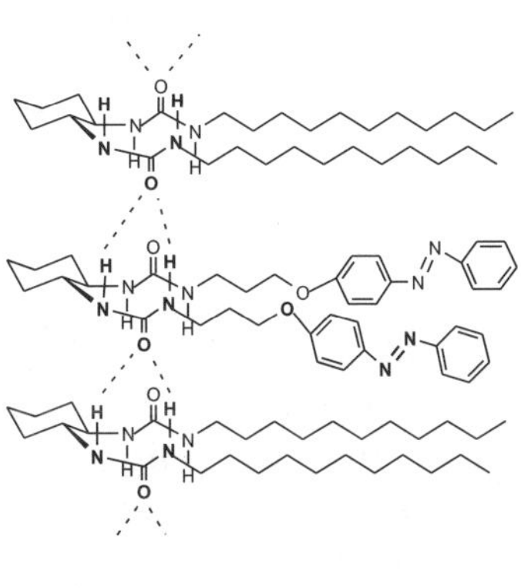

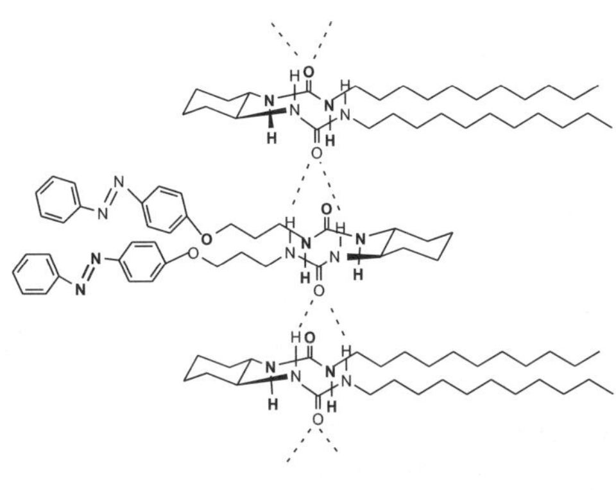

(*R*,*R*)-**162** and (*R*,*R*)-**163** (*S*,*S*)-**162** and (*R*,*R*)-**163**

Fig. 30 Chiral gelators **162** and **163** and representation of different packing in homochiral and heterochiral combinations of **162** and **163**

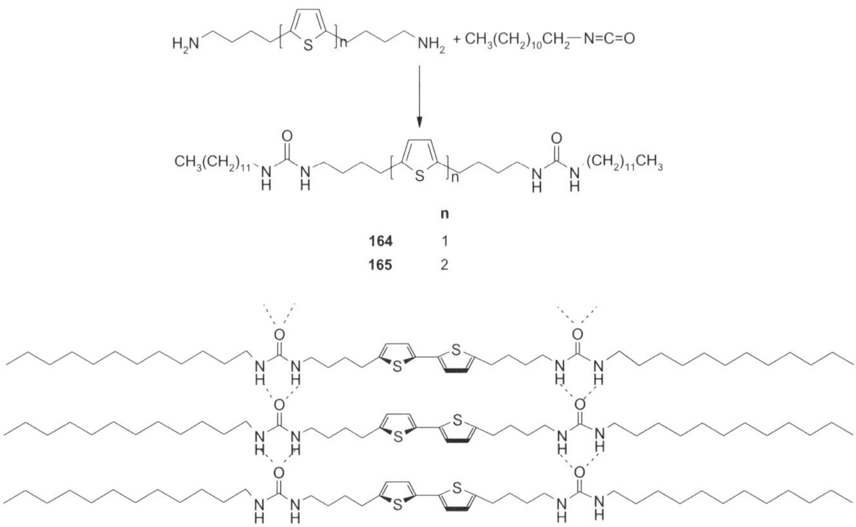

Scheme 23 Preparation of thiophene-containing gelators **164** and **165** and spatial organization by urea–urea hydrogen bonding and stacking of bithiophene units

results from NMR titration studies, CD measurements, and kinetics of *cis–trans* isomerization of the azobenzene groups in the coaggregates consistently point toward different packing of **163** in the homo- and heterochiral coaggregates. In the former, the azobenzene groups are inserted between closely packed alkyl chains and in the latter, the same groups experience a more exposed "solvent-like" environment (Fig. 30).

With the idea of combining the favorable properties of polymers of thiophene as organic semiconducting materials and the ability of bisurea compounds to self-assemble into well-defined lamellar structures, van Esch and Feringa have synthesized gelators **164** and **165** with thiophene and bithiophene groups incorporated in the spacers connecting two urea units [114]. The compounds prepared from 2,5-di(4-aminobutyl)-thiophene and 5,5′-di(4-aminobutyl)-2,2′-bithiophene (Scheme 23) are able to form gels with tetralin and 1,2-dichloroethane.

In gels, the bisurea thiophenes **164** and **165** organize into lamellar ribbons by hydrogen bonding between urea units and stacking interactions between thiophene moieties. The well-defined arrangements of thiophene moieties in the closely packed layers results in efficient charge transport, as measured by the pulse-radiolysis time-resolved microwave conductivity (PR-TRMC) technique, and make these compounds promising candidates for application in charge-transport devices based on organic semiconducting layers. Supramolecular organization in 2D of similar bisurea derivatives containing thiophene, bithiophene, and trithiophene units in the spacers has been studied by STM at the solution/graphite interface [115]. It was found that the compounds form highly ordered physisorbed monolayers organized by hydrogen bonding between urea units and partial overlap of thiophene π systems.

4.2 Urea–Amino Acid Conjugates

The bisurea–amino acid conjugates **166–168**, a new type of low molecular weight organogelators, have been prepared by Hamilton [116]. The compounds were obtained by the reaction of xylylene-1,4-bisisocyanate and *tert*-butyl esters of L-Ala, L-Phe, and L-Val (Scheme 24).

However, only the L-Val derivative **168** formed a gel with ethyl acetate that was stable at room temperature, whereas **166** and **167** gelled acetone and THF at 5 °C and the gels turned into sols at room temperature.

The same group also designed efficient hydrogelators by combining bisurea units, glutamate, and long alkyl chains in the same structure [117]. The hydrogelators **169–176** are able to gel at mgc values below 0.3 wt % in the narrow pH range of 5.9–7.9 and at high ionic strength ($I = 1\ \mathrm{mol\,kg^{-1}}$). Synthesis starts from the commercially available *N*-Boc-protected γ-benzyl ester of L-glutamate which was first condensed with selected long-chain alcohols. The esters were then treated with trifluoroacetic acid (TFA) which removed the Boc protecting groups. The TFA salts were reacted with selected bisisocyanates giving bisurea derivatives, and finally the γ-benzyl esters were removed by hydrogenolysis in the presence of Pd/C catalyst to give the hydrogelators in 45–70% overall yields (Scheme 25).

The combined use of cryo-TEM and X-ray diffraction sheds light on the mechanism of gelation for certain members of the bisurea-glutamate family [118]. Aggregation is governed by intermolecular hydrogen bonding between both urea units, by hydrophobic interactions involving alkyl ester groups, and also by Ca^{2+}–bis(carboxylate) interactions, the latter being in

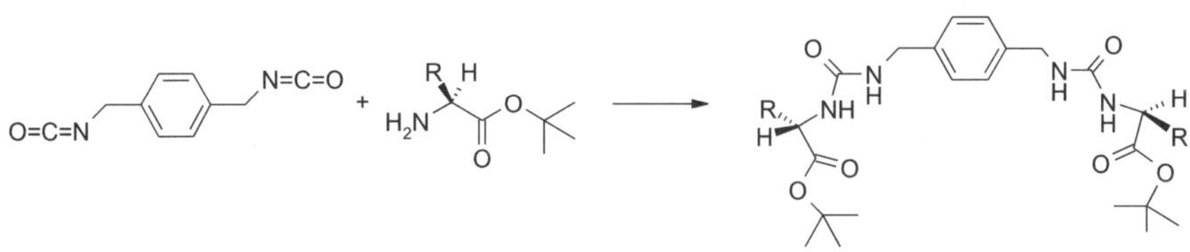

	R
166	CH_3
167	CH_2Ph
168	$CH(CH_3)_2$

Scheme 24 Synthesis of bisurea–amino acid conjugates **166–168**

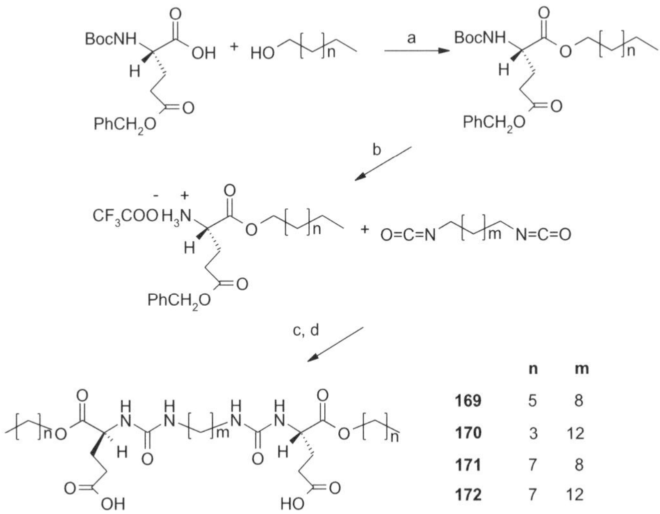

	n	m
169	5	8
170	3	12
171	7	8
172	7	12
173	11	6
174	11	8
175	11	12
176	14	8

Scheme 25 Synthesis of bisurea–glutamate hydrogelators **169–176**: (a) EDCI, DMAP; (b) CF_3COOH; (c) Et_3N, CH_2Cl_2; (d) H_2, Pd/C

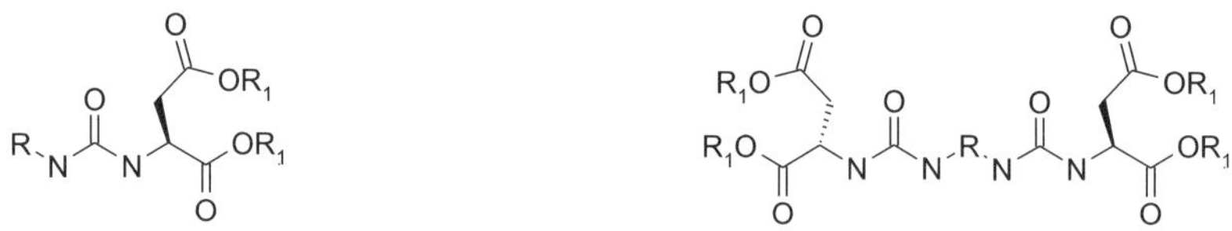

177 **178**

R = $(CH_2)_5CH_3$, p-$CF_3C_6H_4$, Ph R$_1$ = $(CH_2)_2(CF_2)_7CF_3$ R = 1,4-C_6H_4, $(CH_2)_6$, 1,4-xylyl

Fig. 31 Some selected structures of monourea **177** and bisurea **178** types of gelators of sc CO_2

accord with the observation that the presence of Ca^{2+} strongly promotes gelation.

A structurally similar group of compounds based on L-aspartate were designed as supercritical (sc) CO_2 gelators [119]. In addition to two urea and two L-aspartate units, CO_2-philic groups such as fluoroalkyl and fluoroether were incorporated into ester groups or into bridge-connecting urea fragments. Two types of efficient gelators of sc CO_2 were obtained in this way: the monourea **177** and the bisurea **178** (Fig. 31) which are able to gel

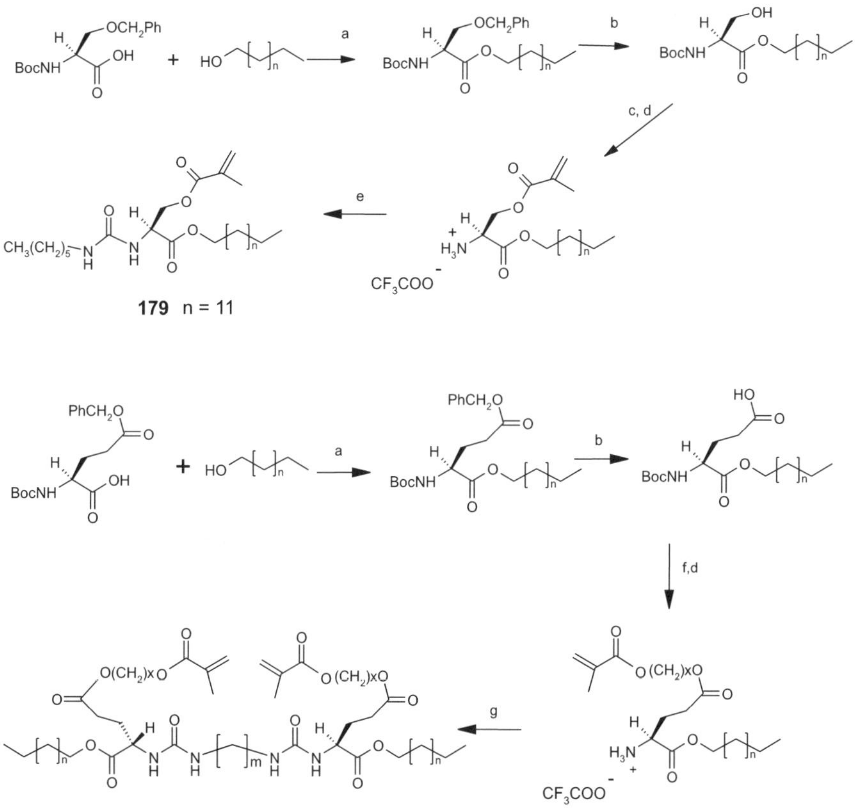

	n	m	X
180	11	12	2
181	11	12	9

Scheme 26 Synthesis of polymerizable monourea L-Ser gelator **179**: (a) EDCI, DMAP; (b) H_2, Pd/C; (c) methacryloyl chloride; (d) TFA, CH_2Cl_2, and bisurea L-Glu derivatives **180** and **181**; (e) TEA, $CH_3(CH_2)_5NCO$ and bisurea L-Glu derivatives **180** and **181**; (f) $HO(CH_2)_xOCOC(CH_3) = CH_2$; (g) $OCN(CH_2)_{12}NCO$

at mgc values between 1.0–6.0 wt %. Impressively, upon removal of CO_2 self-supporting foams with cells possessing an average diameter smaller than 1 μm were obtained, which resulted in a reduction of the bulk density of 97% relative to the parent material. Such gels are of high interest as precursors of microcellular materials which may find numerous applications.

Mono- and bisurea derivatives containing L-Ser and L-Glu alkyl and alkyl methacrylate ester groups were designed as polymerizable organogelators [120]. The L-Ser derivative **179**, and L-Glu derivatives **180** and **181**, (Scheme 26) exhibited efficient gelation of ethylene glycol, hexane, ethyl ac-

	R	n
182	Me	4
183	Me	5
184	Et	4
185	Et	5
186	CH_2Ph	2
187	CH_2Ph	3

	R_1	R_2
188	CH_2Ph	Me
189	CH_2Ph	Et
190	CH_2Ph	CH_2Ph
191	p-$CH_2C_6H_4Br$	Et
192	p-$CH_2C_6H_4Br$	CH_2Ph

Fig. 32 Monourea L-serine hydrogelators **182–192**

etate, acetone, and EtOH at mgc values of 2–10 g L^{-1}. The ethyl acetate gels of **180** and **181** were fully polymerized after 2–4 h of irradiation with a 500 –W Hg lamp, giving gels with impressively higher thermal stabilities; gel melting temperatures were 70 °C higher compared to those of the unpolymerized samples. SEM investigation of polymerized and unpolymerized gels showed practically identical morphologies.

Monourea L-serine derivatives **182–187** containing long *N*-alkyl chains exhibited remarkable hydrogelating properties at mgc values of 0.8–1.5 wt %. Further structural modifications by introduction of *N*-benzyl instead of *N*-

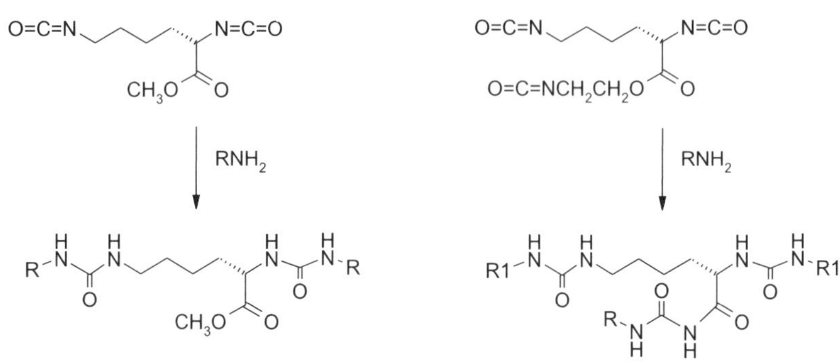

	R
193	C_8H_{17}
194	$C_{12}H_{25}$
195	$C_{18}H_{37}$

	R
196	C_8H_{17}
197	$C_{12}H_{25}$
198	$C_{18}H_{37}$

Scheme 27 Synthesis of di- (**193–195**) and triurea (**196–198**) gelators exhibiting in situ gelation at room temperature

alkyl substituents to give derivatives **188–192** also gave efficient hydrogelators, the best being the *N*-(4-bromobenzyl) ethyl ester derivative **191** capable of forming the hydrogel at a mgc of 0.5 wt % (Fig. 32) [121].

In contrast to the conventional gelation commonly performed by dissolving a gelator in a selected solvent by heating and subsequent cooling of the solution, the example of in situ gelation that occurs by reaction of highly reactive methyl 2,6-diisocyanatohexanoate or 2-isocyanatoethyl 2,6-diisocyanatohexanoate with selected long-chain amines was recently reported by Suzuki (Scheme 27) [122]. In this way, prepared di- and triurea derivatives exhibited gelation of various aromatic solvents, EtOAc, THF, dioxane, DMSO, and DMF at mgc values in the range of 5–40 $g L^{-1}$. Derivatives **196** and **197** failed to gel acetone, ethyl acetate, and acetonitrile by the conventional method due to their insufficient solubility in these solvents even at elevated temperatures. However, in situ gelation performed by reaction of 2-isocyanatoethyl 2,6-diisocyanatohexanoate with octylamine or dodecylamine at room temperature produced gels with the same solvents.

5 Latent Gelators

Recently, a new concept of "latent" low molecular weight gelators was introduced by Weiss and George [123]. They have shown that sols or solutions of various alkylamines can absorb CO_2 giving more strongly gelling ammonium carbamates. Gentle heating of the formed gels results in decarboxylation of the carbamates and bubbling of nitrogen ensures displacement of dissolved CO_2 so that sols or solutions of starting amines are recovered (Scheme 28). The gelation/degelation cycle could be repeated many times without detectable degradation of the components. This example describes the new concept of chemically reversible gels that are also thermoreversible. It shows that gelation could be induced by in situ chemical reaction and that chemically reversible organogels could be prepared via latent gelators.

The described concept has been recently extended to other gases (NO_2, SO_2) and CS_2 (Scheme 29). However, in contrast to the systems with CO_2, none of the adducts **202–204** could be reverted to the starting alkylamines by heating, but the adducts **204** were converted into N,N'-dialkylthiourea derivatives **205** which also exhibited gelation properties [124].

Further extension of the above principle resulted in the highly interesting preparation of polyammonium–polycarbamate alcohol and 1-methyl-2-pyrrolidone gels formed by bubbling CO_2 into polyallylamine solution in these solvents [125]. The reaction of polyallylamine (the latent gelator) with CO_2 resulted in transformation of neutral amine functions into charged ammonium and carbamate groups amenable to strong electrostatic inter-

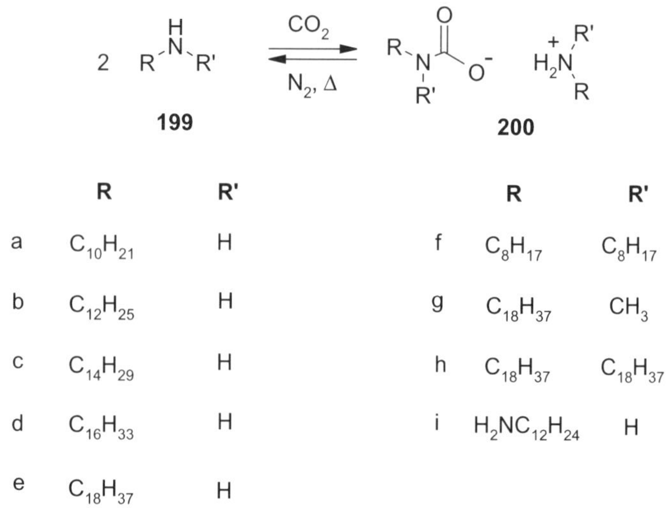

Scheme 28 Synthesis of ammonium carbamate gelators **200** by reaction of alkylamines and CO_2

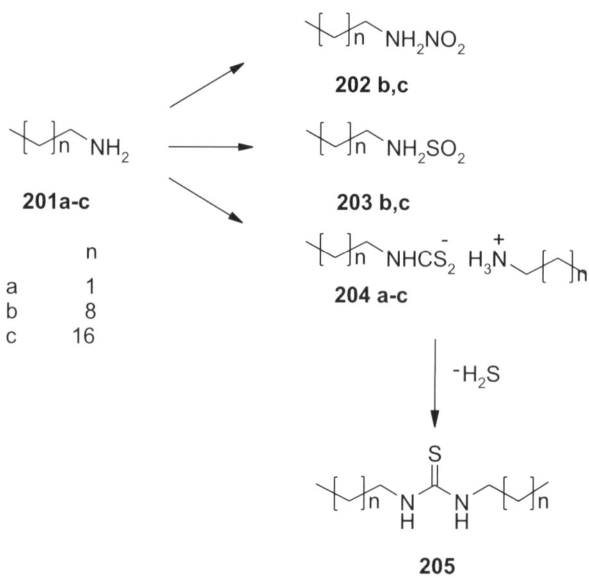

Scheme 29 Synthesis of gelling alkylamine-NO_2 (**202 b,c**), -SO_2 (**203 b,c**), and -CS_2 (**205 a–c**) adducts and transformation of **205** into N,N'-dialkylthiourea gelators

actions which "cross-link" the polymer chains to give a three-dimensional network. However, in contrast to low molecular weight systems which showed reversible gelation/degelation upon absorption and removal of CO_2, respectively, the polyammonium–polycarbamate system exhibited irreversible behavior. The removal of CO_2 by heating of the gel and bubbling of nitrogen resulted in permanent loss of the gel phase despite subsequent retreatment with CO_2.

6
Conclusions

Many efficient gelators of organic solvents and water could be prepared by the structural combination of amidic, carbamate, urea, or oxalamide groups and long aliphatic chains or aromatic groups with a large surface. Such combination of polar hydrogen-bonding groups and highly lipophilic and flexible alkyl chains or large aromatic groups results in diminished solubility and a decreased tendency for crystallization, which are both competitive processes to gelation. In many cases the introduction of chiral centers is favorable for gelation. However, racemates and the centrosymmetric *meso* diastereomers tend to crystallize or show much weaker gelation compared to the respective enantiomers. Some recent work describes exceptions showing that racemates could be more efficient gelators than the pure enantiomers [8, 84].

In most publications the efficiency of gelation of various solvents by different gelator molecules is expressed as minimal gelation concentration (mgc) using weight % (wt %) or $g\,L^{-1}$ units. It should be emphasized that the comparison of gelator efficiencies between structurally different gelator molecules based on mgc values is very uncertain and should be made with reservations. In many papers the precise procedures of mgc determination are not provided. In most cases the mgc values are obtained by repeated additions of small measured volumes of a solvent and repeated heating (sol)–cooling (gel) cycles until finally the cooling step gives a loose gel or viscous solution that restores the fluidity of the system. In such experiments, test tubes of different diameters and different heating–cooling regimes are used. If the experiments are performed in a tube of small diameter, much higher volumes of gelled solvent per weighed amount of the gelator could be measured due to stronger capillary forces compared to experiments performed in a tube with larger diameter; the differences in mgc values in $mol\,dm^{-3}$ could be up to a factor of 10 (see ref. [126]).

Once the prototype gelator structures are discovered, the subsequent synthetic modifications by introduction of various functions give gelators with controllable or advanced properties of interest for numerous possible applications. The mechanical and thermal properties of gels could be improved by photoinduced polymerization of polymerizable functions present in gelator molecules [23, 110, 111, 120]. Introduction of photochromic groups in the gelator molecule gives photoresponsive gels suitable for development of sensing devices [15]. Gel fibers can be used as templates for polymerization of silica or transition metal oxides, and their structures can be transcribed into silica or transition metal oxide nanostructures, such as tubes or helical fibers [24–26, 62]. Some gels exhibited high ionic conductivities and can be used as electrochromic materials [34, 61]. The porphyrin–fullerene gelators could be of interest for the development of photofunctional gels with electron-transfer properties [42]. The bisurea–thiophene gelators could be used in charge-transport devices [114]. Some hydrogels showed pH-responsive volume phase transitions [68] or pH-dependent sol-to-gel transitions [70, 71, 90], both of which are of high interest for development of new drug-delivery systems. Most interestingly, efficient gelators of sc CO_2 were discovered; the gels after removal of CO_2 give self-supporting foams with cells smaller than 1 μm and a reduction of bulk density of 97% [119]. Such gels are of high importance as precursors of microcellular materials with numerous potential applications. The vancomycin–pyrene gelator showed 8–11 times higher activity against VRE than vancomycin itself [101]. The latter opens exciting perspectives for the preparation of drug-containing gelators which may result in improved drug properties.

References

1. Lehn JM (1995) Supramolecular chemistry: concepts and perspectives. VCH, Weinheim
2. Schneider HJ, Yatsimirsky A (1999) Principles and methods in supramolecular chemistry. Wiley, Chichester
3. Doig AJ, Wiliams DH (1992) J Am Chem Soc 114:338
4. Di Capua FM, Swaminathan S, Beveridge DL (1991) J Am Chem Soc 113:6145
5. Beginn U, Sheiko S, Moeller M (2000) Macromol Chem Phys 201:1008
6. George M, Snyder SL, Terech P, Glinka CJ, Weiss RG (2003) J Am Chem Soc 125:10275
7. Mieden-Gundert G, Klein L, Fischer M, Vögtle F, Heuze K, Pozzo JL, Vallier M, Fages F (2001) Angew Chem Int Ed 40:3164
8. Čaplar V, Žinić M, Pozzo JL, Fages F, Mieden-Gundert G, Vögtle F (2004) Eur J Org Chem 4048
9. Luo X, Liu B, Liang Y (2001) Chem Commun 1556
10. Hanabusa K, Kobayashi H, Suzuki M, Kimura M, Shirai H (1998) Colloid Polym Sci 276:252
11. Hanabusa K, Okui K, Karaki K, Kimura M, Shirai H (1997) J Colloid Interface Sci 195:86
12. Bhattacharya S, Ghanashyam Acharya SN, Raju AR (1996) Chem Commun 2101
13. Fuhrhop JH, Schneider P, Rosenberg J, Boekema E (1987) J Am Chem Soc 109:3387
14. Hanabusa K, Maesaka Y, Kimura M, Shirai H (1999) Tetrahedron Lett 40:2385
15. Ahmed SA, Sallenave X, Fages F, Meiden-Gundert G, Müller WM, Müller U, Vögtle F, Pozzo JL (2002) Langmuir 18:7096
16. Maitra U, Potluri VK, Sangeetha NM, Babu P, Raju AR (2001) Tetrahedron Asymmetry 12:477
17. Babu P, Sangeetha NM, Vijaykumar P, Maitra U, Rissanen K, Raju AR (2003) Chem Eur J 9:1922

18. Hashimoto M, Ujiie S, Mori A (2003) Adv Mater 15:797
19. Hanabusa K, Yamada M, Kimura M, Shirai H (1996) Angew Chem Int Ed Engl 35:1949
20. Kato T, Kutsuna T, Hanabusa K, Ukon M (1998) Adv Mater 10:606
21. Kato T, Kondo G, Hanabusa K (1998) Chem Lett 193
22. Hanabusa K, Maesaka Y, Suzuki M, Kimura M, Shirai H (2000) Chem Lett 1168
23. Inoue K, Ono Y, Kanekiyo Y, Hanabusa K, Shinkai S (1999) Chem Lett 429
24. Jung JH, Ono Y, Shinkai S (2002) Chem Eur J 6:4552
25. Jung JH, Ono Y, Hanabusa K, Shinkai S (2000) J Am Chem Soc 122:5008
26. Kobayashi S, Hamasaki N, Suzuki M, Kimura M, Shirai H, Hanabusa K (2002) J Am Chem Soc 124:6550
27. Koshima H, Matsusaka W, Yu H (2003) J Photochem Photobiol A 156:83
28. Sumiyoshi T, Nishimura K, Nakano M, Handa T, Miwa Y, Tomioka K (2003) J Am Chem Soc 125:12137
29. Schmidt R, Adam FB, Michel M, Schmutz M, Decher G, Mesini PJ (2003) Tetrahedron Lett 44:3171
30. Liu P, Shirota Y, Osada Y (2000) Polym Adv Technol 11:512
31. Yasuda Y, Iishi E, Inada H, Shirota Y (1996) Chem Lett 575
32. Hanabusa K, Koto C, Kimura M, Shirai H, Kakehi A (1997) Chem Lett 429
33. Hanabusa K, Kawakami A, Kimura M, Shirai H (1997) Chem Lett 191
34. Yasuda Y, Takebe Y, Fukumoto M, Inada H, Shirota Y (1996) Adv Mater 8:740
35. Kamiyama T, Yasuda Y, Shirota Y (1999) Polym J 31:1165
36. Yasuda Y, Kamiyama T, Shirota Y (2000) Electrochim Acta 45:1537
37. van Gorp JJ, Vekemans JAJM, Meijer EW (2002) J Am Chem Soc 124:14759
38. Green MM, Peterson NC, Sato T, Teramoto A, Cook R, Lifson S (1995) Science 268:1860
39. Palmans ARA, Vekemanms JAJM, Havinga EE, Meijer EW (1997) Angew Chem Int Ed Engl 36:2648
40. Ikeda M, Takeuchi M, Shinkai S (2003) Chem Commun 1354
41. Shirakawa M, Kawano S, Fujita N, Sada K, Shinkai S (2003) J Org Chem 68:5037
42. Shirakawa M, Fujita N, Shinkai S (2003) J Am Chem Soc 125:9902
43. Fuhrhop J-H, König J (1994) Membranes and molecular assemblies: the synkinetic approach. In: Stoddart Fraser J (ed) Monographs in supramolecular chemistry. RSC, Cambridge
44. Hanabusa K, Okui K, Karaki K, Kimura M, Shirai H (1997) J Colloid Interface Sci 195:86
45. Bhattacharya S, Ganashyam Acharya SN (1999) Chem Mater 11:3121
46. Luo X, Liu B, Liang Y (2001) Chem Commun 1556
47. Bhattacharya S, Krishnan-Ghosh Y (2001) Chem Commun 185
48. Couffin-Hoarau A-C, Motulsky A, Delmas P, Leroux J-C (2004) Pharm Res 21:454
49. Ragunathan KG, Bhattacharya S (1995) Chem Phys Lipids 77:13
50. Boettcher C, Schade B, Fuhrhop J-H (2001) Langmuir 17:873
51. Hanabusa K, Nakayama H, Kimura M, Shirai H (2000) Chem Lett 1070
52. Suzuki M, Yumoto M, Kimura M, Shirai H, Hanabusa K (2002) Chem Commun 884
53. Fuhrhop J-H, Spiroski D, Boettcher C (1993) J Am Chem Soc 115:1600
54. Suzuki M, Waraksa CC, Nakayama H, Hanabusa K, Kumura M, Shirai H (2001) Chem Commun 2012
55. Suzuki M, Waraksa CC, Mallouk TE, Nakayama H, Hanabusa K (2002) J Phys Chem B 106:4227
56. Suzuki M, Yumoto M, Kimura M, Shirai H, Hanabusa K (2003) Chem Eur J 9:348
57. Suzuki M, Yumoto M, Kimura M, Shirai H, Hanabusa K (2002) New J Chem 26:817

58. Partridge KS, Smith DK, Dykes GM, McGrail PT (2001) Chem Commun 319
59. Hirst AR, Smith DK, Feiters MC, Geurts HPM, Wright AC (2003) J Am Chem Soc 125:9010
60. Hanabusa K, Tange J, Taguchi Y, Koyama T, Shirai H (1993) J Chem Soc Chem Commun 390
61. Hanabusa K, Hiratsuka K, Kimura M, Shirai H (1999) Chem Mater 11:649
62. Kobayashi S, Hanabusa K, Suzuki M, Kimura M, Shirai H (1999) Chem Lett 1077
63. Hanabusa K, Itoh A, Kimura M, Shirai H (1999) Chem Lett 767
64. Hanabusa K, Kobayashi H, Suzuki M, Kimura M, Shirai H (1998) Colloid Polym Sci 276:252
65. Nakhasima N, Asakuma S, Kunitake T (1985) J Am Chem Soc 107:509
66. Ihara H, Hachisako H, Hirayama C, Zamada K (1992) J Chem Soc Chem Commun 1244
67. Ihara H, Sakurai T, Yamada T, Hashimoto T, Takafuji M, Sagawa T, Hachisako H (2002) Langmuir 18:7120
68. Kiyonaka S, Zhou S-L, Hamachi I (2003) Supramol Chem 15:521
69. Gao P, Zhan C, Liu L, Zhou Y, Liu M (2004) Chem Commun 1174
70. Heeres A, van der Pol C, Stuart M, Friggeri A, Feringa BL, van Esch J (2003) J Am Chem Soc 125:14252
71. van Bommel KJC, van der Pol C, Muizebelt I, Friggeri A, Heeres A, Metsma A, Feringa BL, van Esch J (2004) Angew Chem Int Ed 43:1663
72. de Vries EJ, Kellogg R (1993) J Chem Soc Chem Commun 238
73. Hanabusa K, Matsumoto Y, Miki T, Koyama T, Shirai H (1994) J Chem Soc Chem Commun 1401
74. Hanabusa K, Matsumoto Y, Kimura M, Kakehi A, Shirai H (2000) J Colloid Interface Sci 224:231
75. Becerril J, Burguete MI, Escuder B, Luis SV, Miravet JF, Querol M (2002) Chem Commun 738
76. Becerril J, Burguete MI, Escuder B, Galindo F, Gavara R, Miravet JF, Luis SV, Peris G (2004) Chem Eur J 10:3879
77. Menger FM, Yamasaki Y, Catlin KK, Nishimi T (1995) Angew Chem Int Ed Engl 34:585
78. Menger FM, Caran KL (2000) J Am Chem Soc 122:11679
79. Fuhrhop A-H, Wang T (2004) Chem Rev 104:2901
80. Hanabusa K, Tanaka R, Suzuki M, Kimura M, Shirai H (1997) Adv Mater 9:1095
81. Suzuki M, Owa S, Yumoto M, Kimura M, Shirai H, Hanabusa K (2004) Tetrahedron Lett 45:5399
82. Jokic M, Makarevic J, Zinic M (1995) Chem Commun 1723
83. Makarevic J, Jokic M, Peric B, Tomisic V, Kojic-Prodic B, Zinic M (2001) Chem Eur J 7:3328
84. Makarevic J, Jokic M, Raza Z, Stefanic Z, Kojic-Prodic B, Zinic M (2003) Chem Eur J 9:5567
85. Coe S, Kane JJ, Nguyen TL, Toledo LM, Winiger E, Fowler FW, Lauher JW (1997) J Am Chem Soc 119:86
86. Makarevic J, Jokic M, Frkanec L, Katalenic D, Zinic M (2002) Chem Commun 2238
87. Frkanec L, Jokic M, Makarevic J, Wolsperger K, Zinic M (2002) J Am Chem Soc 124:9716
88. Luo X, Li C, Liang Y (2000) Chem Commun 2091
89. Suzuki M, Nigawara T, Yumoto M, Kimura M, Shirai H, Hanabusa K (2003) Terahedron Lett 44:6841
90. Zhang Y, Gu H, Yang Z, Xu B (2003) J Am Chem Soc 125:13680

91. Maji SK, Malik S, Drew MGB, Nandi AK, Banerjee A (2003) Tetrahedron Lett 44:4103
92. Ariga K, Kikuchi J, Naito M, Koyama E, Yamada N (2000) Langmuir 16:4929
93. Yamada N, Ariga K (2000) Synlett 575
94. Lyon RP, Atkins WM (2001) J Am Chem Soc 123:4408
95. Hanabusa K, Naka Y, Koyama T, Shirai H (1994) Chem Commun 2683
96. Daly WH, Poche D (1988) Tetrahedron Lett 29:5859
97. Angelli A, Bell M, Boden N, Keen JN, Knowles PF, McLeish TCB, Pitkeathly M, Radford SE (1997) Nature 386:259
98. Angelli A, Nyrkova IA, Bell M, Harding R, Carrick L, McLeish TCB, Semenov AN, Boden N (2001) Proc Natl Acad Sci USA 98:11857
99. Aggeli A, Bell M, Boden N, Carrick LM, Strong E (2003) Angew Chem Int Ed 42:5603
100. Xing B, Yu C-W, Chow K-H, Ho P-L, Fu D, Xu B (2002) J Am Chem Soc 124:14846
101. Tiller JC (2003) Angew Chem Int Ed 42:3072
102. Chang Y-L, West M-A, Fowler FW, Lauher FW (1993) J Am Chem Soc 115:5991
103. Kane JJ, Liao RF, Fowler FW (1995) J Am Chem Soc 117:12003
104. Hanabusa K, Shimura K, Hirose K, Kimura M, Shirai H (1996) Chem Lett 885
105. Brinksma J, Feringa BL, Kellogg RM, Vreeker R, van Esch J (2000) Langmuir 16:9249
106. van Esch J, Kellogg RM, Feringa BL (1996) Tetrahedron Lett 38:281
107. van Esch J, De Feyter S, Kellogg RM, De Schryver F, Feringa BL (1997) Chem Eur J 3:1238
108. De Feyter S, Grim PCM, van Esch J, Kellogg RM, Feringa BL, De Schryver F (1998) J Phys Chem B 102:8981
109. De Feyter S, Larsson M, Schuurmans N, Verkuijl B, Zoriniants G, Gesquiere A, Abdel-Mottaleb MM, van Esch J, Feringa BL, van Stam J, De Schryver F (2003) Chem Eur J 9:1198
110. de Loos M, van Esch J, Stokroos I, Kellogg RM, Feringa BL (1997) J Am Chem Soc 119:12675
111. van Esch J, Schoonbeek F, de Loos M, Kooilman H, Spek AL, Kellogg RM, Feringa BL (1999) Chem Eur J 5:937
112. van der Laan S, Feringa BL, Kellogg RM, van Esch J (2002) Langmuir 18:7136
113. de Loos M, van Esch J, Kellogg RM, Feringa BL (2001) Angew Chem Int Ed 40:613
114. Schoonbeek FS, van Esch J, Wegewijs B, Rep DBA, de Haas MP, Klapwijk TM, Kellogg RM, Feringa BL (1999) Angew Chem Int Ed 38:1393
115. Gesquiere A, Abdel-Mottaleb MS, De Feyter S, de Schryver FC, Schoonbeek F, van Esch J, Kellogg RM, Feringa BL, Calderone A, Lazzaroni R, Bredas JL (2000) Langmuir 16:10385
116. Carr AJ, Melendez R, Geib SJ, Hamilton AD (1998) Tetrahedron Lett 39:7447
117. Estroff LA, Hamilton AD (2000) Angew Chem Int Ed 39:3447
118. Estroff LA, Leiserowitz L, Addadi L, Weiner S, Hamilton AD (2003) Adv Mater 15:38
119. Shi C, Huang Z, Kilic S, Xu J, Enick RM, Beckman EJ, Carr AJ, Melendez RE, Hamilton AD (1999) Science 286:1540
120. Wang G, Hamilton AD (2002) Chem Eur J 8:1954
121. Wang G, Hamilton AD (2003) Chem Commun 310
122. Suzuki M, Nakajima Y, Yumoto M, Kimura M, Shirai H, Hanabusa K (2004) Org Biomol Chem 2:1155
123. George M, Weiss RG (2002) Langmuir 18:7124
124. George M, Weiss RG (2003) Langmuir 19:1017
125. Carretti E, Dei L, Baglioni P, Weiss RG (2003) J Am Chem Soc 125:5121
126. Makarevic J, Jokic M, Raza Z, Caplar V, Katalenic D, Stefanic Z, Kojic-Prodic B, Zinic M (2004) Croat Chem Acta 77:403

Top Curr Chem (2005) 256: 133–165
DOI 10.1007/b107173

Published online: 1 July 2005

Nucleobase-Containing Gelators

Koji Araki (✉) · Isao Yoshikawa

Institute of Industrial Science, University of Tokyo, 4-6-1 Komaba, Meguro-ku, 153-8505 Tokyo, Japan
araki@iis.u-tokyo.ac.jp

Abstract Nucleobases are nitrogenous heterocyclic compounds and have high ability to form directionally controlled multiple intermolecular interaction, i.e., in-plane multiple hydrogen-bonding interactions and stacking interactions perpendicular to the plane. Here the recent development of nucleobase-containing low molecular mass gelators in aqueous and organic systems is reviewed, and the self-assemblies of nucleobase-containing small molecules and their arrangement within the gel are discussed from the molecular, and mesoscopic to the macroscopic level in regard to becoming a macroscale three-dimensional network. Brief overview of the nucleoside-containing supramolecular materials are also presented.

Keywords Nucleobase · Hydrogen bond · Stacking interaction · Hydrogel · Organogel

Abbreviations

A adenine
AFM atomic force microscopy
C cytosine

CD	circular dichroism
G	guanine
GMP	guanosine monophosphate
isoG	isoguanosine
LC	liquid crystal
SAXS	Small-angle X-ray scattering
SEM	scanning electron microscopy
SUV	small unilamellar vesicle
TEM	transmission electron microscopy
T_{gel}	sol-gel transition temperature
THN	teterahydronaphthalene
U	uracil
XRD	X-ray diffraction

1
Introduction

Since the DNA double-helix structure was elucidated fifty years ago, non-covalent base-base interactions have been the subject of continuing interest both in biological and synthetic systems. A nucleobase is a nitrogenous heterocyclic compound that is optimized for self-recognition and interaction with other molecules by multiple base-base hydrogen bonds. Since formation of the multiple hydrogen bonds requires a strict spatial arrangement of the nucleobase moieties, recognition of other molecules can be achieved with high specificity. In biological systems, genetic information storage and processing are realized through the complementary base pair formation by the multiple hydrogen bonds. In synthetic systems, highly specific base-base interactions are used to design drugs, molecular sensors, and functional architectures, attracting considerable interest not only in pharmaceutical but also supramolecular and materials chemistry.

Sol-gel transition is the macroscopic process induced by formation of a macro-scale, three-dimensional network of gelator molecules. Therefore, low molecular mass gelator molecules have to be brought together into a macroscopic scale three-dimensional assembly by intermolecular interactions [1, 2]. Because of the high ability to form intermolecular interactions, carbohydrate and amino acid units are frequently found in the molecular structure of the gelators. These units serve as the building blocks of the macroscale three-dimensional structure. However, there have been relatively few reports about nucleobase-containing gelators even though a nucleobase has a suitable molecular structure for intermolecular interactions. It has to be noted that a compound that can be in a gel state is not necessarily classified as a gelator. For example, DNA sediment in aqueous alcoholic solution is known to be in a highly concentrated viscous gel state [3] but DNA can not be regarded as a gelator in this case. In this chapter, the term "gelator" is limited to that which can induce gelation of a system without phase separation, preferably with relatively low concentrations. Since formation of mesoscopic-scale assembly is not sufficient to induce gelation, macroscopic-scale arrangement and network structure of the assemblies in the gel is essential to understand the gelation process. We will review the self-assemblies of nucleobase-containing small molecules from the molecular, and mesoscopic to macroscopic level in regard to becoming a macroscale three-dimensional network.

2
Base-Base Interactions

Nucleobase molecules are substituted pyrimidines and purines. Guanine (G), adenine (A), thymine (T), and cytosine (C) are the four principal nucleobases found in DNA, and uracil (U) instead of T in RNA. There are many other naturally occurring or chemically modified nucleobases that are the constituents of the nucleic acids. Hydrogen-bonding and aromatic stacking are the major base-base interactions contributing to the thermodynamic stability of the DNA double helix. Since the nucleobase molecules have a planar ring structure, hydrogen-bonding interactions are mostly limited within the plane but aromatic stacking operates in a perpendicular direction to the nucleobase plane. These directionally controlled base-base interactions allow construction of the well-defined three-dimensional assemblies. Directionally controlled multiple intermolecular interactions are also suitable for precise molecular recognition, which are the key to the fidelity of transcription and translation of the genetic information. In the following sections, the properties of these noncovalent base-base interactions will be briefly summarized.

2.1
Hydrogen Bond

Nucleobases have multiple hydrogen-bond donor and acceptor sites within their molecular structures. The hydrogen bonding sites of guanine, adenine, thymine and cytosine are shown in Fig. 1. Because of the presence of multiple hydrogen bonding sites, the hydrogen bonding patterns between nucleobases are not limited to the Watson-Crick or Hoogsteen type base pairs. There is a wide variety of hydrogen bonding patterns in the crystal structures of nucleobases, nucleosides, and nucleotides. They form dimer-type base pairs, cyclic oligomers, infinite linear polymers, or more complex structures. Typical examples of the hydrogen bonding patterns for the principal nucleobases are shown in Fig. 2

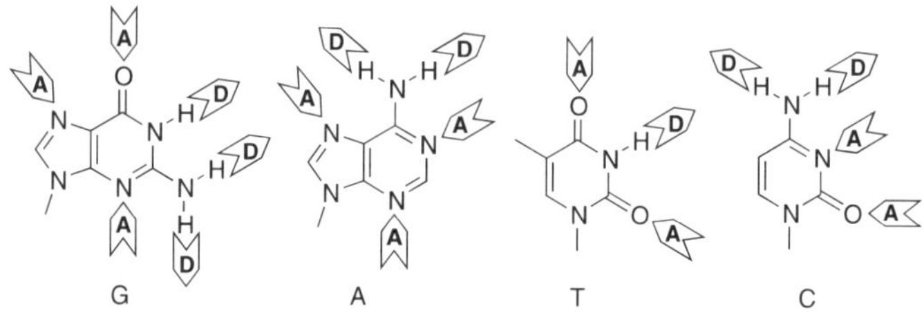

Fig. 1 Hydrogen-bonding donor (D) and acceptor (A) sites of guanine (G), adenine (A), thymine (T) and cytosine (C)

The hydrogen bond is a moderately strong and directional intermolecular interaction among noncovalent interactions. Though the definition of the hydrogen bond is still the subject of heavy debate, it is widely accepted that the electrostatic interaction is the major attractive force [4]. Generally, hydrogen bond energy depends greatly on the type of donor-acceptor sets and surrounding media. In the case of a weak hydrogen bond like C – H – O, hydrogen bond energy is only a few kJ mol^{-1}, while that of a strong hydrogen bond with a charged acceptor like O – H – $O^{(-)}$ is more than a hundred kJ mol^{-1} [5]. The interbase hydrogen bond is a relatively strong hydrogen bond ranging between 20–40 kJ mol^{-1}. Because of the additivity of intermolecular interaction, formation of multiple hydrogen bonds contributes to high stability of their assemblies. In addition, the cooperativity of the hydrogen bond has to be considered when multiple hydrogen bonds are involved. Because of the polarizability or charge transfer nature of the components of the hydrogen bonding chains or networks, the binding energy is larger than the sum of the individual hydrogen bond energies. Recent ab initio calculation estimated that the cooperative contributions provided 31% and 12–16%, respectively, of the stability of the A-T and G-C base pairs [6].

As the electrostatic force is inversely correlated to the relative permittivity (ε) of the solvent, hydrogen bond interaction does not operate effectively in water that has large relative permittivity ($\varepsilon_r = 80.1$). As the major attractive force of the base-base interaction, this characteristic of the hydrogen bond has to be considered especially in water.

2.2 Aromatic Stacking

Though aromatic stacking is the important noncovalent interaction which stabilizes the double helical structure, the nature of this interaction is not well understood. Electrostatic interactions, dispersion forces, and solvophobic effects are the potentially important factors in stabilizing the face-to-face base-base interactions [7]. The electrostatic interactions operate between par-

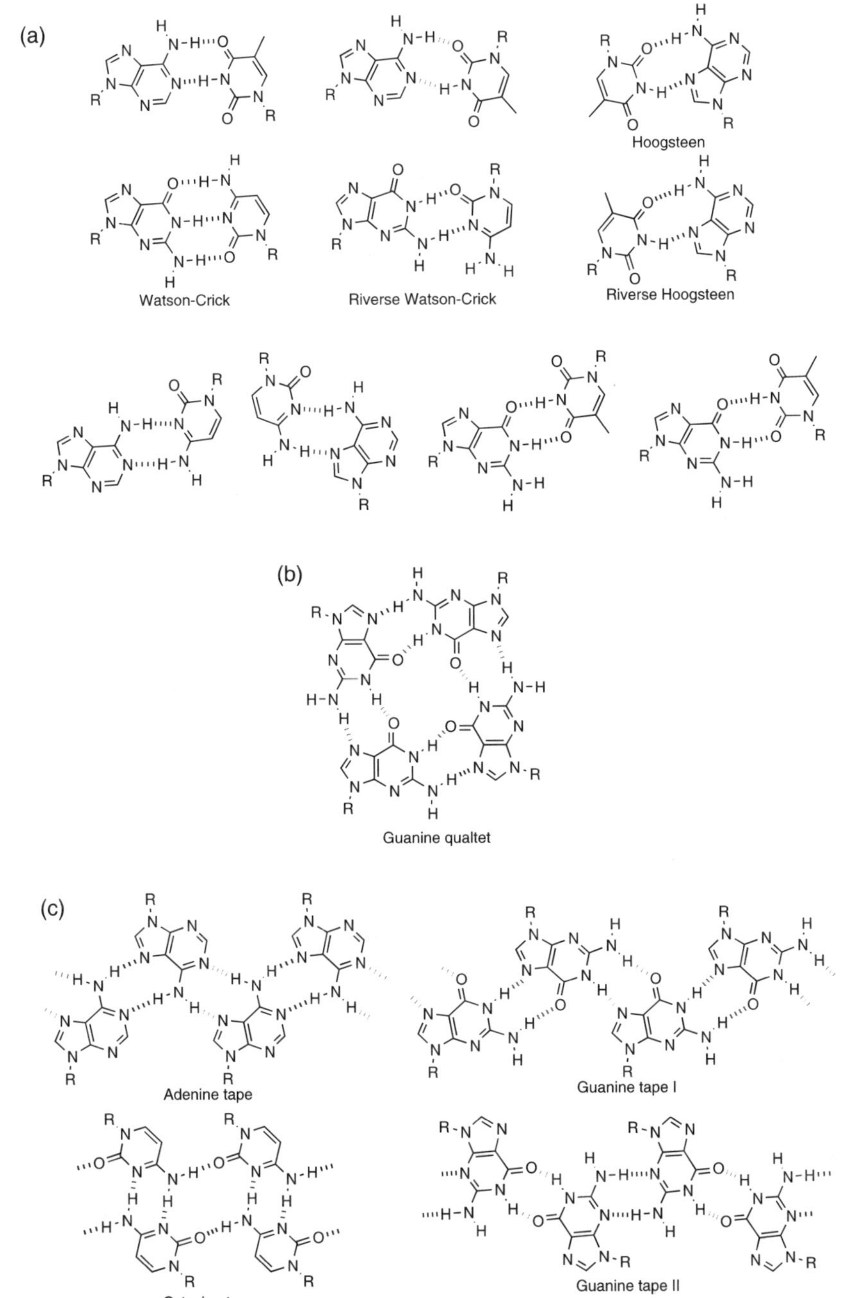

Fig. 2 Typical hydrogen bonding patterns for the principal nucleobases (**a**) base pairs (**b**) cyclic oligomer and (**c**) infinite linear polymers

tial charges localized at specific parts of a given heterocyclic base. Dispersion forces depend on the surface area of contact and on the polarizability of the two species. Finally, solvophobic and other solvent-driven effects depend on the relative energies of the solvation of bases when stacked and unstacked as well as the amount of surface area desolvated on stacking. It has been indicated that the relative stacking ability of the natural bases qualitatively goes in the order purine-purine > pyrimidine-purine > pyrimidine-pyrimidine [8].

3
Molecular Assemblies and Gelation in Aqueous Systems

As discussed above, hydrogen bond interaction does not operate effectively in water that has large relative permittivity. Indeed, nucleobases can not form hydrogen-bonded complementary base pairs in water [9]. As the hydrogen bond is essential to the base-base interaction, an effective approach to the design and construction of stable and predictable nucleobase assemblies is to make the microenvironment of the nucleobase more hydrophobic (lower permittivity). The use of an air/water interface is one of the effective approaches for this purpose.

3.1
Molecular Assemblies at Air/Water Interface

Though the hydrogen bonding interaction between nucleobases (or their mimicry) is usually suppressed in bulk water, base pair formation at the air-water interface was first suggested by Kitano and Ringsdorf [10], and then was extensively studied by Kunitake and his group [11]. Selective association of diaminotriazine-functionalized amphiphile (**1**) monolayers with thymine and thymidine in aqueous solution took place at the air-water interface, and this molecular recognition was indicated to be due to complementary hydrogen bonding (Fig. 3a) [12]. This remarkable feature of the air-water interface is ascribed to the unique properties of water at the interface, where the water structure is different from that in bulk water [13]. Similar base pair formation is also reported for the monolayers of nucleoside-lipid hybrids at the air-water interface [14] or within the monolayer [15].

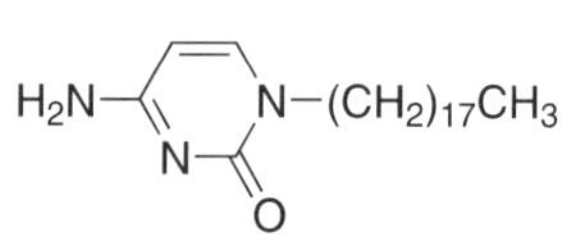

3

Scheme 1

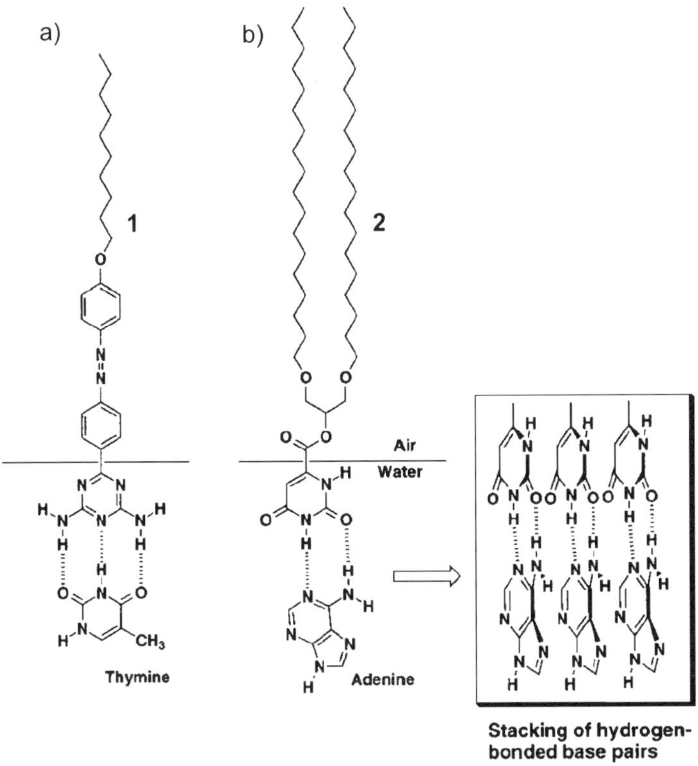

Fig. 3 Recognition of nucleic acid bases at the air-water interface: (**a**) recognition of thymine by a **1** monolayer; (**b**) recognition of adenine by a **2** monolayer. Reprinted with permission from Acc Chem Res 1998, 31, 371

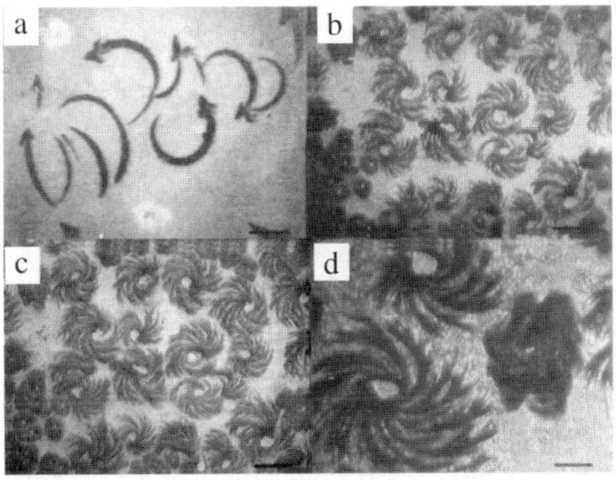

Fig. 4 Fluorescence images of the **3** monolayer at 3 mN m^{-1} (**a**) on pure water subphase observed with a G-filter (546 nm) for octadecylrhodamine B excitation, scale bar = 200 μm, (**b**) on guanosine subphase observed with a G-filter, scale bar 100 μm, (**c**) on guanosine subphase observed with a B-filter (410 nm) for octadecylacridine orange excitation, scale bar 100 μm, and (**d**) close-up image on guanosine subphase, scale bar 25 μm. Reprinted with permission from J Am Chem Soc 1997, 119, 2341

Stacking interaction operates effectively at the interface. Base pairs of cyclic imide-functionalized amphiphile (2) monolayers with aqueous adenine at the air/water interface were suggested to stack along the interface, and this stacking interaction enhanced cooperatively the base pair formation (Fig. 3b) [16]. By using amphiphilic fluorescent probes, Shimomura et al. observed formation of spiral-shaped two-dimensional domains in a compressed alkylated cytosine (3) monolayer spread on an aqueous guanosine or deoxyguanosine solution (Fig. 4b–d) [17]. This macroscale spiral-shaped domain of more than 10^{-1} mm in size was a two-dimensional crystal consisting of laterally stacked cytocine-guanine pairs. This observation demonstrated that the unique properties of the air/water interface allowed formation of macroscale nucleobase assemblies by the complementary hydrogen bond and the base stacking interactions though it was not three-dimensional but two-dimensional.

3.2
Molecular Assemblies in Bulk Aqueous Phase

3.2.1
Small Assemblies

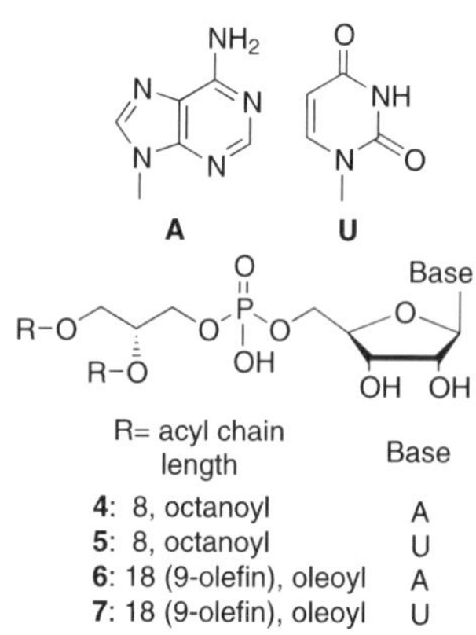

Scheme 2

In the bulk aqueous phase, it is necessary to make the hydrogen bonding interaction effective in order to induce self-assembly of the nucleobase. The first approach is to use the hydrophobic interior of amphiphile aggregates, micelles, and the formation of a complementary A-T base pair was observed at the micellar core [18–20]. However, the surface region of micelles and vesicles also provides an effective microenvironment for the base-base interactions. Phosphatidylnucleosides, a group of phospholipids that have a nucleoside at their head group, have been synthesized by enzymatic reaction, and base-base interaction was studied in their micellar and vesicular aggregates. The phosphatidylnucleosides having short and saturated dioctanoyl chains, **4** and **5**, formed quasi-spherical micelles composed of 28–45 monomers in the pH range of 3.5–7.5 [21], and the micelles were in a thermodynamically equilibrated state due to the fast lipid exchange. Though nucleobases were anchored at the micellar surface in a highly aqueous environment, both base-base stacking and hydrogen bonding interactions were observed for the **4** and **5** 1 : 1 mixture by spectroscopic and other analytical methods [22]. NMR studies confirmed formation of the Watson-Crick type adduct at the hydrophilic micellar surface. Similar results are also observed for the micelles of monoalkyl-phosphoryl nucleosides [23].

In the case of vesicles, the lipid exchange was not sufficiently fast. Small unilamellar vesicles (SUV) of approximately 33 nm in radius were prepared from **6** and **7**, phosphatidylnucleosides having A and U, respectively, with dioleoyl chains [24, 25]. When the **6** and **7** vesicles prepared separately were mixed together (1 : 1), no base-base interaction was observed immediately. Only after aging for nearly two weeks were formation of hydrogen-bonded complementary base pairs and aromatic stacking completed. This slow development of base-base interaction was attributed to the slow lipid exchange rate in the case of SUV having high curvature and tension. The results showed that the hydrogen bonding and stacking interactions operated effectively at the hydrophilic vesicular surface.

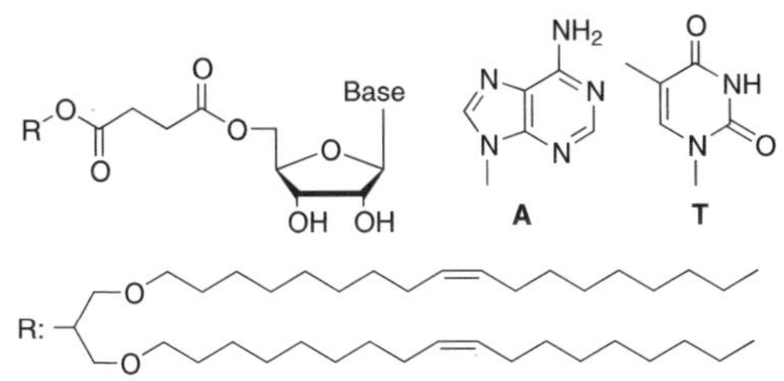

8: Base=A, **9**: Base=T

Scheme 3

The A-T base pair formation at the vesicular surface was also shown to induce hemifusion of bilayer membranes of giant unilamellar vesicles (10–150 μm in diameter) which consisted of dioleylphosphatidylcholine containing deoxyadenosine (**8**) or deoxythimine (**9**)-functionalized lipids (5 w/w %), which gives some insight into the growing process of the assemblies [26].

For small molecular aggregates, self-assembled aggregates are usually in stable thermodynamic equilibrium, and can be discussed mostly from the viewpoint of static intermolecular interactions. However, as it is shown in the above examples, mesoscopic- to macroscopic-scale assemblies that are responsible for macroscale gelation are frequently in metastable or kinetically frozen states, and their morphologies are quite sensitive to many factors, especially to coexistent components, temperature, and the method of preparation.

3.2.2
Hydrogels

It has been known for a long time that guanosine derivatives yield gels in aqueous solutions at relatively high concentration, and the presence of an alkali metal is essential for formation of the homogeneous gel. Some examples of the gel-forming guanosine derivatives in the presence of 0.1 M KCl and 0.1 M nucleoside are guanosine, guanosine 2′-, 3′-, and 5′-monophosphate (2′-GMP, 3′-GMP and 5′-GMP), isoguanosine (isoG), and oligoguanylic acids [27]. The assemblies in the gel were shown to be the cyclic tetramer (G-quartet, Fig. 2b) of the guanine moieties in a 3′-GMP gel (Fig. 5a) and a continuous helix with 15 bases (Fig. 5b) in a 5′-GMP gel determined from their X-ray diffraction (XRD) patterns [28, 29]. The multiple hydrogen-bond donor and acceptor sites within the guanine units allowed the guanosine derivatives to form the hydrogen-bonded G-quartet or continuous helix in the presence of alkali metal ions as a template. The structure of the metal-induced G-quartet was extensively studied, and high size-selectivity toward metal ions was demonstrated [30]. Na^+ and K^+ ions were effective in inducing the G-quartet formation but Li^+ and Cs^+ were not effective. Large hypochromicities and an increase of circular dichroism (CD) indicated that the G-quartets were assembled into columnar stacks in the gel. Formation of the columnar stacks was not only due to the stacking interactions between the cyclic tetramers, but also the alkali metal ions contributed to forming colum-

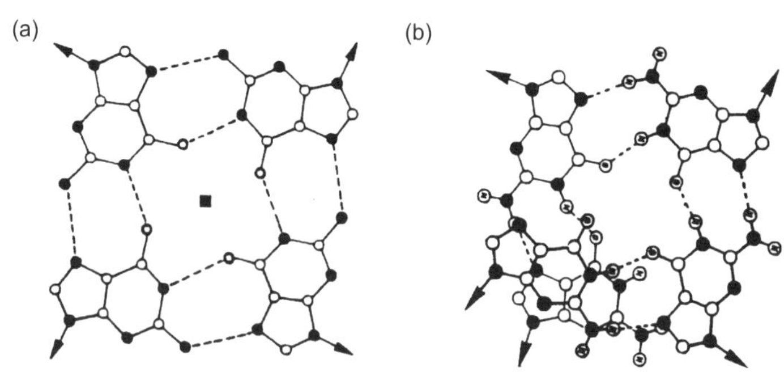

Fig. 5 (**a**) Tetrameric arrangement of guanine bases for the Guo-3′-P gel and (**b**) the continuous helical arrangement in the Guo-5′-P gel. Reprinted with permission from J Biomol Struct Dyn 1990, 8, 491

nar assemblies. The arrangement of the columnar or helix assemblies on a macroscopic scale was further studied in detail in the liquid crystalline (LC) phase [31]. In the cholesteric liquid crystalline phase of 5′-GMP, the arrangements of the rod-like columnar assemblies are shown in Fig. 6b. At higher concentrations, arrangement of the columnar assemblies became a hexagonal (Fig. 6c).

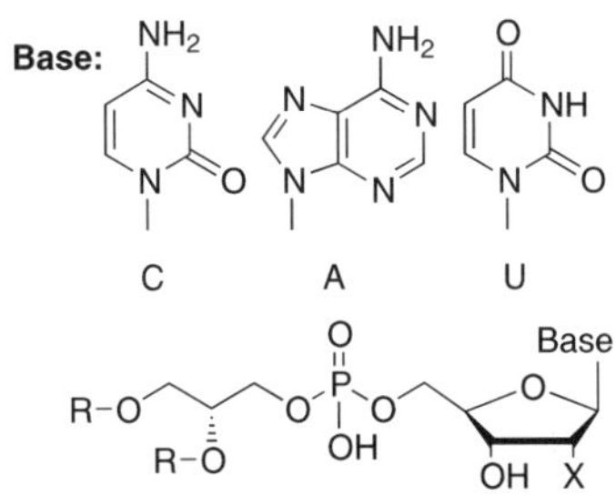

	R= acyl chain length	X	Base
10 :	16, palmitoyl	-H	C
11a:	14, myristoyl	-OH	A
11b:	12, lauroyl	-OH	A
12 :	14, myristoyl	-OH	U

Scheme 4

In 1988, Yanagawa et al. reported that phosphatidylnucleosides self-assembled into circular and linear helical strands in aqueous salt solution [32]. In the case of phosphatidyldeoxycytidine with dipalmitoyl chains

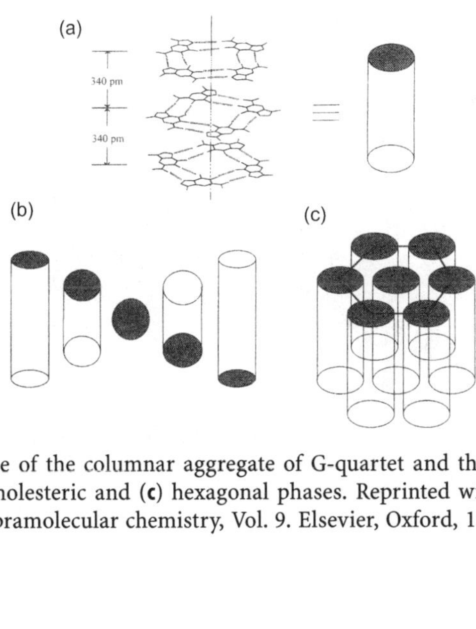

Fig. 6 (**a**) A scheme of the columnar aggregate of G-quartet and their relative arrangements in the (**b**) cholesteric and (**c**) hexagonal phases. Reprinted with permission from Comprehensive supramolecular chemistry, Vol. 9. Elsevier, Oxford, 1996, p 483

(**10**) [33], a sonicated aqueous solution of the lipid in a Tris-HCl buffer (pH 8.0) at 50 °C for 45 min formed vesicles of approximately 20–160 nm in size, and the vesicles were slowly transformed into superhelical strands after aging at 25 °C for 1 day. The strands consisted of a right-handed duplex with a diameter of approximately 11 nm and a helical pitch of 24 nm or double duplex in the case of the thick strands, and the duplex or double duplex further formed a superhelical structure with a helical pitch of 95–110 nm. Development of the superhelical strands on a mesoscopic scale was clearly shown by transmission electron microscopy (TEM) (Fig. 7). After aging for a week, the superhelical structures were converted into gels and the gels contained a number of thick strands consisting of linearly extended helical strands (Fig. 7d). A greatly increased positive Cotton effect, presumably due to the helical assemblies was observed in the CD spectra below the transition temperature of the gel-liquid crystal of the lipid. Since this large positive Cotton effect was quite similar to that of poly(C) in a nonprotonated single-stranded helical structure, the helical assemblies were suggested to be stabilized by the base-base stacking interactions and the hydrophobic interactions at the long alkyl chain moieties. Formation of different hybrid helical strands was also reported when the 1 : 1 mixture of the lipid having complementary bases (**11a** and **12**) were used, suggesting the importance of the interbase hydrogen bonding interactions [34].

Similar giant polymer-like assemblies of phosphatidylnucleosides (**11b**) were also observed in solution by light scattering and small-angel neutron-

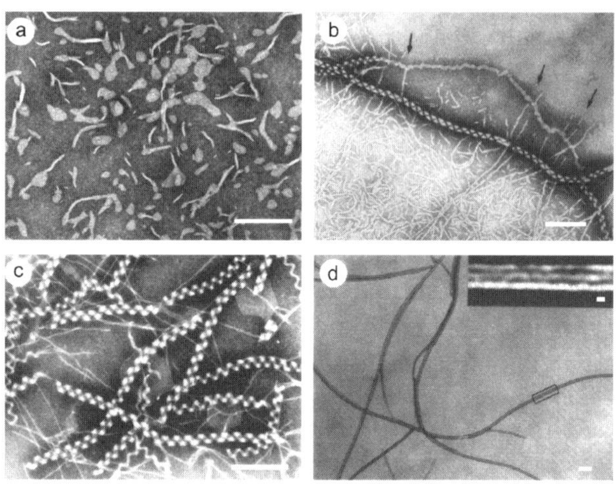

Fig. 7 Formation process of superhelical strands of **10**. The electron micrographs were taken after aging for (**a**) 5 h (**b**) 15 h (**c**) 1 day and (**d**) 1 week respectively. All scale bars represent 2000 Å. *Arrows* in (**b**) represent positions where strands are twisting. *Inset* in (**d**) in the Fourier-transformed image (scale bar, 100 Å) of the segment boxed by a rectangle. Reprinted with permission from J Am Chem Soc 1989, 111, 4567

scattering methods [35]. The results indicated micellar growth to giant worm-like assemblies that entangle to form a transient network with a response to mechanical stress, which is similar to that of polymer solutions in the semi-dilute range. TEM observation supported the formation of a polymer-like network. However, no gelation of the system was reported.

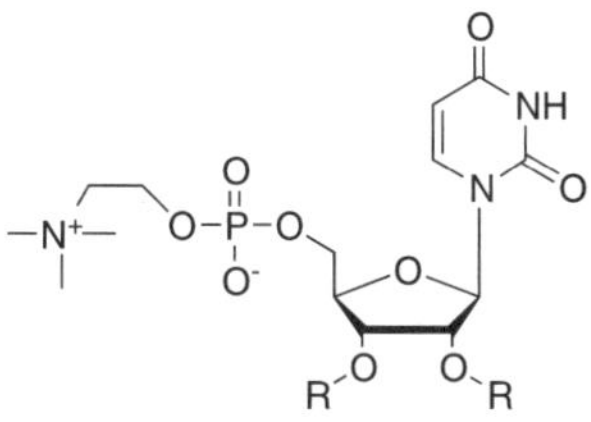

R= acyl chain length
13a: 14, myristoyl
13b: 16, palmitoyl
13c: 18, stearoyl
13d: 20, arachidoyl
13e: 18 (9-olefin), oleoyl

Scheme 5

Recently, Moreau et al. reported gel formation of a different type of lipid-nucleoside hybrids [36]. Structures of diacyluridinophosphocholine with different acyl chains (**13**) were prepared. All the derivatives formed bilayer vesicles in aqueous solution. In addition, **13b–d** showed the ability to form

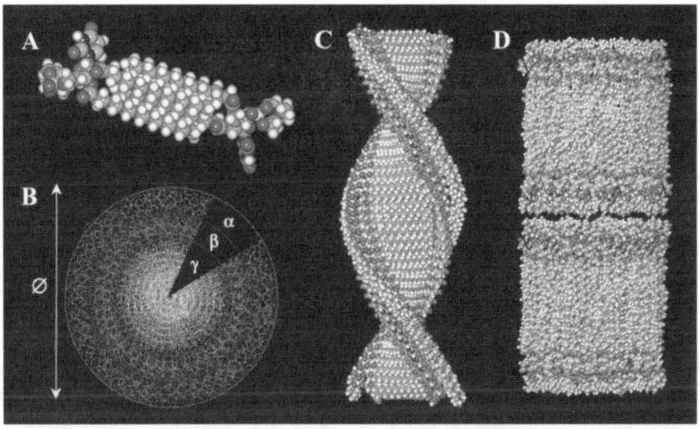

Fig. 8 Proposed model illustrating DNA-like helical fiber and multilamellar organizations in water formed by **13b**. (**A**) Two molecules of **13b** forming the basic repeat unit of the helical structure. (**B**) Top view of the strand presented in (**C**) with a diameter Φ of 4.2 ± 0.5 nm: α is hydrophilic domain (phosphocholines), β corresponds to uridine packing, and γ is the hydrophobic core. (**C**) Drawing of the fiber (for helical pitch) 7.0 ± 0.5 nm. (**D**) Drawing of a multilamellar, L_α, organization. Reprinted with permission from J Am Chem Soc 2004, 126, 7533

stable opaque hydrogels in aqueous solution of more than 6 w/w % below their melting points (mp). Scanning electron microscopy (SEM) showed an entangled fiber network throughout the hydrogel and magnified TEM images showed aligned helical nanofibers of approximately 4.2 nm with a helical pitch of 7.0 nm for **13b**. Small-angle X-ray scattering (SAXS) measurements showed formation of nanofibers below mp and a lamellar-like structure of 4.6 ± 0.2 nm in repeat distance above mp. Based on the above observation, the observed hypochromic effect and CD spectra, the molecular assemblies of nanofibers and lamellae were determined, which are shown in Fig. 8. The temperature-dependent transition between nanofibers and bilayer assemblies took place, and the phase diagram of this transition is shown as a function of the acyl chain length Fig. 9).

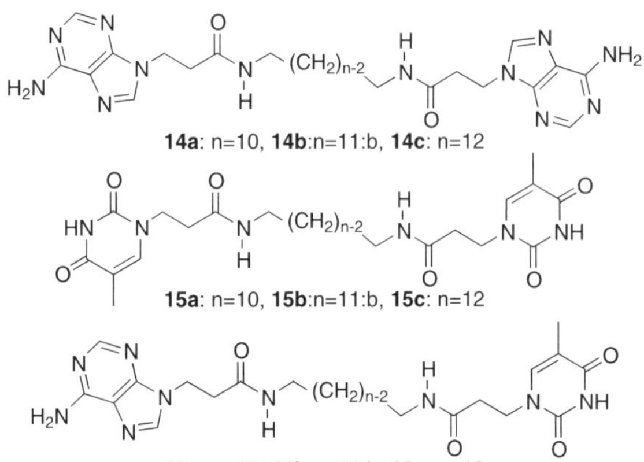

14a: n=10, **14b**:n=11:b, **14c**: n=12

15a: n=10, **15b**:n=11:b, **15c**: n=12

16a: n=10, **16b**:n=11:b, **16c**: n=12

Scheme 6

Shimizu and co-workers have been actively studying molecular assemblies of a new family of amphiphiles, bolaamphiphiles having two hydrophilic head groups at both ends of a hydrophobic alkyl chain [37], and reported the self-assemblies of the bolaamphiphiles having nucleobases at their head groups. In the case of amide-connected bolaamphiphiles, **14–16**, an equimolar mixture of the complementary **14a** and **15a** components in 10% ethanolic aqueous solution yielded an amorphous gel [38]. Heteroditopic **16c** also formed an amorphous gel in 50% ethanolic aqueous solution. In the **14a/15a** gel, a network of nanometer-sized fibers with uniform width ranging from 15 to 30 nm was observed by TEM. A network of nanofibers 20–100 nm in width was also found in the **16c** gel. Hydrogen bonding interactions were noticed between complementary A-T base pairs and the amide units. Base stacking was also indicated from the observed large hypochromicity. From these results and X-ray diffraction patterns, the structure of the nanofiber was elucidated (Fig. 10).

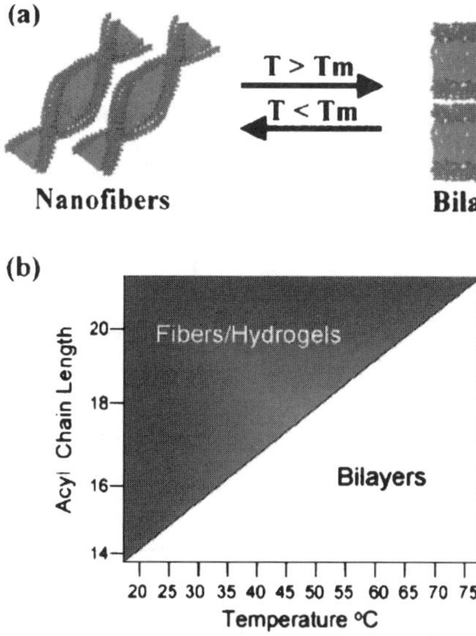

Fig. 9 (**a**) Illustration of the thermoreversible helical-bilayer transition and (**b**) phase diagram at 1 atm for the transformation between fibers and bilayers formed by **13**. Reprinted with permission from J Am Chem Soc 2004, 126, 7533

It is interesting to note that **15a** alone in 10% ethanolic aqueous solution formed a double helical rope of 1–2 μm in width and several hundred μm in length. Approximately equal amounts of the right-handed and left-handed double helical ropes were observed. Detailed NMR and model studies indicated that the presence of small amounts of a natural light-induced [2 + 2] photodimer between adjacent thymine units was responsible for formation of the ropes. The presence of complementary base, A, suppressed formation of the photodimer, leading to the formation of the nanofibers described above.

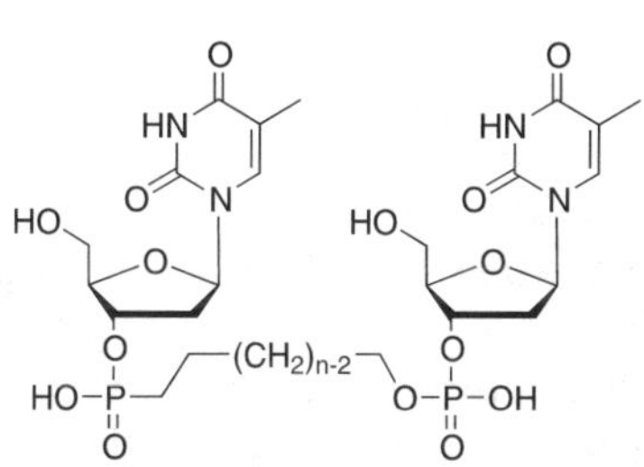

17a: n=12, **17b**: n=13, **17c**: n=14,
17d: n=15, **17e**: n=16, **17f**: n=18,
17g: n=20

Scheme 7

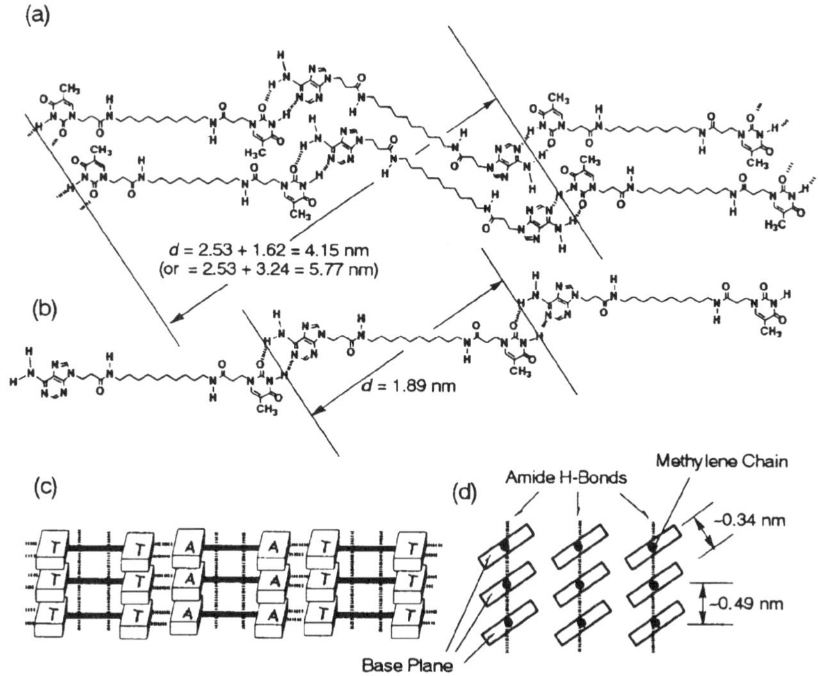

Fig. 10 A possible molecular packing and hydrogen bond scheme for (**a**) the heteroassembly formed from an equimolar mixture of **14a** and **15a** and (**b**) the homoassembly from **16a**. (**a**, **b**) Top view of a layered structure composed of linear polymolecular arrays ("reversed" Hoogsteen base pair configuration is employed here for the thymine-adenine heteroassociation). (**c**) Front view showing 2-D complementary and 1-D amide hydrogen bond network. (**d**) Side view of the polymolecular arrays. In (**d**), the one-dimensional amide hydrogen bond chain contributes to the stabilization of the base stacking and the formation of complementary hydrogen bonds. Reprinted with permission from J Am Chem Soc 2001, 123, 5947

Nucleotide bolaamphiphiles, **17**, in which two deoxythymidine 3′-monophosphate were connected by C12 – C20 oligomethylene spacers, showed good solubility in aqueous media at weakly acidic to neutral pH, and the longer homologues with the C18 (**17f**) and C20 (**17g**) chains effectively caused gelation [39]. The **17g** showed excellent gelation ability, requiring only 0.2 wt % to induce gelation, and the translucent gel was stable for over a week under neutral or mild alkaline conditions. This is actually the first report on the designed nucleobase-containing low molecular mass "hydrogelator" inducing effective gelation at low concentrations. Sol-gel transition took place at approximately 85 °C. Microscopic observations showed formation of nanofibers of 10–30 nm in diameter and several μm in length, which eventually formed well-developed three-dimensional fibrous networks. An atomic force microscopic (AFM) image showed the typical fiber to be a ribbon-type with a 10-nm thickness and 80-nm width. It is interesting to note that slightly shorter **17f** formed stable giant vesicles of 1–2 μm in a neutral buffer solution, and hydrogel was formed only in deionized water. Hydrogen bonds involving the 5′-hydroxyl group, hydrophobic interaction between the long oligomethylene chains, and π–π stacking of the thymine residues were indicated to be responsible for the network formation, leading to the effective hydrogel formation.

When a series of oligoadenylic acids $d(A)_n$ ($n = 2$–40) were present (total $A/T = 1$) in the **17g** solution, hydrogels were formed after several days of incubation [40]. In the case of shorter $d(A)_n$ ($n = 2$–8), three-dimensionally intertwined nanofibers similar to those without a template were formed. For longer $d(A)_n$ ($n = 10$–40), however, formation of discrete helical nanofibers of 7–8 nm in width and several hundred nm in length were indicated, showing the template effect of the oligoadenylic acids. Formation of the complemen-

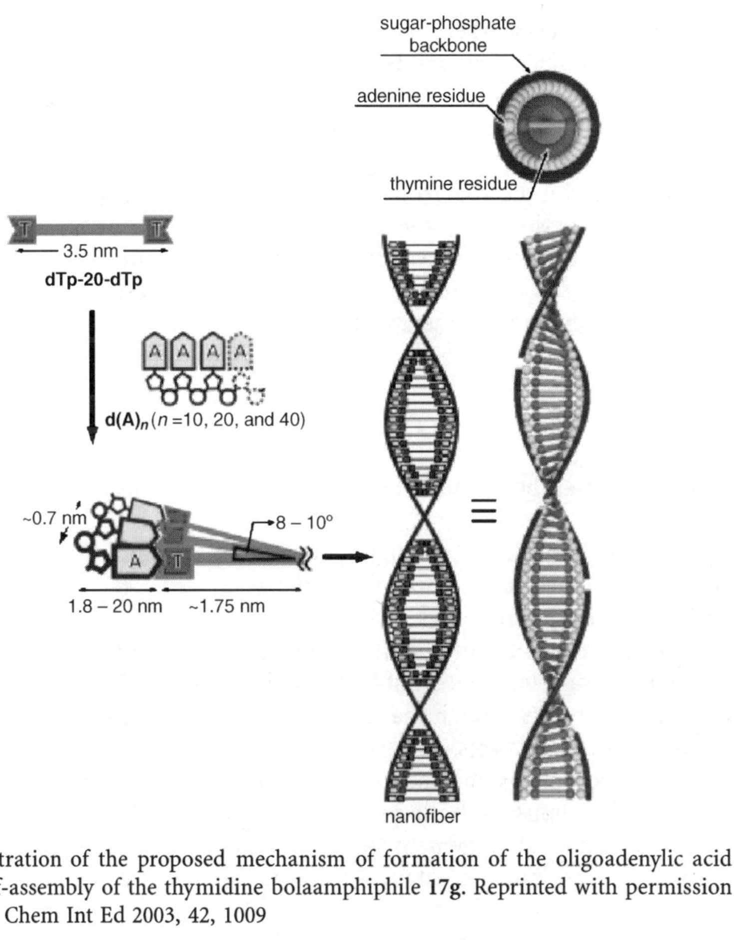

Fig. 11 Illustration of the proposed mechanism of formation of the oligoadenylic acid template self-assembly of the thymidine bolaamphiphile **17g**. Reprinted with permission from Angew Chem Int Ed 2003, 42, 1009

tary A-T base pairs was confirmed by infra-red (IR) and mass spectrometry [41]. The structure of the helical nanofibers was proposed to be the double strands of the $d(A)_n$ twisted around the vertically stacked rods of the bolaamphiphiles (Fig. 11).

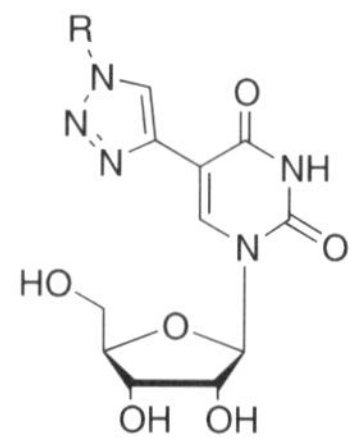

18a : R = Benzyl
18b : R = *o* - Methyl benzyl
18c : R = *m* - Methyl benzyl
18d : R = *p* - Methyl benzyl
18e : R = *o* - Ethyl benzyl
18f : R = *p* - Ethyl benzyl
18g : R = *p* - Butyl benzyl

Scheme 8

Kim and his co-workers applied a different gelator design [42]. They introduced several benzyltriazole derivatives to the C5 position of 2′-deoxyuridine (**18**) in order to control the hydrophilic/hydrophobic balance and to increase the aggregation ability. Most of them showed excellent gelation ability in less than 1 wt % in water, and formation of fibrous or lamellar assemblies were indicated from their SEM images. In these gels, hydrogen bonding interactions between hydroxyl groups of the deoxyribose moieties, and additional interbase hydrogen bonding interactions were suggested to be important for formation of the fibrous assemblies. However, the stacking interaction between the attached benzyltriazole moieties was not clearly discussed.

4
Molecular Assemblies and Gelation in Organic Systems

Since nucleobases are only slightly soluble in organic solvents, it is necessary to introduce some lipophilic moieties in order to dissolve nucleobase derivatives. In organic solvents having low permittivity, base-base hydrogen bonding interaction operates effectively, and this interaction plays a primary role in organogel formation but hydrophobic interaction is much less effective in organic solvents though the solvophobic effect is expected to some extent in polar organic solvents.

Lipophilic guanosine-based organogelators were reported by Gottarelli's and Araki's groups. In both cases, structures of the gelator assemblies and their macroscale arrangement in the gel states were clarified, and, interestingly, they were not the entangled fibrous networks generally observed for low molecular mass gelators.

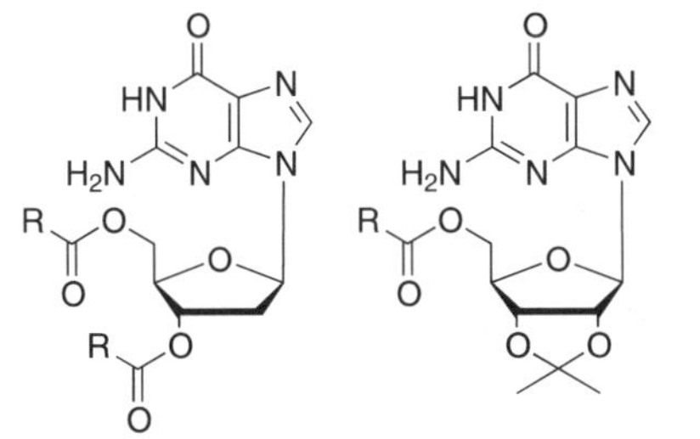

Scheme 9

Gottarelli and his group have been studying various supramolecular structures formed by guanosine derivatives [31, 43]. Formation of a gel-like liquid crystalline-phase was observed for a lipophilic didecanoyl deoxyguanosine **19** in hydrocarbon solvents [44]. However, no fibrillar structure was observed, and instead, a liquid crystalline phase composed of a planar tape-like assembly of the hydrogen bonded guanine moieties was suggested. Detailed examination of the tape-like assembly was further performed [45, 46]. The lipophilic guanosine derivatives **19–22** formed a gel-like phase in chloroform, toluene and hexadecane. The gel-like phase was birefringent, indicating an ordered structure. The critical concentrations at which birefringence appeared are shown in Table 1. Based on the in-depth NMR study of **19** in chloroform, the tape-like assembly was indicated to be a linear polymer connected by double interbase hydrogen bonds shown as tape II in Fig. 2. It is interesting to note that the mode of the hydrogen bond networks of **19** immediately after dissolution in solution was not the tape II but tape I, indicating that the hydrogen-bonded linear polymer in the solid phase was in tape I. IR spectra of the gel-like structures indicated that the hydrogen-bonded tape present in hexadecane (and possibly in toluene) must be similar to tape I and different from the tape present in chloroform. X-Ray diffraction measurements for the gel phase of **19** and **20** in hexadecane gave narrow Bragg

Table 1 Critical concentration (wt%) of **19–22** for the formation of the gel-like phases [46]

Compound	$CHCl_3$	Toluene	Hexadecane
19	23%	4%	8%
20	22%	4%	5%
21	4%	5%	$*^a$
22	4%	5%	$*^a$

$*^a$ = no birefringent phase observed.

reflections at small-angle regions. The peak reciprocal spacings were in the ratio $1 : \sqrt{2} : \sqrt{4}$..., indicative of two-dimensional square packing. The absence of additional small-angle peaks indicated that no correlation existed in the direction perpendicular to the two-dimensional square cell. Interestingly, the unit cell parameters continuously increase as a function of the solvent content, indicating a continuous swelling of the two-dimensional cell. From in-depth analysis of the XRD data, it was confirmed that the tapes contained the guanine residues in the extended hydrogen-bonded configuration, while the hydrocarbon chains, together with the hydrocarbon solvent in which they were dissolved, filled the lateral gap between the tapes (Fig. 12).

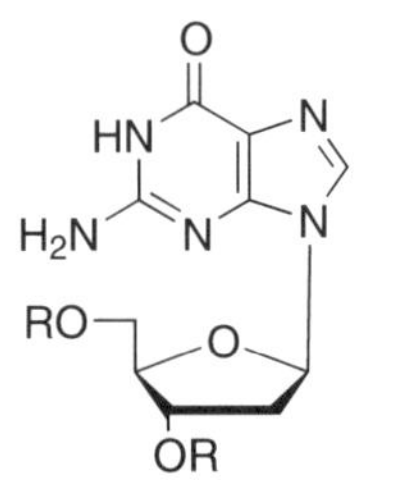

23a: R = $Si(iPr)_2C_8H_{17}$
23b: R=$SiPh_2tBu$
23c: R=$Si(iPr)_3$

Scheme 10

Araki and his group introduced nonpolar and flexible alkylsilyl groups to deoxyguanosine, and alkylsilylated guanosine derivative **23a** was shown to

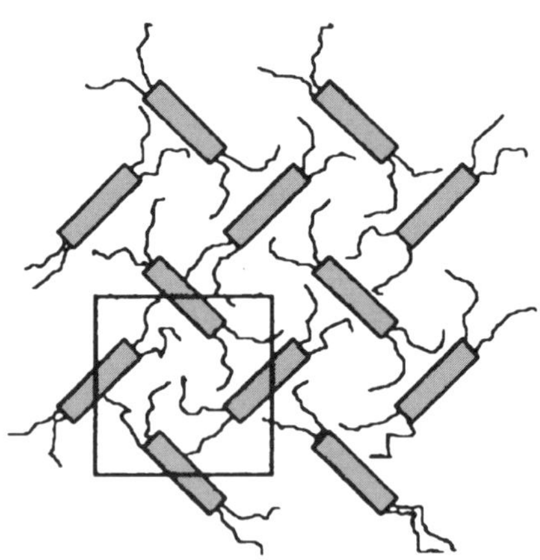

Fig. 12 A model for gel-like phases of **19** and **20** in hydrocarbon. Reprinted with permission from Chem Eur J 2002, 8, 2143

be a good organogelator for alkanes such as hexane (minimum gelation concentration 1.0 wt %), dodecane (0.6 wt %) and cyclohexane (4.3 wt %) [47]. In the AFM image of the gel transferred onto a silicon wafer (Fig. 13), mesoscopic-scale sheet-like structures instead of fibrous structures were clearly observed. The depth profile along the line in the image showed that the height of the sheets was 2.5 ± 0.2 nm but their width was as large as 50–300 nm. A concentrated gel (41 wt %) showed further extended and piled sheet-like assemblies, but the height of the sheets was practically unchanged. By comparing the IR spectra of reference compounds **23b** and **23c**, whose crystal structures were established by X-ray crystallography, the hydrogen bond pattern of the **23a**-dodecane system was examined. The reference compound **23b** has the hydrogen-bonded tape I shown in Fig. 2, but reference compound **23c** has additional double inter-tape hydrogen bonds between 2-NH_2 and 3-N of the two guanine-bases located at the adjacent tapes, leading to formation of two-dimensional hydrogen-bonded sheets (Fig. 14a). Partial formation of the inter-tape hydrogen bonds in the **23a**–dodecane system in the gel state was indicated from the IR spectrum of

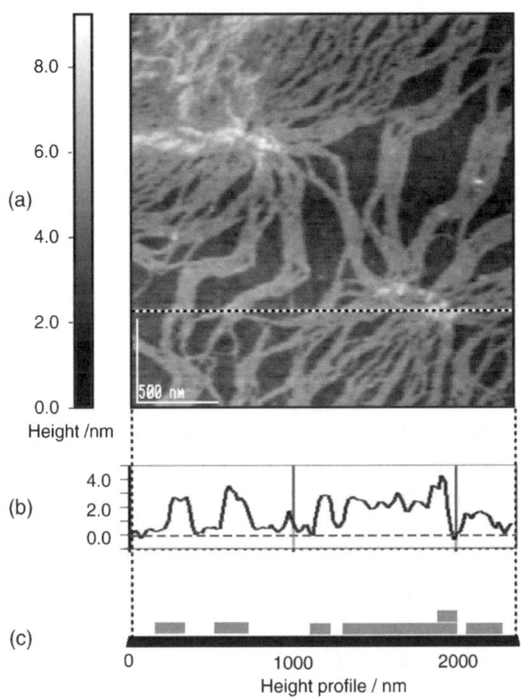

Fig. 13 (**a**) AFM view of the **23a**–dodecane (5 wt %) sample. (**b**) Depth profile along the line in the image. Height from the silicone surface is shown in nm. (**c**) Schematic view of the depth profile. Reprinted with permission from J Mater Chem 2001, 11, 3018

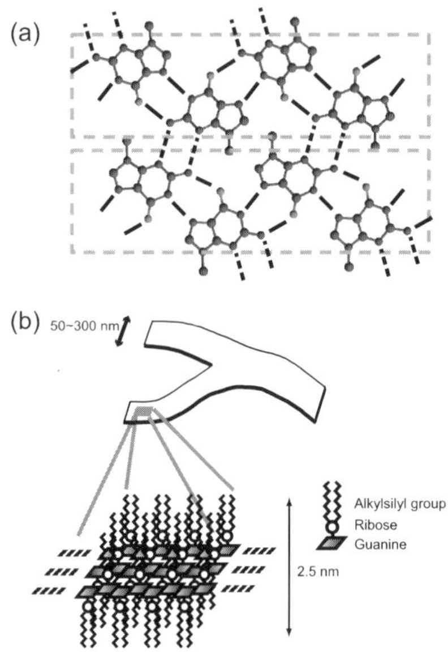

Fig. 14 (**a**) Two-dimensional hydrogen-bonded sheet of guanine moieties and (**b**) schematic view of the mesoscopic-scale sheet-like assembly of the **23a**–dodecane gel. Reprinted with permission from J Mater Chem 2001, 11, 3018

the gel. Furthermore, the gel to LC phase transition by heating was observed for the **23a**–dodecane system at high concentration (37–63 wt %), and this transition was shown to be due to the selective cleavage of the inter-tape hydrogen bonds. The results indicated that formation of the sheet assemblies by the inter-tape hydrogen bonds was essential for the gelation process.

The observed thickness of 2.5 nm in the AFM image was comparable to that of the sheet-like structure of **23a**, while spacings of 4.0 and 3.2 nm were obtained from XRD patterns of the **23a**–dodecane gel at 41 and 60 wt %, respectively. The observed spacing of the gel was larger than the thickness of the sheet assembly. Since a higher dodecane content increased the spacing, the gel was suggested to have a layer structure composed of the hydrogen-bonded sheet-like assemblies and dodecane molecules were entrapped in between the sheet assemblies. A schematic view of the mesoscopic-scale sheet-like assembly in the gel is given in Fig. 14b. The splitting and stitching of the sheet assemblies, clearly shown in the AFM image, led to formation of well-developed macroscale three-dimensional networks of the sheet assemblies, offering a molecular-level understanding of the network formation.

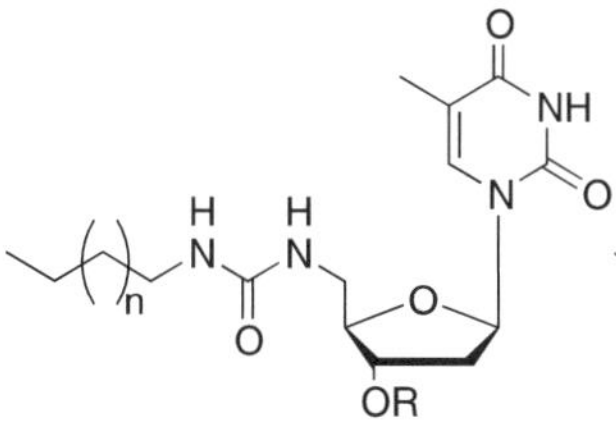

24a: n=5
24b: n=9
24c: n=15

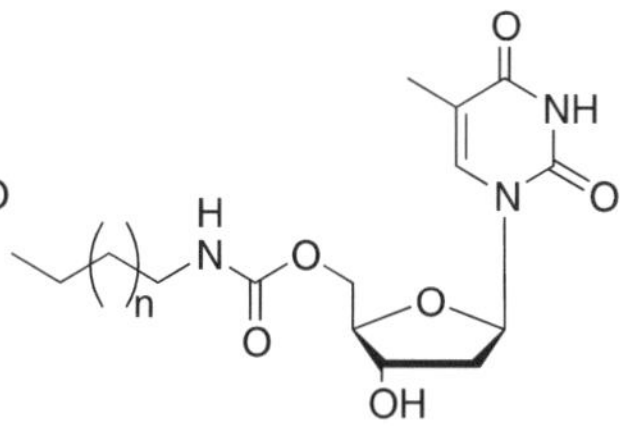

25a: n=5
25b: n=9
25c: n=15

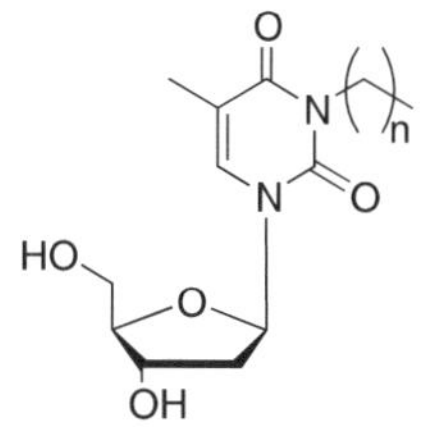

26a: n=11
26b: n=12
26c: n=13
26d: n=14

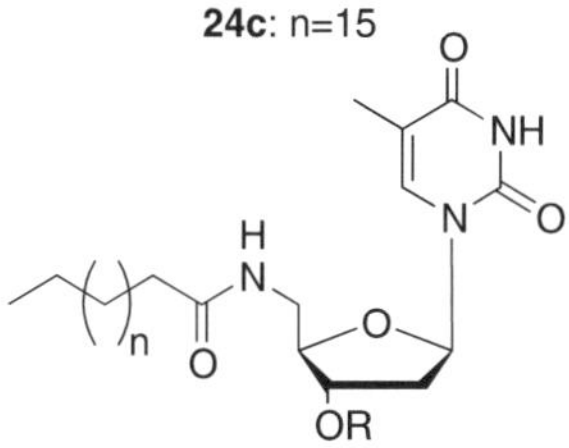

27a: n=8
27b: n=10
27c: n=12

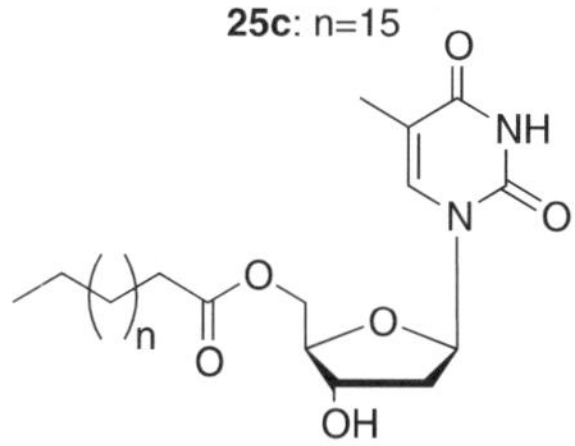

28a: n=8
28b: n=10
28c: n=12

Scheme 11

Kim et al. tested gelation abilities of a series of lipophilic thymidine derivatives having alkyl chains (**24–28**) [48]. All the derivatives (2 w/w %) showed gelation ability in organic solvents (Table 2). The urea- (**24**) and amide-linked (**25**) derivatives gave robust and translucent gels in alkanes, while the carbamate- (**27**) and ester-linked (**28**) derivatives yielded opaque gels in toluene and tetrahydronaphthalene (THN). The *N*-alkylated derivatives (**26**) formed opaque gels in benzene, toluene, and THN. Gelation was not observed in methanol, chloroform, tetrahydrofuran, acetone, and 2-octanol. SEM images of the dried samples showed three types of structures, i.e., fibrous structure (**24**/cyclohexane and **26**/CCl_4), woven structure (**27**/cyclohexane) and lamellar structure (**28**/cyclohexane and **26**/touluene, benzene, and THN) (Fig. 15). Though the pattern of the hydrogen bonding interaction was not clarified in these gels, it is obvious that the hydrogen bonding interactions, including the base moieties, deoxyribose moieties and the linkage units, were the major driving force for gelation. For the *N*-alkylated derivatives, hydrogen bond formation of the hydroxyl groups at the deoxyribose unit was observed by IR spectra. The solvent effect on the gelation ability and morphology of the assemblies indicated that van der Waals interaction and π–π stacking in addition to the hydrogen-bonding interaction were responsible for the organogel formation.

To change the structure and properties of the gel, complementary poly(A) was added to the **28c**/benzene gel system [49]. Since poly(A) was virtually insoluble in organic solvents, the lipohilic poly(A)/cationic lipid complex was used instead. In the presence of an equal amount of the

Table 2 Gelation test of **24–28** (2 wt %)[a] [48]

Solvent	**24a**	**24b**	**24c**	**25a**	**25b**	**25c**	**26a**	**26b**	**26c**	**26d**	**27a**	**27b**	**27c**	**28a**	**28b**	**28c**
Benzene	P	P	–	Ins	Ins	Ins	P	G	G	G	S	S	S	G	pG	G
Toluene	P	P	Ins	Ins	pG	G	P	G	G	–	S	S	S	G	G	G
CCl_4	S	P	–	Ins	Ins	Ins	G	–	–	–	–	–	–	Ins	G	G
THN	–	–	–	G	G	G	G	pG	pG	O	–	–	–	pG	G	G
n-Hexane	G	G	G	Ins	Ins	Ins	–	–	–	–	G	G	G	Ins	pG	G
Cyclohexane	G	G	G	Ins	Ins	Ins	–	–	–	–	G	G	G	Ins	pG	G
n-Pentane	G	G	G	–	–	–	–	–	–	–	G	G	G	–	–	–
n-Heptane	pG	G	G	–	–	–	–	–	–	–	G	G	G	–	–	–
n-Octane	P	pG	G	–	–	–	–	–	–	–	G	G	G	–	–	–

[a]Ins = insoluble; pG = partial gel; G = gel; P = precipitation; S = soluble; O = oilic

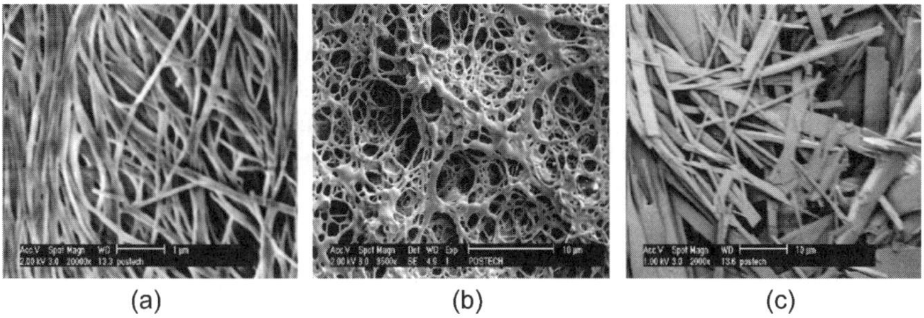

Fig. 15 SEM images of **26** and **27**; (**a**) **26a**/CCl_4 system, (**b**) **27c**/octane system, (**c**) **26c**/toluene system. The different gelling abilities are attributed to differences in hydrogen-bonding interactions. Reprinted with permission from Chem Commun 2003, 254

poly(A)/lipid complex, the opaque **28c**/benzene gel became transparent (Fig. 16a) though the noncomplementary poly(C)/lipid showed no effect at all. Formation of the complementary A-T hydrogen-bonding pair in **28c** with the poly(A)/lipid gel was confirmed by IR measurement. SEM images indicated that an addition of the poly(A)/lipid complex altered a plate-like structure of the **28c**/benzene gel into an entangled fiber structure (Fig. 16c). Based on the X-ray diffraction pattern, it was suggested that formation of the complementary A-T hydrogen bonds with poly(A) prevented the bilayer assemblies of **28c** from developing into a multi-layered lamellar structure, inducing morphological change from a plate-like to a fiber structure.

Some of the uridine-phosphocholine hybrids, which are discussed as hydrogelators in the preceding section, also caused gelation of nonaqueous solutions [36]. For example, **13b** formed a clear organogel in cyclohexane (6 w/w % above 42 °C). The longer chain derivatives (**13c–e**) did not form organogels, and this is a likely consequence of their increased solubility.

Fig. 16 (**a**) Photograph of **28c** gel (*left*) and **28c** with poly(A)/lipid gel (*right*); SEM images of (**b**) **28c** gel, (**c**) **28c** with poly(A)/lipid gel and (**d**) **28c** with poly(C)/lipid gel, prepared from benzene, bar length = 5 mm, [A or C]/[T] = 1.0. Reprinted with permission from Chem Commun **2004**, 1996

A SEM image of the **13b** organogel in cyclohexane showed formation of multilamellar assemblies. SAXS measurements confirmed the presence of repeat periods of 4.6 ± 0.2 nm, which was consistent with the lamellar structure found in the aqueous system. The IR spectrum of the organogel indicated formation of the U-U hydrogen bonding in the **13a** supramolecular assembly.

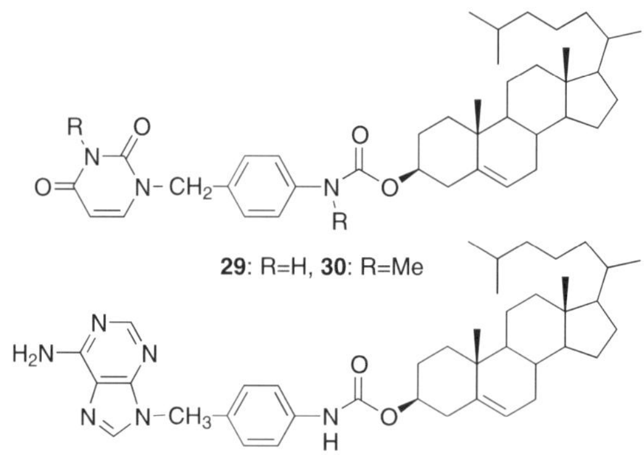

29: R=H, **30**: R=Me

31

Scheme 12

All the organogelators discussed above use base-base interaction as the major driving force for assembling the gelators. Instead, Shinkai and coworkers used a steroid unit as the major gelation unit, and a nucleobase moiety was introduced as an additional unit to control the mode of assembly. Steroid derivatives that have extended planar surfaces with molecular chirality are known to serve as an effective mesogenic unit for inducing the columnar cholesteric mesophase. Many types of low molecular mass organogelators having steroid structures have been reported [1]. In these gels, the steroid moieties are stacked in a helical fashion to form a central core. Shinkai et al. have been actively studying a variety of self-assemblies formed by cholesterol derivatives [50], and have found that functional groups introduced at the C3 position of cholesterol stick outwards like a spiral staircase from the core [51]. They introduced a uracil unit to this position [52, 53]. The uracil-appended derivative **29** showed excellent gelation ability in cyclohexane (minimum gelation 0.25 wt %), and the solvents that were gelated by **29** at 3 wt % were *n*-octane, benzene, toluene, *p*-xylene and decaline. The stability of the gel was estimated from the sol-gel transition temperature T_{gel}. Methyl substitution of the uracil NH of the gelator (**30**) resulted in decreased gel stability, which suggested that the intergelator hydrogen-bonding interaction at the uracil moiety was also important for gel formation in addition to the columnar stacking of the steroid units. The presence of lipophilic nucleosides was found to destabilize the gel system presumably due to suppression of the intergelator hydrogen bonds, further supporting the importance of the hydrogen bonds for the gel formation. The SEM picture of the xerogel prepared from **29** showed a well-developed network structure consisting of fibrils. The diameter of the fibrils was about 20–40 nm and the fiber structure showed clear right-handed helicity. Since no helical structure was observed for the xerogel of **30**, hydrogen-bonding interaction was suggested to be essential to induce neat molecular packing in the gel phase, leading to the helical superstructure.

Though the presence of the adenosine-appended cholesterol derivative **31** in cyclohexane apparently decreased the T_{gel} of **29**, an increase of the molar fraction of **31** from 0.3 to 0.8 did not result in a monotonous decrease of the T_{gel}. Instead, the highest T_{gel} within this range was observed when the molar fraction of **31** was 0.5, suggesting a contribution of the complementary A-T hydrogen-bonding interaction to the gel stability. In a SEM picture of the xerogel, a change in the morphology of the assemblies was evident, which showed a mixture of both sheet and fiber structures. The presence of a polynucleotide also affected the gelation ability of **29** in *n*-butanol [54]. As shown in Table 3, **29**/poly(A) and **29**/poly(C) complexes showed an increased gelation ability of *n*-butanol compared to that of **29** alone. A TEM image of the **29**/poly(A) complex showed a well-developed fibrous network structure, and a close-up picture showed that most tape-like fibers of 50–100 nm in diameter were twisted in a right-handed helical fashion with 600–700 nm pitches. The **29**/poly(C) complex also showed development of the fibrous network but no helical structure was observed even in a magnified picture. These results indicated the importance of the interbase hydrogen-bonding interaction for compact packing, affecting the morphology of the assemblies.

Table 3 Gelation test and T_{gel} values[a] of **29**/polynecleotide complexes [54]

[**29**] wt %	[adenine base]/[**29**] molar ratio					
	0	0.05	0.10	0.30	0.50	0.70
3 wt %	S			pG		
5 wt %	S	pG	G (70)	G (76)	G (72)	G (70)
7 wt %	G (80)			G (82)		
[**29**] wt%	[cytosine base]/[**29**] molar ratio					
	0	0.05	0.10	0.30	0.50	0.70
3 wt %	S			pG		
5 wt %	S	pG	G (72)	G (69)	G (70)	G (70)

[a] G = gel, pG = partial gel, S = solution, °C in parenthesis

5 Design of Nucleobase-Containing Molecules for Supramolecular Materials

A nucleobase has high ability to form directionally controlled multiple intermolecular interactions, i.e., in-plane multiple hydrogen bonding interaction and stacking interaction perpendicular to the plane. As a result, molecular assemblies of the nucleobases have geometrically well-defined structures like helical or lamellar motif discussed above rather than randomly arranged amorphous states. Molecular assemblies to form macroscale networks leading to the amorphous gel state are not necessary to have a well-defined, ordered structure. Therefore, the nucleobase assemblies are more suitable for designing and constructing supramolecular materials with highly organized structures. As concluding remarks to show the high ability of the nucleobase-containing molecules in supramolecular chemistry, a brief overview of the nucleobase-containing supramolecular materials and their potential applications will be presented.

5.1 Base Pairs

Base pair formation is the principal interbase hydrogen-bonding motif, and has been used for specific recognition of receptors [55], as the mesogenic unit for liquid crystals [56], the cross-linkage of polymer systems [57–59], and the connecting unit for supramolecular polymers [60]. Among them, inter-chain cross-linking of polymers by the base pairs at their side chains offers a convenient approach to a tunable gel, though they can not be included in the category of low molecular mass gelators. As shown in Fig. 17a, association of a self-complementary nucleobase analog (ureidopyrimidinone) appended to a polyolefin chain caused reversible cross-link formation, leading to gelation of toluene [57]. The bulk polymer showed unique elastomeric properties due to the reversible network formation. Instead of direct self-association, bi-functional cross-linkers can be used (Fig. 17b and c) to induce inter-chain cross-linking [58, 59]. In the case of Fig. 17c, two DNA strands (SA1 and SA2) appended to polyacrylamide chains were cross-linked by the third DNA strand L2, leading to reversible hydrogel formation. In addition, Guan et al. reported the unique mechanical properties of polymeric materials when cross-links are formed within the same chain [61].

Supramolecular polymers, in which monomer units are connected by noncovalent interactions, have been attracting considerable interest [60], and there are many reports using base-base interactions as connecting units [62–64]. Among them, an interesting approach was reported by Meijer and his co-workers, polymeric assembly of monofunctional and bifunctional

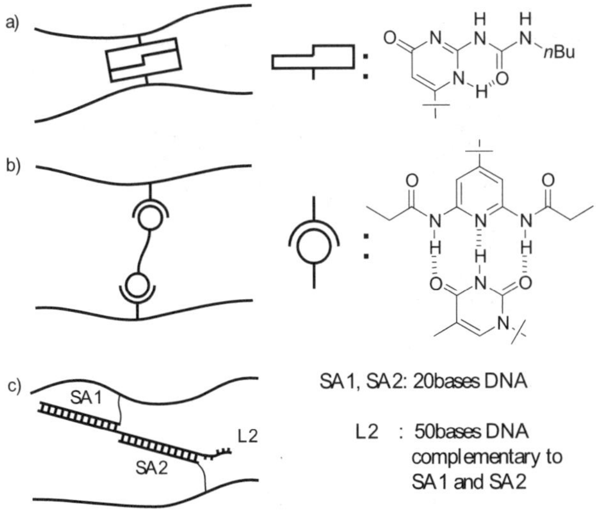

Fig. 17 Cross-linkage of polymer systems by use of (**a**) self-complementary sidechains [57] (**b**) a bifunctional cross-linker [58] and (**c**) a DNA cross-linker [59]

building blocks having self-complementary nucleobase analogs, ureidotriazine derivatives, as the functional units [65]. They formed columnar stacks by the interbase hydrogen bonds and the stacking interaction.

5.2 Cyclic Oligomers

Cyclic oligomers are observed mostly for the guanosine derivatives because they have sufficient hydrogen bond donor and acceptor sites within the guanine ring [30]. As discussed in Sect. 3.2, the highly size-selective cation binding ability of the cyclic oligomers of G and isoG as self-assembled supramolecular ionophores has attracted considerable interest, and their ionophoric properties have been studied [66–68]. Their use as ion-channels has also been examined [69, 70]. In addition, the liquid crystal properties of the columnar stacks of guanosine and guanine-like folic acid derivatives have been studied [71–73].

5.3 Linear Polymers

The linear polymer formed by multiple interbase hydrogen bonds has a rigid structure with sufficient stability, and is suitable as a component of liquid crystals [31, 44–46]. The use of linear polymers as photoconductive and electronic materials was also reported [74, 75].

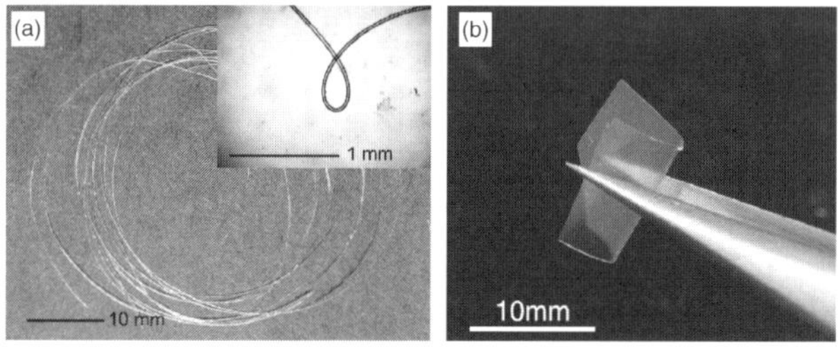

Fig. 18 Guanosine-based supramolecular (**a**) fiber and (**b**) film. Reprinted with permission from Chem Commun **2001**, 1826 and Angew Chem Int Ed **2004**, 43, 100

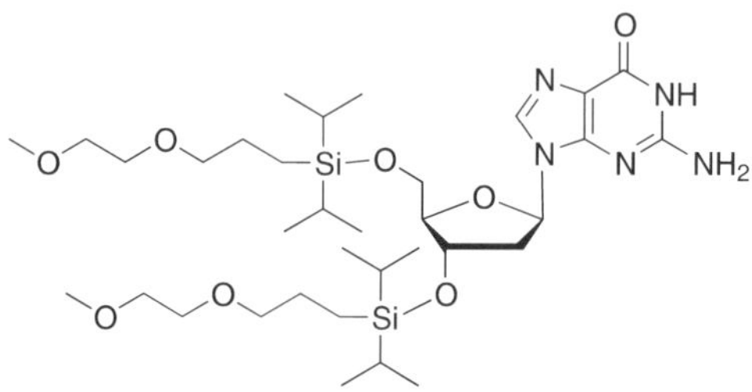

32

Scheme 13

Furthermore, supramolecular fibers and films were fabricated from these assemblies. The molecular design to wrap the hydrogen-bonded linear polymer chains of the guanosine derivatives by nonpolar and flexible alkylsilyl side chains allowed melt spinning to give sufficiently tough and flexible macro-scale fibers of 0.02 to 0.21 mm in diameter (ϕ) and infinite length (Fig. 18a). High orientation of the hydrogen-bonded polymer chains along the fiber axis was indicated [76]. When the two-dimensional hydrogen-bonded sheet assemblies of the guanosine derivative **32** was sandwiched by nonpolar alkylsilyl groups having an oxyethylene unit at the chain end, self-standing and flexible supramolecular film was obtained by the solvent-cast method [77]. Unlike guanosine derivative **23a**, the presence of inter-sheet interaction allowed fabrication of the supramolecular film instead of an organogel (Fig. 18b).

6
General Conclusion

Heterocyclic nucleobase molecules have high ability to form in-plane multiple hydrogen bonding and base stacking perpendicular to the plane. These directionally controlled multiple base-base interactions make the nucleobase moiety a highly useful structural element for design and construction of macroscale helical, lamellar, and other geometrically well-defined three-dimensional assemblies. Though it requires further study, rapid progress of the nucleobase-related supramolecular and materials chemistry in recent years, which is briefly reviewed here, will open the way to a variety of novel functional materials that consist of geometrically well-defined three-dimensional assemblies of nucleobase-containing molecules.

Acknowledgements This work was partly supported by a Grant-in-Aid for Scientific Research (grant nos. 14045211 and 14350482) from the Ministry of Education, Science, Sports and Culture, Japan

References

1. Terech P, Weiss RG (1997) Chem Rev 97:3133
2. Estroff LA, Hamilton AD (2004) Chem Rev 104:1201
3. Saenger W (1984) Principles of Nucleic Acid Structure. Springer-Verlag, Berlin Heidelberg New York
4. Steiner T, Saenger W (1993) J Am Chem Soc 115:4540
5. Braga D, Maini L, Polito M, Grepioni F (2004) In: Michael D, Mingos P (eds) Structure and bonding 111. Springer, Berlin Heidelberg New York, p 1
6. Asensio A, Kobko N, Dannenberg JJ (2003) J Phys Chem A 107:6441
7. Guckian KM, Schweitzer BA, Ren RX-F, Sheils CJ, Tahmassebi DC, Kool ET (2000) J Am Chem Soc 122:2213
8. Mutai K, Gruber BA, Leonard NJ (1975) J Am Chem Soc 97:4095
9. Ts'o POP (1974) Basic principles in nucleic acid chemistry. Academic Press, New York
10. Kitano H, Ringsdorf H (1985) Bull Chem Soc Jpn 58:2826
11. Ariga K, Kunitake T (1998) Acc Chem Res 31:371
12. Kurihara K, Ohto K, Honda Y, Kunitake T (1991) J Am Chem Soc 113:5077
13. Ebara Y, Itakura K, Okahata Y (1996) Langmuir 12:5165
14. Miao WG, Du XZ, Liang YQ (2003) Langmuir 19:5389
15. Berti D, Franchi L, Baglioni P, Luisi PL (1997) Langmuir 13:3438
16. Kawahara T, Kurihara K, Kunitake T (1992) Chem Lett 1839
17. Shimomura M, Nakamura F, Ijiro K, Taketsuna H, Tanaka M, Nakamura H, Hasebe K (1997) J Am Chem Soc 119:2341
18. Nowick JS, Chen JS (1992) J Am Chem Soc 114:1107
19. Nowick JS, Chen JS, Noronha G (1993) J Am Chem Soc 115:7636
20. Nowick JS, Cao T, Noronha G (1994) J Am Chem Soc 116:3285
21. Berti D, Pini F, Baglioni P, Teixeira J (1999) J Phys Chem B 103:1738
22. Berti D, Barbaro P, Bucci I, Baglioni P (1999) J Phys Chem B 103:4916
23. Zandomeneghi G, Luisi PL, Mannina L, Segre A (2001) Helv Chim Acta 84:3710
24. Berti D, Baglioni P, Bonaccio S, Barsacchi-Bo G, Luisi PL (1998) J Phys Chem B 102:303
25. Berti D, Luisi PL, Baglioni P (2000) Colloids Surf A 167:95
26. Pincet F, Lebeau L, Cribier S (2001) Eur Biophys J 30:91
27. Guschlbauer W, Chantot JF, Thiele D (1990) J Biomol Struct Dyn 8:491
28. Gellert M, Lipsett MN, Davies DR (1962) Proc Natl Acad Sci USA 48:2013
29. Sasisekharan V, Zimmerman S, Davies DR (1975) J Mol Biol 92:171

30. Davis JT (2004) Angew Chem Int Ed 43:668
31. Gottarelli G, Spada GP, Garbesi A (1996) In: Sauvage JP, Hosseini MW (eds) Comprehensive supramolecular chemistry, vol. 9. Elsevier, Oxford, p 483 and references therein
32. Yanagawa H, Ogawa Y, Furuta H, Tsuno K (1988) Chem Lett 269
33. Yanagawa H, Ogawa Y, Furuta H, Tsuno K (1989) J Am Chem Soc 111:4567
34. Itojima Y, Ogawa Y, Tusno K, Handa N, Yangawa H (1992) Biochem 31:4757
35. Bombelli FB, Berti D, Keiderling U, Baglioni P (2002) J Phys Chem B 106:11613
36. Moreau L, Barthélémy P, Maataoui MEI, Grinstaff MW (2004) J Am Chem Soc 126:7533
37. Shimizu T (2002) Macromol Rapid Commun 23:311
38. Shimizu T, Iwaura R, Masuda M, Hanada T, Yase K (2001) J Am Chem Soc 123:5947
39. Iwaura R, Yoshida K, Masuda M, Yase K, Shimizu T (2002) Chem Mater 14:3047
40. Iwaura R, Yoshida K, Masuda M, Ohnishi-Kameyama M, Yoshida M, Shimizu T (2003) Angew Chem Int Ed 42:1009
41. Iwaura R, Ohnishi-Kameyama M, Yoshida M, Shimizu T (2002) Chem Commun 2658
42. Park SM, Lee YS, Kim BH (2003) Chem Commun 2912
43. Spada GP, Gottarelli G (2004) Synlett 4:596
44. Gottarelli G, Masiero S, Mezzina E, Spada GP, Mariani P, Recanatini M (1998) Helv Chim Acta 81:2078
45. Gottarelli G, Masiero S, Mezzina E, Pieraccini S, Spada GP, Mariani P (1999) Liq Cryst 26:965
46. Giorgi T, Grepioni F, Manet I, Mariani P, Masiero S, Mezzina E, Pieraccini S, Saturni L, Spada GP, Gottarelli G (2002) Chem Eur J 8:2143
47. Sato T, Seko M, Takasawa R, Yoshikawa I, Araki K (2001) J Mater Chem 11:3018
48. Yun YJ, Park SM, Kim BH (2003) Chem Commun 254
49. Sugiyasu K, Numata M, Fujita N, Park SM, Yun YJ, Kim BH, Shinkai S (2004) Chem Commun 1996
50. Shinkai S, Murata K (1998) J Mater Chem 8:485
51. Murata K, Aoki M, Suzuki T, Harada T, Kawabata H, Komori T, Ohseto F, Ueda K, Shinkai S (1994) J Am Chem Soc 116:6664
52. Snip E, Shinkai S, Reinhoudt DN (2001) Tetrahedron Lett 42:2153
53. Snip E, Koumoto K, Shinkai S (2002) Tetrahedron 58:8863
54. Numata M, Shinkai S (2003) Chem Lett 32:308
55. Zimmerman SC, Corbin PS (2000) In: Structure and Bonding 96. Springer, Berlin Heidelberg New York, p 63
56. Kato T (2000) In: Structure and Bonding 96. Springer, Berlin Heidelberg New York, p 95
57. Rieth LR, Eaton RF, Coates GW (2001) Angew Chem Int Ed 40:2153
58. Thibault RJ, Hotchkiss PJ, Gray M, Rotello VM (2003) J Am Chem Soc 125:11249
59. Lin DC, Yurke B, Langrana NA (2004) J Biomechanical Eng 126:104
60. Brunsveld L, Folmer BJB, Meijer EW, Sijbesma RP (2001) Chem Rev 101:4071
61. Guan Z, Roland JT, Bai JZ, Ma SX, Mclntire TM, Nguyen M (2004) J Am Chem Soc 126:2058
62. Fouquey C, Lehn J-M, Levelut A-M (1990) Adv Mater 2:254
63. Sivakova S, Rowan SJ (2003) Chem Commun 2428
64. Folmer BJB, Sijbesma RP, Meijer EW (2001) J Am Chem Soc 123:2093
65. Hirschberg JHKK, Koevoets RA, Sijbesma RP, Meijer EW (2003) Chem Eur J 9:4222
66. Marlow AL, Mezzina E, Spada GP, Masiero S, Davis JT, Gottarelli G (1999) J Org Chem 64:5116

67. Tirumala S, Davis JT (1997) J Am Chem Soc 119:2769
68. Lee SC, Lamb JD, Cai MM, Davis JT (2001) J Inclusion Phenom Macrocyclic Chem 40:51
69. Forman SL, Fettinger JC, Pieraccine S, Gottarelli G, Davis JT (2000) J Am Chem Soc 122:4060
70. Sidorov V, Kotch FW, El-Kouedi M, Davis JT (2000) Chem Commun 2369
71. Pieraccini S, Gottarelli G, Mariani P, Masiero S, Saturni L, Spada GP (2001) Chirality 13:7
72. Giorgi T, Lena S, Mariani P, Cremonini MA, Masiero S, Pieraccini S, Rabe JP, Samori P, Spada GP, Gottarelli G (2003) J Am Chem Soc 125:14741
73. Kanie K, Nishii M, Yasuda T, Taki T, Ujiie S, Kato T (2001) J Mater Chem 11:2875
74. Rinaldi R, Branca E, Cingolani R, Masiero S, Spada GP, Gottarelli G (2001) Appl Phys Lett 78:3541
75. Maruccio G, Visconti P, Arima V, D'Amico S, Biasco A, D'Amone E, Cingolani R, Rinaldi R, Masiero S, Giorgi T, Gottarelli G (2003) Nanoletters 3:479
76. Araki K, Takasawa R, Yoshikawa I (2001) Chem Commun 1826
77. Yoshikawa I, Li J, Sakata Y, Araki K (2004) Angew Chem Int Ed 43:100

Top Curr Chem (2005) 256: 167–218
DOI 10.1007/b107174

Published online: 1 July 2005

Chirality Effects in Self-assembled Fibrillar Networks

Aurélie Brizard · Reiko Oda · Ivan Huc (✉)

Institut Européen de Chimie et Biologie, 2 rue Robert Escarpit, 33607 Pessac, France
r.oda@iecb.u-bordeaux.fr, i.huc@iecb.u-bordeaux.fr

Abstract Chirality seems to be intimately associated with the growth and stability of self-assembled fibrillar networks and with the most common macroscopic property of these networks, which is the thermoreversible gelation of the solvent. The presence and the relative configurations of stereogenic centers in the structure of a small molecule gelator are generally (but not always) observed to be critical to its ability to form gels. Symmetry considerations of chiral molecular packing provide thermodynamic and kinetic arguments that may explain why chirality favors fiber growth. Additionally, molecular chirality is sometimes expressed at a scale of nanometers or micrometers and gives rise to twisted or coiled fiber structures that are readily observable by microscopic techniques. These chiral fiber morphologies have already found some applications as templates for helical protein crystallization or for the growth of chiral inorganic replicas. The chiroptical properties of assembled chiral molecules, e.g., circular dichroism, allow monitoring of aggregation and may sometimes give insights into molecular packing. But determining chiral molecular arrangements in the fibers remains a challenge and requires the use of multiple techniques.

Keywords Chirality · Gels · Self-assembly · Helices · Fibers

Abbreviations

AFM	Atomic force microscopy
CD	Circular dichroism
DIC	Differential interference contrast
HRSEM	High-resolution scanning electron microscopy
NMR	Nuclear magnetic resonance
SEM	Scanning electron microscopy
STM	Scanning tunneling microscopy
TEM	Transmission electron microscopy
VCD	Vibrational circular dichroism
WAXS	Wide-angle X-ray scattering

1 Introduction

Chirality seems to be intimately associated with the growth and stability of self-assembled fibrillar networks of small organic molecules and with the most common macroscopic property of these networks, which is the thermoreversible gelation of the solvent. The importance of chirality is apparent both at the molecular scale and at the scale of the self-assembled fibers, which typically ranges from nanometers to micrometers. This chapter attempts to review literature data about both these aspects.

As is discussed in the second section of this chapter, many research groups have reported that the presence of at least one stereogenic center in the structure of a small molecule gelator determines its ability to form gels in organic solvents or in water. Molecular chirality thus seems to promote the growth of assemblies with high aspect ratio that entrap the solvent in which they form. When the molecule possesses several stereogenic centers, their relative configuration is also critical to the gelling properties. This has led to the simple empirical (though not general) rule that a molecule has a better chance to be a good gelator if it is chiral. But these observations remain for a large part difficult to explain.

On the other hand, molecular chirality is sometimes expressed at a much larger scale in the morphology of self-assembled fibers. Elongated, fibrous objects such as rods, tapes, or tubes may be helically twisted, coiled, or wound around one another, and therefore exist in a left-handed or a right-handed form. These intriguing shapes can often be simply visualized by microscopic techniques and have fascinated both chemists and physicists. The third section of this chapter aims to illustrate the structural variety of these objects and the extent to which their formation can be altered and their shapes tuned upon changing the experimental conditions. It also gives a perspective of the relations that can sometimes be drawn between on the one hand, chirality of small molecular components and, on the other hand, supramolecular chirality within their assemblies.

The fourth section of the chapter focuses on the various approaches to experimentally address chirality in self-assembled fibers. The unique chiroptical properties of chiral molecules can be studied by circular dichroism and give access to valuable information about the conformations and relative positions of the molecules in the fibers. This section also gives an overview of the techniques that have been used to observe and assign fiber handedness.

One of the reasons why gels and self-assembled fibers attract so much attention is their very high potential for applications. The last section of this chapter addresses applications which specifically make use of chirality, such as those based on enantioselective molecular recognition, or those based on the use of chiral fibers as templates for the generation of helical arrays of proteins or of chiral inorganic materials.

This chapter was constructed so as to limit overlap with previous reviews on the topic of chirality in self-assembled fibers. Excellent papers by Fuhrhop and Helfrich [1], Schnur [2], and Kunitake [3] were published at the beginning of the 1990s. More recently, an important book chapter appeared about chiral molecular self-assembly of amphiphilic molecules in water or other protic solvents [4]. The present chapter should complement this review in several ways. It extends the field covered to chiral assembly and to gelation in organic solvents. It presents short sections on the tools to investigate chirality in self-assembled fibers and on applications of these chiral fibers. Additionally, the systems are not classified here according to the family of molecules to which they belong, but rather according to the type of properties that they exhibit. This was partly facilitated by the way the field tends to be divided. Indeed, molecular chirality as a factor of gelling ability on the one hand, and chirality expressed in the morphology of the fibers on the other hand, often belong to the same systems. However, many publications tend to focus on one aspect only. Publications dealing primarily with gelling ability describe various aspects which can be considered from the perspective of chirality: the solvent range of gel formation, gel transition temperatures as a function of concentration, possible molecular arrangements of the gelators in the gels etc. These papers often show a few images of the network of fibers, but rarely give detailed accounts of the morphology of these fibers. Conversely, extensive studies about helical or twisted fiber morphology often do not even make mention that the fibers that are studied form a gel.

To further define the scope of this chapter, a line should be drawn between solid fibers and fluid micellar rods [1]. The solid fibers of which organogels and hydrogels are made and which interest us here are generally produced by fast precipitation after having elevated the temperature above the solubility limit of the gelator. However, fluid-elongated micelles can be produced by simply suspending an amphiphilic molecule in water or, more rarely, in other solvents. Fluid micellar fibers thus exist at temperatures above the solubility limit of the amphiphile (Krafft temperature). Compared to solid fibers, the effect of chirality is rare in fluid fibers [5]. However, the border between these two types of fibers is not always easy to establish. Both may lead to a large increase of viscosity and the visual aspect of the samples can be similar. Moreover, some systems seem to combine both solid and fluid features, for example, the thixotropic gels which spontaneously repair after physical damage.

Formally, physical organogels or hydrogels consist of a network of solid fibers. In other words, connections between the fibers should be necessary to the physical integrity of the gels. But again, this aspect is often not presented in detail in the literature. Micrographs of the samples do not always show clear differences between connected and entangled fibers. Rheological studies should provide definite answers about the importance of connections in a gel, but they are rarely undertaken.

Even with these limitations, the very large body of literature pertaining to the topics presented here could not be covered exhaustively in the context of this chapter. We have selected examples which we thought were illustrative. As a consequence, this chapter is not perfectly representative of the scientific production of the various contributors to this field, and we apologize for any important omissions.

2 Chirality and Gelation

2.1 Most Small Molecule Gelators are Chiral

The number of reports on low molecular mass gelators of organic liquids and water has grown steadily in the last ten years. From this large body of data has emerged an empirical view of the structural requirements for a molecule to gel a liquid. Yet, one is rarely able to predict whether a given molecule will be a gelator. Most new gelator families are still being discovered by serendipity and not by de novo design. A quick look at the literature on small molecule gelators and, to start with, at review articles on this topic makes it clear that the ability to form gels is often associated with the presence of stereogenic centers in the gelator molecular structure [6, 7]. The fact that chirality enhances gelling ability has been persistently (although hardly conclusively) reported. It has become a widely accepted fact and seems to be true for most families of gelators no matter how different they may be from each other. On the other hand, no broad explanation has yet been proposed. In this section, we have tried to give a clear perspective on the extent to which chirality matters in gelation. We have gathered and organized data from a number of publications, and derived from them hypotheses on the role of chirality in gelation phenomena. Given the large body of literature that pertains to this subject, we did not intend to make an exhaustive presentation and have selected relevant examples. It should be added that our sampling of chiral gelators is not necessarily representative of the numerous families of chiral gelators. Several of these families are derived from steroid derivatives or from complex sugars [6]. In these cases, the synthetic chiral precursors of the gelators are obtained from the chiral pool of natural products, and are often available only as a single enantiomeric form. Neither the racemate nor the other enantiomeric form is readily available. Because of this, the gelling properties of these compounds have not always been addressed from the perspective of chirality and are not accounted for here, even though these gelators are chiral as well. The chiral gelators that are mentioned here are limited to those for which the property of gelation has been addressed in one way or another from the perspective of chirality.

As stated above, the vast majority of gelators possess at least one stereogenic center in their structure. On the contrary, Scheme 1 shows 11 examples of gelators that possess none: chirality is a common but not a universal feature of organic gelators [8–18]. These nonchiral compounds were discovered by different authors in different contexts. Nevertheless, several of them are structurally related. These 11 examples may in fact be reduced to a smaller number of subclasses, such as compounds **1–4**, which all consist of two alkylamido or alkylurea units connected directly or by an aliphatic spacer, C_3 symmetrical tris-amides such as **5** and **7**, and alkoxy-aromatics such as compounds **8** and **9**. In short, the number of families of achiral gelators may be limited to just a few. It is interesting to note that compounds **1–11** gel organic liquids and that none of them gels water. A recent review on water gelation by small organic molecules [7] actually shows that almost all hydrogelators are chiral. As mentioned in the Introduction, a clear distinction should be made between genuine gels, which consist of solid-like fiber networks, and the highly viscous solutions of cylindrical micellar aggregates of amphiphilic molecules [1]. The latter may have a gel-like aspect because of their high viscosity and viscoelasticity but they differ fundamentally from a gel in that they exist at temperatures above the Krafft temperature, and in that they can flow, albeit slowly. This distinction is important because cylindrical micelles seem to form regardless of the presence of stereogenic centers in the amphiphile, and most of the studied micellar cylinders are made of achiral molecules (e.g., cetyltrimethylammonium salicylate [19], or many classes of gemini surfactants [20]).

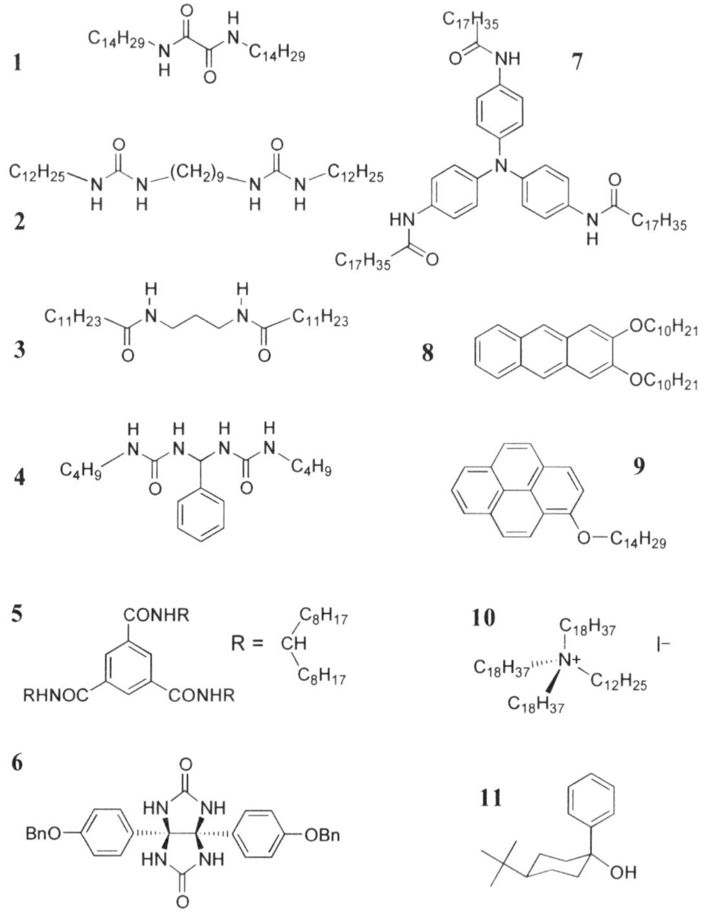

Scheme 1 Representative examples of nonchiral gelators. Compounds **1–11** are described in [8–18], respectively.

2.2 Diastereomers Show Different Gelation Behaviors

When a small molecule gelator contains more than one stereogenic center, the various diastereomers may show completely different gelation behaviors. Changing the configuration of a single stereocenter in such a molecule may have dramatic effects on its solubility properties, the range of solvents that can be gelled, the stiffness of the gels it forms, its critical gelation concentration, and the temperature at which its gels melt. The most compelling example of such effects is displayed by 11 isomers of methyl 4,6-*O*-benzylidene monosaccharides described by Shinkai et al. (Scheme 2) [21,22]. Among them, only **12a, 12b, 12f, 12i,** and **12j** are good gelators. A closer look at

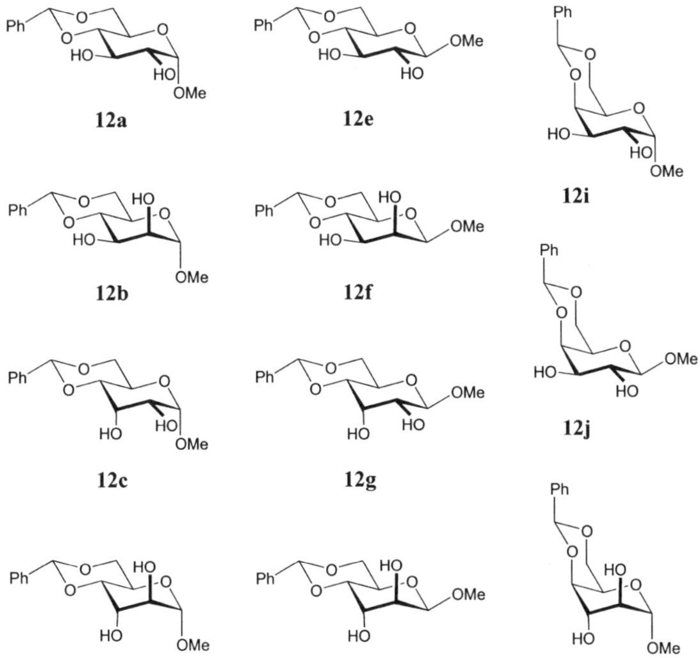

Scheme 2 Methyl 4, 6-*O*-benzylidene monosaccharides as gelators [21, 22]

these structures allows one to conclude that efficient gelation—though not necessarily in the same solvents—is possible for both *α* and *β* anomers, with equatorial benzylidene groups (*gluco*-**12a** and *manno*-**12b**) and also with axial benzylidene groups (*galacto*-**12i**). No particular configuration appears to prevent gelation by itself, except maybe the axial configuration of the 3-OH group in the *allo*- (**12c, 12f**), *altro*- (**12d, 12h**), and *ido*-pyrranosides (**12k**). The nongelators either yield poorly soluble material which precipitates instead of forming gels, or very soluble material which remains soluble at concentrations much above the critical gel concentrations of the gelators. These variations are difficult to predict but can to some extent be rationalized on the basis of the X-ray crystallographic structures of some of these compounds. In the solid state, gelators tend to show one-dimensional arrays of intermolecularly hydrogen bonded molecules. Compounds that are poorly soluble show two-dimensional arrays of intermolecularly hydrogen bonded molecules. And, in one example, exclusively intramolecular hydrogen bonds seem to correlate with high solubility.

In the specific case of *meso* compounds, the gelling properties of chiral and achiral diastereomers can be generalized. Many gelators have a C_2 symmetrical structure and contain two chiral units with the same stereochemistry (Schemes 3 and 4). It is not clear whether such C_2 symmetrical structures have higher chances to be good gelators, or whether more of them have been found because they are generally easier to prepare in a convergent fashion. Among this rather large family of chiral gelators, we could not find a single example where the achiral *meso* diastereomer is also a gelator. Whenever the *meso* compounds are mentioned in the literature, they invariably tend to form precipitates or crystallize.

2.3
Enantiomerically Pure Gelators Overrun their Racemates

The fact that the presence of a stereogenic center is essential to the gelation properties of a small organic molecule is often characterized by the observation that pure enantiomers of a given gelator have a better ability to form gels than their corresponding racemate. For example, Scheme 3 shows structures which, taken as a single enantiomer, are found to be good gelators, whereas their racemates are reported not to form any gel at all, or at best to form weak, partial, or unstable gels, or to form gels in a restricted range of solvents [23–31]. Again, it should be kept in mind that these examples are not representative of the numerous chiral gelators, but are cases that have been studied from the perspective of chirality. Because they are poor gelators, the properties of the racemates are often considered to be of lesser importance. Few reports described them in detail, and most reports do not mention anything about them. When information about the nature of the aggregates formed by the racemates is available, they are reported to crystallize, or to precipitate as flakes or platelets. Thus, when the two enantiomers of a gelator are mixed two phenomena result. First, they generally combine in a single aggregate. Second, this aggregate differs significantly in size and aspect ratio from the gel fibers of the single enantiomer, and does not lead to a gel.

The tendency of enantiomeric gelators to coaggregate is nicely illustrated by the results of van Esch et al. [29] who showed that the probe (*S*, *S*)-**19b** with its azobenzene chromophores (Scheme 3) has a marked preference to be incorporated in the gel fibers of (*R*, *R*)-**19a**, which possesses an opposite chirality, rather than in the gel fibers of (*S*, *S*)-**19a**, which has the same chirality. Other interesting examples were provided by Fuhrhop et al., who showed that the two enantiomers of gluconamide **18** [28] coaggregate in platelets even when they possess alkyl chains of different lengths [32].

Coaggregation of enantiomers thus appears to be a common behavior of racemic mixtures of chiral gelators. Yet, there are also rare examples where a racemic gelator forms a conglomerate: the two enantiomers spontaneously resolve into gel fibers consisting of a single enantiomer. The fibers obtained from the racemates and, presumably, from a mixture of the two enantiomers in any proportions are then of an identical nature to the fibers obtained from a pure enantiomer. What varies is only the proportion of fibers containing one enantiomer and of fibers containing the other enantiomer. This was demon-

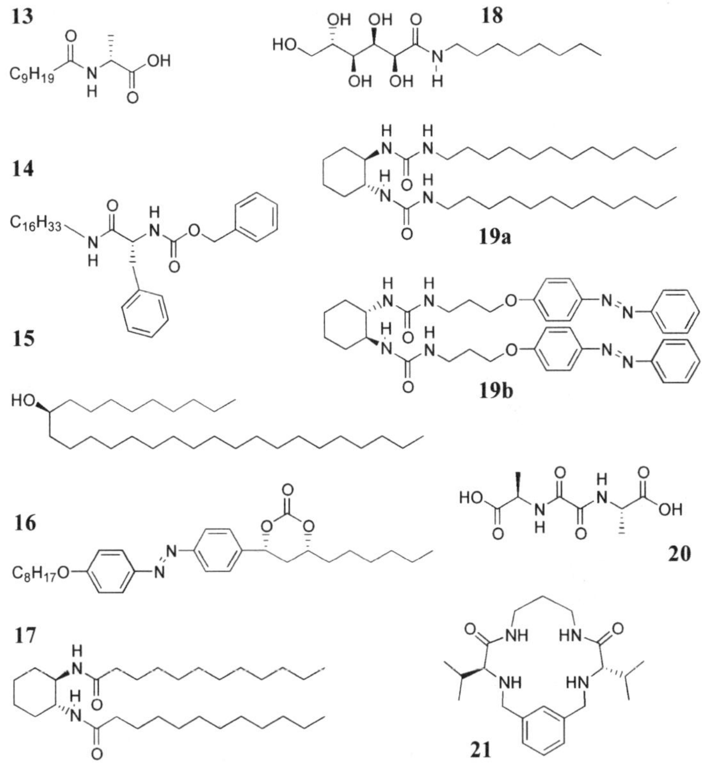

Scheme 3 Gelators that gel as a single enantiomer. Compounds **13–21** are described in references [23–31], respectively

strated unambiguously by calorimetric studies and a phase diagram of mixtures of the two enantiomers of diacetylenic phospholipid **22** (Scheme 4) [33]. It is also supported by circular dichroism studies [34]. A similar phase separation was also proposed in the case of the mixture **27a+27b** (Scheme 4) [35]. Right-handed helical fibers are observed in gels of L-**27a**+L-**27b**, left-handed helical fibers form in gels of D-**27a**+D-**27b**, and both right- and left-handed fibers are observed in *rac*-(**27a+27b**). Similar results were reported for compound **28** [36]. When such separation between the enantiomers occurs, the racemate forms gels similar to those of the individual enantiomers. But it has also been proposed that even when a conglomerate forms, the presence of the other enantiomer may slow down the formation of the fibers and possibly alter their shapes and thus the properties of the gel [31].

Compounds **23–26** (Scheme 4) exhibit an even less conventional behavior. As for compounds **13–21**, they are good gelators as single enantiomers and their racemic solutions produce racemic aggregates. However, unlike compounds **13–21**, the racemic aggregates of **23–26** have equal or better gelling properties than the single enantiomers. Because of this unusual property,

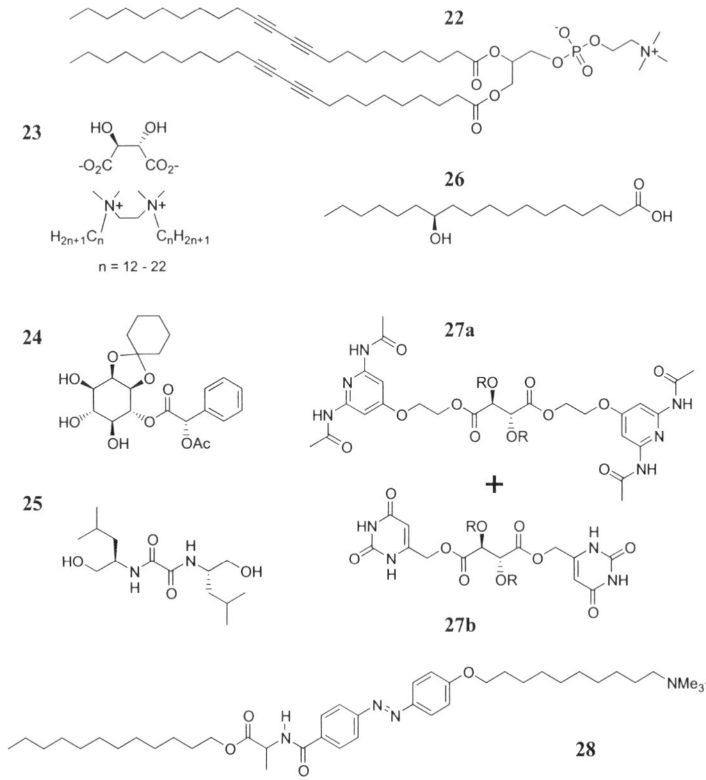

Scheme 4 Gelators that gel as racemates, conglomerates, or solid solutions

these racemates have been the object of detailed studies. For example, Terech et al. clearly established that the racemic gels of 12-hydroxyoctadecanoic acid **26** [37] are comparable in every respect to the gels of the single enantiomer: the aspect of the samples; the concentration range and the range of solvents in which the gels form; the longevity of the gels; the viscosity and viscoelasticity of the gels; the shape of the gel fibers; and the molecular packing within the fibers. In toluene, gelator **24** [38] is able to gel a volume three times as large as a racemate than as a pure enantiomer. Similarly, gelator **25** [39] gels larger volumes (up to ten times) of a variety of solvents as a racemate, and even as a mixture of diastereomers, than as a single enantiomer. For both **24** and **25**, transmission electron microscopy (TEM) shows that the fibers produced by the racemates are thinner, more numerous, and more connected than those produced by the single enantiomer. This is consistent with their ability to gel larger volumes of solvent and contrasts strongly with the large aggregates normally produced by racemates. In a recent publication, Žinić et al. systematically compared the gelation properties of the single enantiomers and of the racemates of eight gelators derived from various amino acids [40]. As expected, most of the racemates were found to be poor gelators. However, some proved able to gel significantly (up to 16 times) larger volumes of solvents, as for **24** and **25**. Thus, the fact that racemates are poorer gelators cannot be taken as a general rule.

Gelator **23** [41] shows a slightly different behavior. In chlorinated and organic solvents, its racemate forms a crystalline precipitate of large aggregates as for compounds **13–21**. In water, however, its racemate forms stiffer gels than the single enantiomer. TEM studies show that the racemate consists of large platelets, as do most racemates, but these still have a high aspect ratio, and are thicker, wider, longer, and have larger junction zones than the fibers formed by a single enantiomer. Another unusual aspect of this gelator is that the two enantiomers mix in the same gel fibers in any proportions [42], as in a solid solution (a crystal in which the two enantiomers may differently occupy the same crystallographic position [43]). The shape and size of the fibers vary with the enantiomeric excess, but a hydrogel is formed in all cases.

2.4
Comparing Racemic or Homochiral Gel Fibers to Racemic or Homochiral Crystals

The general tendency of both enantiomers of a chiral gelator to aggregate together in racemic fibers is an illustration of their crystalline nature. Indeed, about 90–95% of racemic solutions produce racemic crystals, and at most 5–10% spontaneously resolve in crystals containing exclusively one or the other enantiomer (conglomerate) [43]. In this respect at least, gel fibers seem to behave similar to crystals. The actual physical origin for the preferential formation of crystalline racemates is not yet settled and is still being debated. As early as 1885, it was observed that racemic crystals were generally denser [44, 45]. This higher density was presented as arising from lower values of crystal network free energy. Statistical arguments have also been invoked involving the larger number of nonchiral space groups and the additional symmetry elements that they offer. Racemates have more possibilities of packing than conglomerates. The chances are thus higher that the most stable arrangement lies among racemic packing. More specifically, it can be observed that crystal packing of organic molecules is often guided so as to avoid energetically unfavorable mutual orientations of juxtaposed polar groups of neighboring molecules related by symmetry elements. Thus, a glide reflection is preferable to a pure reflection, a screw rotation is preferable to a pure rotation, and the combination of glide reflection and screw rotation is preferable to either individually [45, 46]. This is reflected by the symmetry properties of the most common achiral space groups ($P2_1/c$, $C2/c$, and $P1$ account for about 85% of the space groups of crystalline racemates) [43] and also by the most common chiral space groups ($P2_1$ and $P2_12_12_1$ account for about 85% of the space groups of organic crystalline enantiomers) [43].

Achiral space groups offer the symmetry elements available in chiral space groups and also additional possibilities (glide reflections and inversion centers), which may lead to a preference for racemic crystals over conglomerates. Finally, kinetic arguments may be involved in the process of forming a viable crystal nucleus from a racemic solution. For a racemic crystal, all molecules (both enantiomers) may find a suitable binding site on the surface of the nucleus, whereas for a conglomerate half of the molecules may scroll on a wrong subcritical cluster belonging to the other enantiomer and inhibit its growth.

These arguments were developed for crystals, but their validity probably extends largely to gel fibers. Though they consist in metastable structures, gel fibers most often give powder diffraction data that reveal a long-range crystal-like order. The various factors presented above may thus shed some light on the fact that racemic gelators tend to coaggregate. However, it remains to explain why these racemates generally do not lead to a gel. Gel fibers arise upon a strong preference for anisotropic unidirectional growth. Instead, racemates aggregate as platelets or crystals, which result from a preferential bidirectional growth or from three-dimensional growth, respectively. Again, it is likely that symmetry considerations come into play in this process. As mentioned above, crystal packing of organic molecules is often guided so as to avoid energetically unfavorable mutual orientations of juxtaposed polar groups of neighboring molecules related by symmetry elements. Since racemic crystals possess additional symmetry elements to limit these repulsions (hence their stability), one can conceive that crystal growth along the various (*hkl*) crystallographic directions has more chances to be even. On the contrary, crystals of a single enantiomer lack some of these symmetry elements, and attractive or repulsive interactions have more chances to be unevenly distributed along the crystallographic directions.

The few examples of chiral gelators that have been characterized by single-crystal X-ray crystallography tend to support this view. This is the case for the example of gelator **25** [38] (Scheme 4), the crystal of which reveals strong directional interactions along the *b* plane and much weaker interactions along the *a* and *c* planes, suggesting an anisotropic growth into fibers. As pointed out by the authors, this rationale relates to the theory of crystal growth proposed by Hartman and Bennema which states that "the relative growth rate of a crystal face increases with E_{att}, which is the energy per molecule released when one slice of thickness d_{hkl} crystallizes onto a crystal face (*hkl*)." Other examples of crystal structures of chiral gelators (or closely related analogues) also show distinct unidirectional arrangements of the molecules. This includes a chiral analogue of **9** [16] (Scheme 1); 4,6-*O*-benzylidene monosaccharides [21, 22] (Scheme 2); a derivative of **16** [26] having shorter alkyl chains (Scheme 3); the bis(valine) analogue of bis(alanine) derivative **20** [30, 47] (Scheme 3); analogues of gelator **21** (Scheme 3) [31], di(*p*-toluyl)-L-cystine [48], sugar-based bolaamphiphiles [49], or a tetralin derivative [50].

It should be added that the most common chiral space groups ($P2_1$ and $P2_12_12_1$) feature twofold screw axes which, by construction, represent columnar arrangements. The crystal structures referred to above belong to these space groups. The unidirectional arrangements of the molecules do not always match a twofold-symmetry screw axis: weak intermolecular interactions may be observed along the 2_1 axis whereas much stronger interactions are observed along another direction. However, some families of molecules have been identified where the 2_1 axis does indeed represent the direction where the strongest intermolecular interactions take place in the crystal, thus defining this crystallographic direction as the direction of most favorable growth. This is the case, for example, in salts of chiral primary amines and achiral carboxylic acids [51].

When considering gel fiber growth, kinetic factors should also be invoked. Despite their crystalline nature, gel fibers are produced out of equilibrium, and are not true, thermodynamically stable crystals. The typical conditions of formation of a gel consist of dissolving the gelator at high temperature and letting it cool upon standing, or upon cooling it actively to accelerate the temperature drop. It has been noted that very slow cooling of the gelator solution may promote crystallization or precipitation instead of gel fiber growth [52]. Conversely, some racemic solutions of gelators may produce gels when cooled very rapidly, whereas simply standing at room temperature leads to weak gels or precipitates [31]. Clearly, the change in temperature has to be commensurate with the energetic factors which promote growth of a gel fiber in a particular direction.

Finally, the fact that gel fibers are not at equilibrium raises the question of what prevents their rearrangement into crystals. Again, it is possible that chirality comes into play. Fuhrhop et al. proposed a rationale to explain the stability of chiral fibers called the "chiral bilayer effect." They predict that the kinetic barrier to rearrange a chiral micellar cylinder into an enantiopolar crystal should be larger than the barrier to rearrange an achiral micellar cylinder into an achiral crystal having a bilayered structure [28]. The generality of this concept has been questioned and it was suspected that its application is limited to amphiphilic structures in aqueous medium [37]. It seems to us that the enantiopolar lamellar structure proposed for crystals of single enantiomers is far from being general. Some chiral gelators show distinct bilayer structures in the solid state [38].

Perhaps an important factor that contributes to the stability of fibers of chiral gelators is their ability to adopt morphologies that express chirality at a mesoscopic scale, giving rise to fibers possessing complex superstructures, such as helical cylinders or multiple helices. As is presented in more detail in the next section, the width of these structures cannot grow without perturbing the molecular packing, thus stabilizing the high aspect ratio of the fibers. Moreover, twisted or helical fibers are less prone than flat objects to forming thick bundles that would eventually precipitate, because the contact surface involved in a stack of helices is limited compared that of a stack of flat objects.

3 Chirality and Fiber Morphology

When chiral molecules self-assemble into fibrillar structures, chirality is sometimes expressed in the morphology of the fiber at a supramolecular scale of nanometers or micrometers. Elongated objects may be coiled or twisted and exist as a left-handed or a right-handed form. The two enantiomers of the chiral molecule individually generate fibers with opposite handedness but for rare exceptions, or unless they are mixed. This section presents in detail the diverse morphologies of chiral fibers that can be generated, as well as the various means by which these morphologies can be tuned. The last part of this section briefly discusses chiral molecular organization in self-assembled fibers and the mechanisms through which chirality may be expressed at a mesoscopic scale.

As mentioned in the Introduction, literature reports concerning chirality in self-assembled fibers often do not make mention of whether the fibers studied belong to true physical gels or to fibrous precipitates as, for example, in references [2, 53–60]. However, this distinction is not essential when discussing fiber morphology. Gel formation is often a consequence of inhibited crystallization or precipitation. It is generally a strongly history-dependent process and the same solution may produce a fibrous precipitate or a gel depending on how it is treated. However, even though the aspect of the samples may seem different, it is often assumed that the local supramolecular organizations in gel fibers and in fibrous precipitates are very similar.

3.1 Various Chiral Morphologies of Fibrillar Aggregates

Self-assembled fibrillar objects, such as rods, tapes, or tubes, may be helically twisted, coiled, or wound around one another to give multiple helices or even coiled coils. These structures are intrinsically chiral and thus possess a right- or left-handedness. The length, the diameter, and the chiral pitches of the fibers are highly variable: from ~10 nm to 1 mm. The observation that aggregates of chiral molecules may exhibit supramolecular chirality was first reported with natural lipids such as sodium deoxycholate [61], deoxycholic acid [62], and hydroxystearic acid **26** (Scheme 4) and its salts [37, 58, 59]. The D and L enantiomers of hydroxystearic acid were used to provide the first evidence that the handedness of the helical fibers is determined by the configuration of the molecular component. The L enantiomer gives right-handed twisted fibers, whilst the D enantiomer gives left-handed twisted fibers. Almost 15 years later, the first examples of synthetic chiral molecules forming chiral aggregates were reported [53, 55, 63].

Among the variety of chiral fibrillar structures encountered in the literature, probably the most commonly described are chiral ribbons of single or multiple bilayer membranes. Such ribbons can be roughly classified into two different morphologies: helical and twisted shapes. The distinction between helical and twisted ribbons is shown in Fig. 1. Helical ribbons have a cylindrical curvature and can be precursors of tubules (see below). By comparison, twisted ribbons have Gaussian or saddle-like curvature. Typical examples of helical ribbons are those formed by diacetylenic lipids such as **22** (Scheme 4) [2, 53, 54, 60], by *N*-octyl-L-galactonamide [64], by cytidine myristoylphosphatidyl conjugates [65], or by some chiral single-chain ammonium amphiphiles [66]. Typical examples of twisted morphologies are those of hydrostearic acid [58, 59], of double-chain ammonium amphiphiles derived from glutamic acid [63], or of diammonium cationic gemini surfactants **23** (Scheme 4) [41, 43].

The distinction between twisted and helical chiral morphologies is not always straightforward, and ribbons having both a twisted and a helical component can also form. Additionally, authors do not always use the same nomenclature and term as "helical" ribbons that are actually twisted and vice versa. Why some ribbons have a cylindrical curvature whereas others have a saddle-like shape is still a matter of debate. A theoretical model treating the energy differences in the two types of curvatures has been introduced and it was suggested that for fluid membranes, twisted ribbons are favored, whereas crystalline molecular organization imposes a helical curvature [43, 67]. But this may not be general and is not always verified at the experimental level [67, 68].

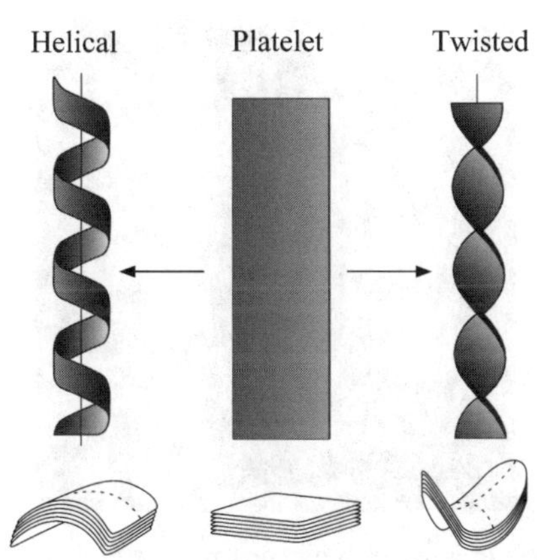

Fig. 1 Schematic representation of helical and twisted ribbons

The expression of molecular chirality at the level of the fiber and the emergence of helical or twisted shapes presumably play a role in the high aspect ratio of the fibers and in the ability of chiral compounds to form gels (see Sect. 2). For example, whilst the width of a flat ribbon may grow infinitely, the width of a helical ribbon is limited to the point when the ribbon transforms into a tubule (see below). The growth in width of a twisted ribbon is also predicted to be limited. For a given pitch of a twisted structure, the tear stress of the edges increases with the width of the ribbon. Despite its edges exposed to the solvent, a twisted ribbon is predicted to reach a finite width that is inversely proportional to the strength of the chiral effect [42].

Helical and twisted ribbons formed from self-assembled chiral amphiphilic molecules may have a single or a multiple bilayer structure. Some have been shown to consist of interdigitated bilayers [69, 70]. Interdigitation is considered to have a consolidation effect for such membrane structures as it results in highly ordered packing. In a few cases, helical ribbons have been shown to wind around one another and generate well-organized multiple helices. This is the case, for example, for gluconamide **18** [28] (Scheme 3). After numerical treatment, the microscope images of the fibers show a spectacular helix comprising four ribbons (Fig. 2) [71, 72].

The various ribbons presented above consist of amphiphilic molecules arranged in bilayers. The long axis of the molecules is perpendicular to the ribbon plane or slightly tilted from this direction. But ribbons or tapes can also be formed from the assembly of molecular rods oriented with their long axis parallel to the width of the ribbon. This is the case for some peptides that form extended β-sheet tapes which stack to form chiral superstructures (Fig. 3) [73]. It is also the case for numerous gelators consisting of a central aromatic core and chiral cholesteryl saccharidic moieties on the sides, such as the porphyrin derivative shown in Fig. 4 [74]. Chirality effects in these

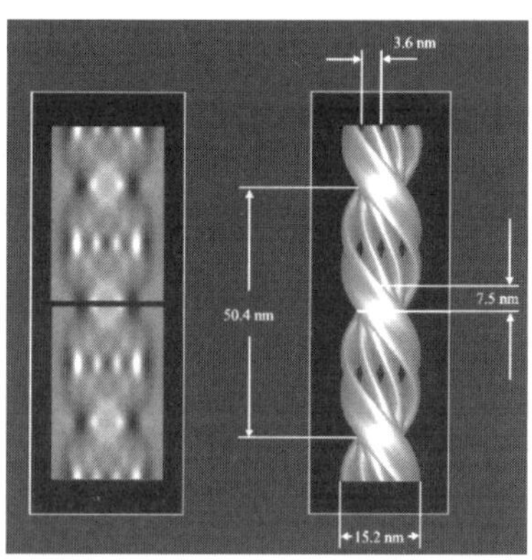

Fig. 2 Electron micrograph of *N*-octyl-D-gluconamide self-assembled fibers and 3D model of four entwined ribbons [71, 72]. Image kindly provided by Dr. Böttcher

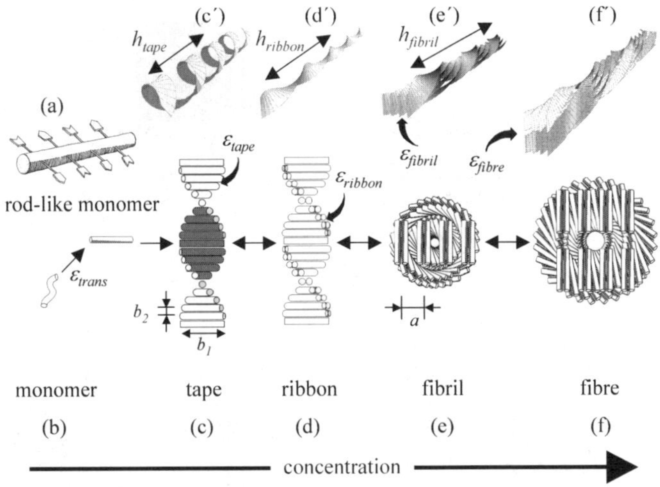

Fig. 3 Model of hierarchical self-assembly of chiral rodlike units. Local arrangements (c–f) and the corresponding global equilibrium conformations (c′–f′) for the hierarchical self-assembling structures formed in solutions of chiral molecules (a), which have complementary donor and acceptor groups, shown by arrows, through which they interact and align to form tapes (c). The black and white surfaces of the rod (a) are reflected in the sides of the helical tape (c), which is chosen to curl toward the black side (c′). The outer sides of the twisted ribbon (d), of the fibril (e), and of the fiber (f) are all white. One of the fibrils in the fiber (f′) is drawn in a darker shade for clarity. (e) and (f) show front views of the edges of fibrils and fibers, respectively. Reprinted with permission from [73]. Copyright 2001 National Academy of Sciences USA

tapes leads to a tilt between each consecutive rod which eventually produces a long-range twist.

Although a cylinder possesses no apparent chirality, chirality may play a role in the formation of cylindrical (tubular) fibers. As seen in Fig. 5, helical ribbons may spontaneously evolve into tubules upon an increase of their width. Tubules that are formed through this pathway often bear helical scars, where the edges of the ribbons have merged, as an indication of their chiral origin. Depending on the experimental conditions, this may be the case for the tubules of diacetylenic lipid **22** (Scheme 4) [2]. Some tubes, however, are produced from vesicles and show no apparent chirality even though the amphiphiles they are made of are chiral [75]. Sometimes, tubular structures form upon rolling up a sheet to form a cigar-like object [64]. As rolling does not occur parallel to the edge of the sheet, the cigar has a right- or left-handedness which may reflect the chirality of the molecular components. Rolling up a sheet is one of the very few possibilities to produce fibers—and potentially gels—from molecules that have no spontaneous tendency to assemble into elongated structures.

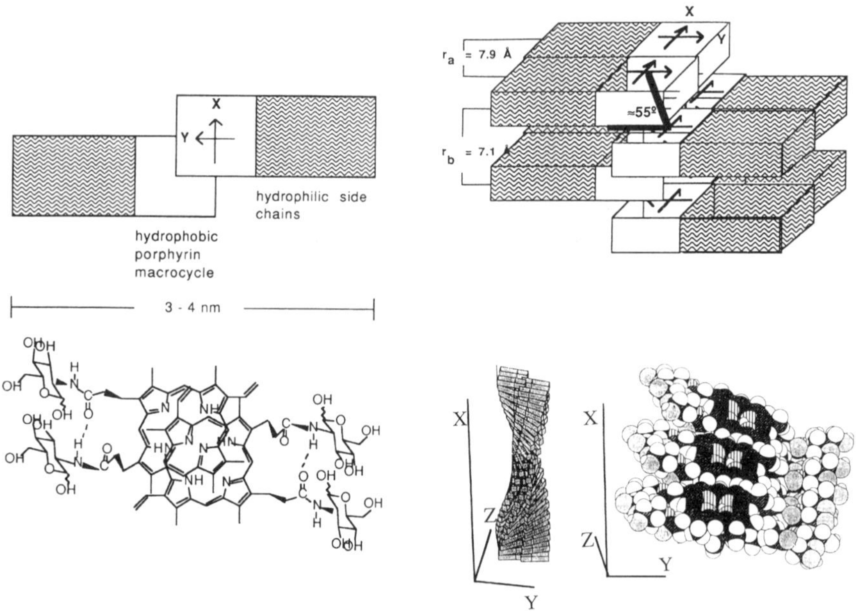

Fig. 4 Proposed twisted ribbons consisting of octametric units of porphyrins bearing aminoglycosamide functions. Reprinted with permission from [74]. Copyright 1992 American Chemical Society

Whilst tubules are hollow fibers, rodlike fibers with no hollow may also reveal the chirality of their molecular components at a mesoscopic scale. However, rodlike fibers are often very thin and structurally more difficult to

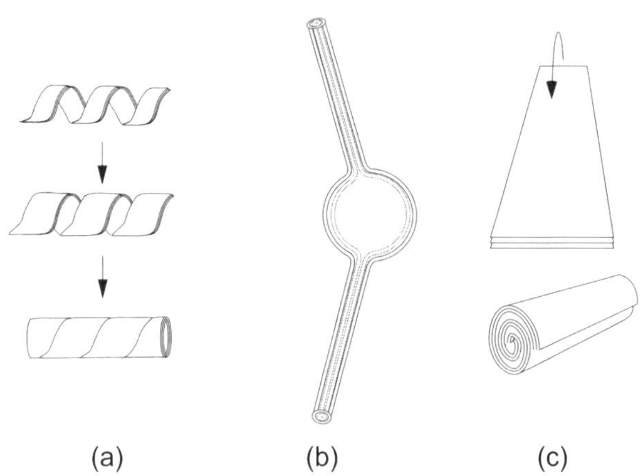

Fig. 5 **a** Helical bilayer growing to form tubules, **b** tubules growing from vesicles, and **c** a planar bilayer sheet rolling up directly to form a cigar-like structure

characterize than fibers derived from ribbons or tapes. Thin fibers tend to wind into bundles in which a global helicity is detected. Some of these fibers supposedly have a cylindrical micellar-type structure with a chiral (helical) solid-like molecular packing (Fig. 6) [64, 65, 76], in contrast to the more usual fluid nature of cylindrical micelles [19, 20]. Fibers may also be formed from stacks of disklike objects such as large aromatic groups which pile up to form a rod. A clockwise or counterclockwise staggering of the disks in the fibers may then give rise to a left-handed or right-handed twist [77]. Both chiral micellar fibers and chiral stacks of disks can wind together to form multiple helices (Fig. 6) and thus generate a second level of helicity (coiled coil). The handedness of such a multiple helix is a function of the handedness of the individual chiral fibers and of various parameters, such as the inclination and depth of the grooves of the chiral fibers. In practice, it is often observed that the multiple helix has a handedness opposite to that of the chiral fibers it is made from [77], as in the coiled coils of some proteins, e.g., keratin. Many other chiral rodlike fibers have dimensions much larger than the diameter of micelles (a few tens of nanometers), and the exact nature of molecular packing in these is far from obvious.

The various chiral fiber morphologies presented above are characterized by long-range helicities which are directly observable by microscopic techniques. These techniques have their own detection limits, and the fact that chirality is not observable in micrographs does not mean that there is no chiral twist within the molecular packing. Indeed, chiral packing can be revealed by spectroscopic methods or even by X-ray crystallography, without necessarily being visible by microscopy. Numerous fibers show intense circular dichroism bands that are not observed in the sol states and that can be interpreted as arising from chiral packing of the molecules (see Sect. 4), even

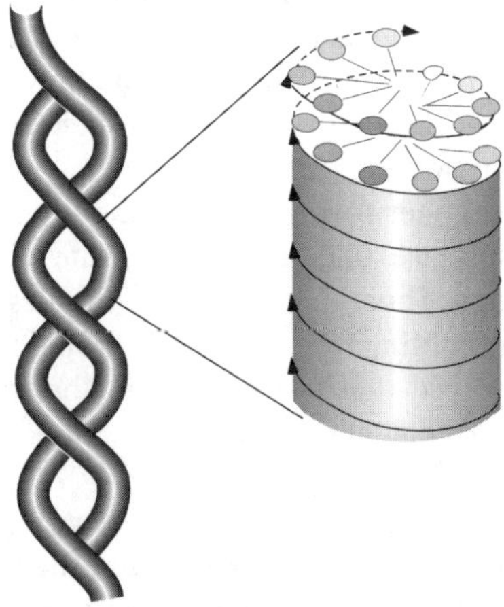

Fig. 6 Micellar fibers with chiral packing of amphiphilic molecules

though no specific chiral morphology is observed. For example, intense induced circular dichroism was observed for L-glutamic acid derivatives [78], various cholesteryl derivatives [79, 80], or oxidized glutathione [81]. Examples of helical organizations revealed by X-ray crystallography have been mentioned in Sect. 2.4 (e.g., references [48] and [50]). Note that in crystals, helical organizations are most commonly simple twofold screw axes. The scale of the helical pitch is then on the order of a few molecules, which is very small compared to the pitches that can be reached in the chiral fibers described above. It is interesting to observe that some fibers may possess both a local crystal-like order and a long-range helical order.

3.2 Morphology Control of Chiral Fibers

It is not easy to predict what parameters cause a chiral gelator to assemble into one type of chiral fiber or another. Very often, the balance between these parameters is subtle, and seemingly slight changes in the experimental conditions may lead to important changes of chiral fiber morphology. For instance, the nature of the solvent, the cooling rate of the sample, or the age of the sample may strongly affect fiber morphology. In addition to these environmental parameters, the most critical is the molecular structure of the amphiphile, within which very slight changes may lead to completely different aggregate morphologies.

The effect of the solvent on chiral fiber morphology can be very strong, but is unfortunately very difficult to rationalize. In some systems, a gelator may assemble into fibers of identical shapes in solvents as different as chloroform or toluene, and water [23, 41, 82]. In other systems, slight solvent changes may be reflected in fiber morphology. For example, diacetylenic phospholipid **22** (Scheme 4) gives single-walled tubules in methanol/water, whereas in ethanol/water or water, the same molecule forms multiple bilayers. Some gelators form twisted or helical ribbons in one solvent whereas they form achiral fibrils in other solvents; the size of the helices can also vary significantly depending on the solvent [83–85].

As mentioned above, fiber formation is generally a kinetically controlled event. It is thus no surprise that the cooling rate of the samples strongly affects fiber morphology. Thus, the size of the tubules formed by **22** was reported to vary with the cooling rate [86]. With azobenzene-linked cholesterol derivatives, it was observed that fast cooling of the solution in cyclohexane leads to a mixture of right-handed and left-handed helices, whereas slow cooling gives homochiral helices [87].

Morphology evolution with time has been reported. As explained in Fig. 5, tubular morphology may be derived from the growth of a helical ribbon, and such transitions have often been observed experimentally [65, 88]. Going from tubules to ribbons is also possible. In a mixture of diacetylenic phospholipid **22** and a short-chain lipid (1,2-bis(dinonanoyl)-*sn*-glycero-3-phosphocholine), a gradual transformation of nanotubules into a lipid gel phase consisting of interconnected "helical" ribbons (i.e., twisted ribbons as defined in Fig. 1) was observed at room temperature upon polymerization of the diacetylenic functions [89].

Small changes of the chemical structures also lead to strong changes of chiral aggregate morphology. Typical molecular parameters that induce significant morphology variation are the length [55, 87, 90, 91] and the number of unsaturations [68] of alkyl chains. In the case of bolaamphiphiles [92–94], not only the length of the spacer alkyl group matters but also whether it has an odd or an even number of carbon atoms, as revealed by IR spectroscopy (Fig. 7). With the salts of L- and D-12-hydroxystearic acid **26** (Scheme 4) mentioned above [37, 58, 59], the nature of the cation affects helix handedness: Rb^+ and Cs^+ salts give a handedness opposite to that of Li^+ salts, and Na^+ and K^+ salts give a mixture of both right- and left-handed helices [95]. As an example of the delicate balance that determines whether twisted or helical ribbons are formed, the former were obtained from a long-chain glutamate lipid with an oligosarcosine head group, whereas the lipids with an oligoproline head group gave the latter [96]. It is interesting to note that these changes in fiber morphology do not necessarily affect significantly the gelling properties (gel aspect, gel stiffness, or gelling temperature).

Diastereomers can have highly variable and unpredictable behaviors as demonstrated by the n-alkyl aldonamide-derived lipids [32, 64]. An alkyl galactonamide derivative forms helical ribbons whereas the mannonamide derivative forms a rolled-up sheet, and the gluconamide derivative forms multiple helices. Mixtures of these three aldonamides with various chain lengths showed that they may coaggregate or aggregate separately depending on their absolute and relative configurations, and on the difference in alkyl chain length. The functionalization of these alkyl aldonamides also has strong effects on the morphology of their aggregates [84, 97].

The various relations that can be established between molecular chirality and fiber handedness are worth a detailed presentation. The general rule is that the handedness of a chiral self-assembled fiber is controlled by the stereochemistry of the molecule. One enantiomer gives a right-handed fiber and the other enantiomer a left-handed fiber. However, there are some rare cases where a pure enantiomer of a chiral molecule assembles into a mixture of right- and left-handed helices. This is the case for the phosphonate analogues of diacetylenic lipid **22** (Fig. 8) [98–100], for cholesteryl anthryloxybutanoate [83], or for a mixture of a bile salt, a phosphatidylcholine, and cholesterol (Fig. 9) [101]. In the latter case, in addition to the fact that both right- and left-handed helical ribbons are observed, two or three different and well-defined helical pitches coexist (Fig. 9) [101].

Sometimes, even achiral molecules can form helical fibers: achiral isomers of diacetylenic lipid **22** (Fig. 10) [102], cyanurate derivatives [103],

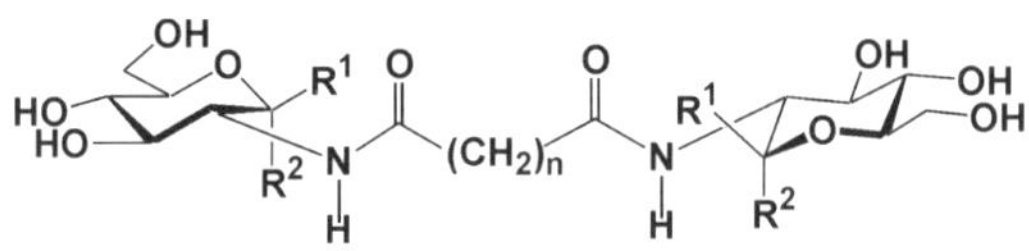

(n = 9, 10, 11, 12, 13, 14, 16, and 18)

1(n) : R^1 = H, R^2 = OH (α-anomer)
R^1 = OH, R^2 = H (β-anomer)

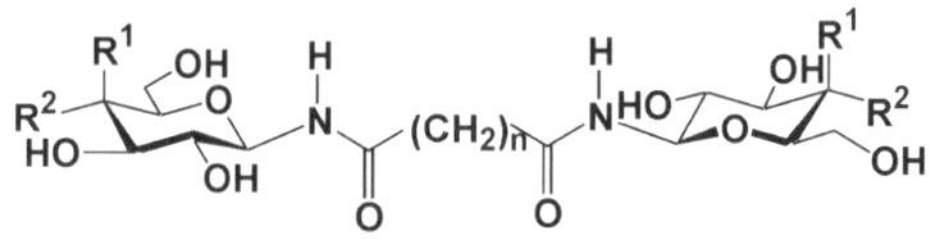

2(n) : R^1 = H, R^2 = OH; n = 6, 9, 10, 11, 12, 13, and 14
3(n) : R^1 = OH, R^2 = H; n = 10, 11, and 12

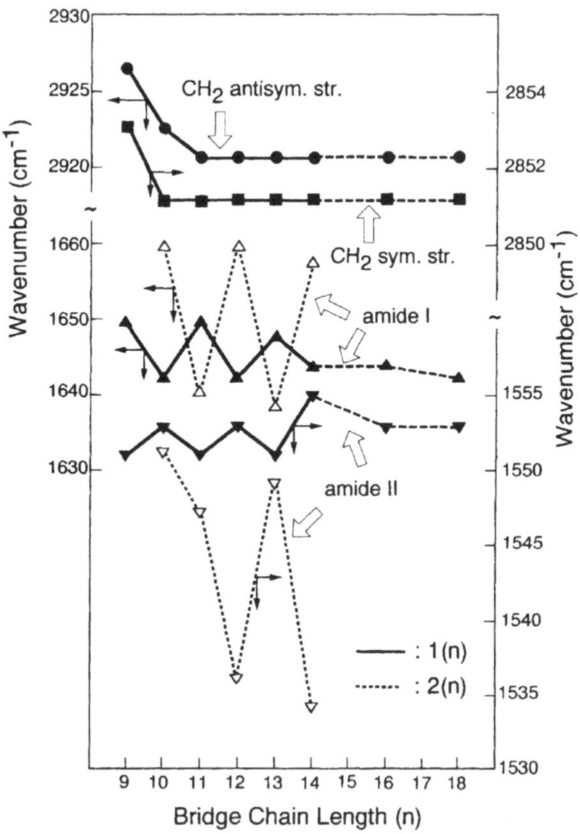

Fig. 7 Dependence of IR band frequencies on the spacer length (n) of glucosamide bolaamphiphiles. Reprinted with permission from [93]. Copyright 1999 American Chemical Society

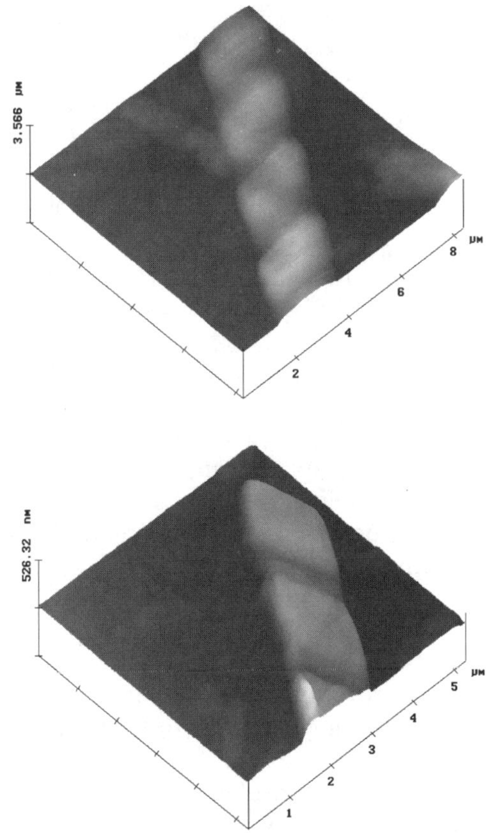

Fig. 8 Contact-mode AFM images of diacetylenic phosphonate lipids showing the presence of both right- and left-handed helices with a pure enantiomer of the lipid. Reprinted with permission from [98]. Copyright 1998 American Chemical Society

some cyanine dyes [104, 105], perfluorinated lipids [106, 107], and various bolaforms [108–110]. In the case of a family of nucleobase-derived bolaamphiphiles, nonchiral bis-thymine bolaamphiphiles form double-helical ropes (Fig. 11) whereas bis-adenine bolaamphiphiles only form microcrystalline solids [110]. A mixture of bis-thymine and bis-adenine or thymine–adenine bolaamphiphiles yield nonchiral nanofibers. It was suggested that bis-thymine bolaamphiphiles may photodimerize leading to chiral molecules which may, in turn, pack in a chiral way. Assembly of the same molecules in the dark did not yield helical structures.

To finish this section, additives or mixtures in various proportions have also been used to tune chiral aggregate morphologies. For example, a transition from twisted ribbons to helical ribbons has been observed in mixtures of cardanyl glucosides with saturated and unsaturated alkyl chains [68]. When

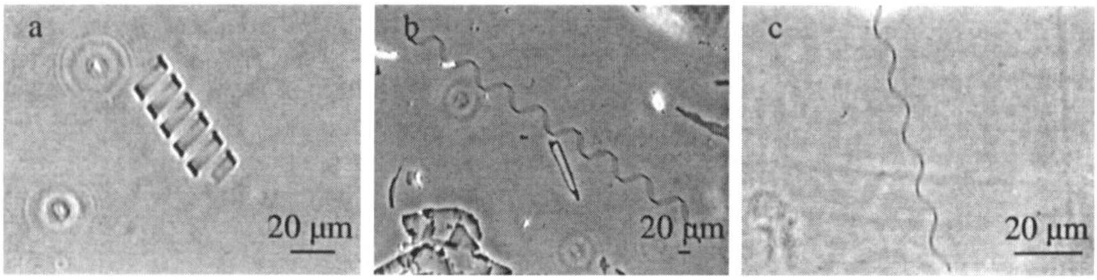

Fig. 9 Three different helical pitches observed with a mixture of a bile salt, phosphatidylcholine, and cholesterol. The pitch angle is **a** $11 \pm 2\,°$, **b** $54 \pm 2\,°$, and **c** $40.8 \pm 2\,°$. Reprinted with permission from [101]. Copyright 1999 National Academy of Sciences USA

the relative proportion of the unsaturated derivatives is increased, twisted ribbons change into helical ribbons and then tubules. The diameter of the tubules formed by diacetylenic lipid **22** can be reduced by a factor of 10 upon adding a short-chain phospholipid [89]. For the same lipid, it was observed that the presence of a few percent of negatively charged lipids yields tubules with three different pitch angles, 14, 28, and 42°, and with diameters of 2000–2500, 1000–1500, and 700–1000 nm, respectively [111]. These results can be compared to what was observed with the bile system presented above [101].

As discussed in Sects. 2.3 and 2.4, mixing the two enantiomers generally results in important changes in fiber morphology. Racemates tend to form platelets that do not express any chirality. In a couple of cases [35, 36] as, for example, for diacetylenic lipid **22** [33, 34], a mixture of right-handed and left-handed helices is observed instead of platelets. Data are rarely available concerning the behavior of mixtures of enantiomers other than 1:1 racemic mixtures. In the case of diacetylenic lipid **22**, the phase separation between the enantiomers should lead to tubules of opposite handedness, the propor-

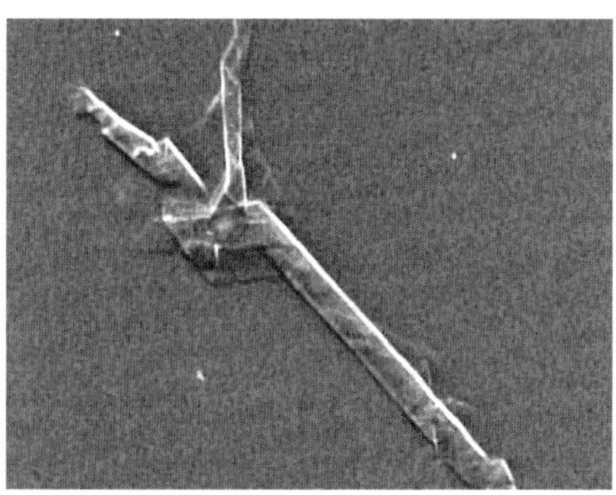

Fig. 10 Chiral tubular structure formed by an achiral β-lecithin molecule. No handedness preference is shown. The width of the image is about 8 µm. Reprinted with permission from [102]. Copyright 2003 National Academy of Sciences USA

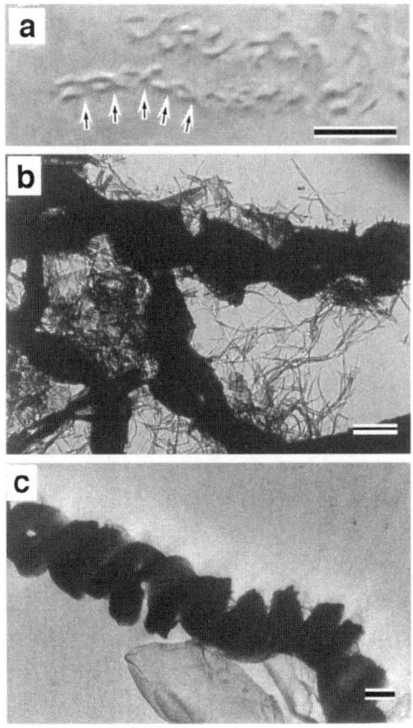

Fig. 11 Double-helical ropes of achiral bis-thymine bolaamphiphiles. Reprinted with permission from [110]. Copyright 2001 American Chemical Society

tions of which reflect the overall proportion of enantiomers (Fig. 12). In the case of cationic bis-quaternary ammonium gemini amphiphiles with tartrate counterions **23** (Scheme 4), studies of aggregate morphology as a function

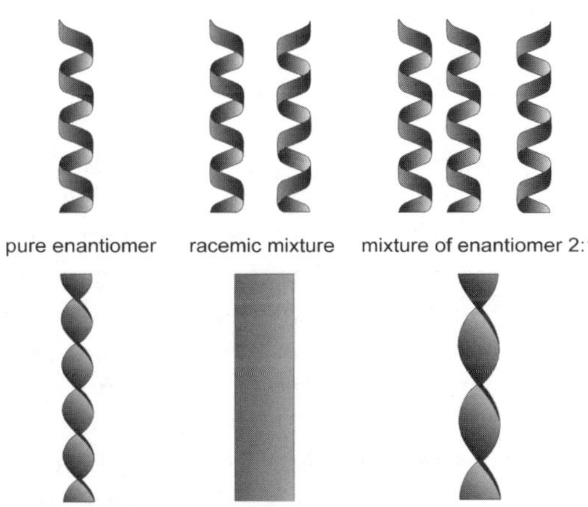

Fig. 12 Schematic representation of the effect of the proportion of enantiomers (*top*), and the case of continuous mixing of the enantiomers (*bottom*) as in a solid solution [43]

of enantiomer proportions showed that increasing the enantiomeric excess from 0 (racemic) to 1 (pure enantiomer) results in a continuous decrease of the pitch of the twisted fibers from infinite (flat ribbons) to about 200 nm (Fig. 12) [43]. This behavior may be due to the unique features of this system; the chirality belongs to the counterion and not to the amphiphile itself, but a conformationally labile chirality can be induced in the structure of the amphiphile [112]. Up to now, such continuous pitch variation has not been observed in any other system.

3.3 From Molecular Chirality to Supramolecular Chirality: Molecular Organization in Chiral Fibers

The helical pitch of twisted or helical fibers or ribbons typically ranges from ten nanometers to microns. This means that the number of molecules needed to span one turn of a helix lies between hundreds to tens of thousands. In other words, the twist between a molecule and its neighbor associated with fiber helicity will be at most one degree and, very often, much less than that. This change of scale of chirality and the cooperative effects that are involved are not easily understood. A number of theoretical models have been proposed to explain the formation of chiral fibers, among which those about cylindrical tubules and helical and twisted bilayer ribbons are probably the most elaborate. These models have been presented in detail in several reviews and we will simply refer the reader to these [4, 67, 113]. Theories are successful in describing some chiral fiber morphologies by introducing chirality as a global parameter. However, they treat the system as a continuous medium instead of an ensemble of discrete chiral molecular objects having a specific shape. Thus, they do not allow prediction of the behavior of a given molecular system under given experimental conditions.

From an empirical perspective, several factors have been pointed to as necessary for designing molecules that self-assemble into fibers with chiral morphologies, besides the presence of stereogenic centers. Such molecules are most often "amphiphilic". This term is understood here in its broadest sense. It implies that the molecules are composed of two antagonist parts. One is solvophobic and provides the driving force for aggregation; the other is solvophilic and assures the solubilization of the amphiphile and the stability of the aggregates in suspension. Additionally, the molecules possess some chemical functions capable of directed noncovalent interactions and/or rigid segments which determine the pseudocrystalline organization of the molecules in the fibrous structures. Thus, small molecules that assemble into chiral fibers often possess hydrogen-bonding functions and may combine amino acid or amide functions [23, 36, 76, 108, 114], sugar groups [21, 22, 69, 74, 115, 116], or both as in gluconamides [28, 32, 84, 97, 117, 118], as well as urea functions [29, 119–121]. Replacing a hydrogen-bonding function by a non-hydrogen-bonding one, e.g., exchanging an amide for an ester [122], results in a loss of directional interactions and no fibrous aggregate forms. Aromatic $\pi - \pi$ stacking is also commonly involved in fibrous chiral aggregates [16, 26, 36, 65, 69, 74, 123–126]. Among the rare precursors to chiral fibers that possess neither hydrogen-bonding functions nor aromatic groups are diacetylenic phospholipids such as **22**, the helices and tubules of which are among the most extensively studied aggregates [2, 32, 34, 53, 54, 60]. To explain why such molecules form helical ribbons or tubules while other phospholipids do not, it was proposed that diacetylene groups add rigidity and a kink to the alkyl chains, which might impose some steric hindrance when the molecules pack parallel to each other. A chiral twist of the packing allows alleviation of this hindrance [34].

The nature of molecular organization within the aggregates is very important to determining whether and how chirality is transferred from individual molecules to their aggregate structures. But accurate data are still very difficult to obtain. Formally, in order to correlate structural information about packing at the molecular level and information about the morphology at the fiber level, both types of data should be collected simultaneously from the solvated fiber. Correlating the orientation of the molecules with respect to the fiber coordinates in the presence of the solvent is complicated and, in practice, rarely achieved [37, 127]. To date, most studies on molecular organization within fibrous aggregates (chiral or not) have been performed on desolvated fibers, assuming that fibrous aggregates are in a pseudocrystalline state and that molecular organization is unchanged when the gels or the fibers are dried. This assumption is necessary as gels are frequently not amenable to detailed determination of molecular packing, but its validity must be questioned in each case. Thus, wide-angle X-ray powder diffraction or wide-angle X-ray scattering (WAXS) [116, 128], solid NMR [129], and infrared spectroscopy [130] were performed on the desolvated fibers and the results extrapolated to the gel structure. Such measurements can provide information about molecular organization at close to atomic resolution if (and only if) the molecules have crystalline order. Even with this, it remains difficult to propose a correct orientation of the molecules within an individual fiber. Techniques which allow the direct observation of isolated fibers (i.e., electron microscopy and AFM) do not yet have the necessary resolution to observe molecular organization. Moreover, some limitation may originate in the nature of the samples. For instance, some aggregates are labile in solution, and the molecules within the aggregate may possess significant mobility, especially when there is no other driving force for molecular assembly than hydrophobic effects which impose little directional constraint and allow a high degree of conformational freedom.

An assumption that should also be made carefully is that the packing within the gel fibers is similar to the packing of the gelators or of their analogues in single crystals (see Sect. 2.4). Single-crystal structures determined by X-ray crystallography provide very accurate information about molecular packing [16, 26, 30, 31, 38, 47–50, 116]. However, while some single-crystal X-ray structures have been validated by comparison with powder diffraction patterns of the gels, this control was not performed in many cases. Occasionally, it was even observed that a single-crystal structure and the corresponding gel structure do not match [129]. Furthermore, even with accurate information at the molecular level, it still is not easy to understand precisely how a given molecular packing gives rise to supramolecular chirality in a fiber.

Sometimes attempts can be made to observe fiber orientation by one technique, and to study molecular orientation in the same fiber by another. In most gels, fibers are at or below the limit of detection of optical microscopy, and are isotropically oriented in the sample. Studies of the properties of a single fiber have been accomplished in very few cases, and only under special circumstances where the fibrous structure was observable by optical microscopy (wavelength in the visible range) and when molecular packing within this fiber could be identified from single-crystal structures [127] or by infrared spectroscopy [130]. However, in none of these cases was chirality involved. A more attractive alternative to determine the orientation of the fibers without having to isolate one is to apply an external force or field that induces the aggregates to orient at the moment of their formation. There are many examples of such orientation induction on elongated supra/macromolecular structures in the literature using electric fields [131–136], magnetic fields [137, 138], or by applying a shear field. These have been applied extensively to flexible objects such as wormlike micelles, disklike assemblies, or polymer solutions. Because the fibers in gels are often mechanically fragile, imposing an orientation to the fibers by an external force or field is not trivial. However, rigid tubules have been aligned through such methods [139–142], which bodes well for the possibility of exploring chiral packing of molecules directly in oriented fibrillar structures.

4 Methods for Probing Chirality in Self-assembled Fibers

4.1 Methods for Probing Chirality at the Molecular Level

Various tools may be used to probe intramolecular and intermolecular interactions and to study molecular packing in self-assembled fibers: infrared or UV/Vis absorption spectroscopy; fluorescence spectroscopy; NMR spectroscopy; X-ray diffraction and X-ray scattering etc. In this section we focus on the techniques that may provide information related to chirality.

Single-crystal X-ray crystallography gives the most complete and accurate description of the conformations and packing of molecules in the solid state. As mentioned in Sects. 2.4 and 3.3, this also applies to chiral gelators whose X-ray structures often show linear arrays of tightly associated molecules suggesting a preferred direction of fiber growth, and 2_1 chiral screw rotation symmetry axes. However, the use of crystallography to study chiral fibers is limited on several grounds. First of all, crystals of gelators suitable for X-ray diffraction are very difficult to obtain, precisely because these molecules tend to aggregate into gel fibers, or at best into thin needles instead of crystals of sizable dimensions. Crystallographic studies are often performed on analogous molecules that possess poor or no gelling abilities. Second, it is very difficult to establish that the linear arrays seen in the crystals define the direction of preferred fiber growth. Moreover, crystals possess a long-range order and by definition show no long-range twist or coiling. Thus, molecular packing in a crystal does not provide much information about the origin of chiral fiber morphologies such as those described in Sect. 3. Wide-angle X-ray scattering may be measured directly on gel fibers. Although the information obtained by this technique does not generally allow an accurate description of molecular packing, it gives some distinctive periodicities that may be compared to and eventually validate an arrangement measured in the crystal.

NMR has sometimes been of some use to assess chiral interactions of small molecule gelators. For example, a selective interaction between chiral tartrate anions and achiral dicationic surfactant head groups in gelator **23** (Scheme 4) results in diastereotopic NMR patterns in the signal of the protons of the head groups [112]. Cross-polarization magic-angle spinning (CPMAS) ^{13}C NMR solid-state spectroscopy may allow characterization of molecular conformations and disorder in the molecular arrangements directly in gel fibers, provided that complementary solution NMR data and solid-state spectra of crystals of related molecules are also available. Such studies have been accomplished on *N*-octylhexonamide supramolecular assemblies [129].

The most versatile tool for studying assemblies of chiral molecules is circular dichroism (CD) spectroscopy. CD measures the differential absorption of right- and left-handed circularly polarized light. CD spectra are generally recorded in the UV/Vis range, but measurements in the short-UV and infrared (see below) are also possible. A CD spectrum amounts to the subtraction of two slightly different spectra and, consequently, CD bands are very weak compared to those of the corresponding absorption spectra. However, the intensity and sign of CD bands are extremely sensitive to changes in the conformation of the molecules in the vicinity of the chromophores to which they are allied. Thus, upon aggregation of chiral molecules into fibrous objects, large changes in CD spectra are commonly observed [143], although the changes are sometimes limited to a simple increase of band intensity [144]. As illustrated in the following, these changes are very practical to determine the onset of aggregation. However, they do not necessarily correspond to dramatic changes in the conformation of the molecules and interpretation of CD data should always be made cautiously [145].

CD signals may be induced in nonchiral chromophores (e.g., aromatic groups) provided that a chiral group lies nearby. The magnitude of such induced CD is inversely proportional to the third power of the distance between the chiral group and the chromophore [146] and should thus decline very fast as this distance increases. However, in a self-assembled fiber, a chiral environment may be provided by the chiral packing of the molecules. Induced CD may thus be observed in chromophores remote from any stereogenic centers when the overall packing has a chiral twist. For example, strong CD bands allied to the electronic transitions of diazobenzene moieties are observed in the CD spectrum of **19b** in gel fibers (Scheme 3) [29]. CD bands are also observed in chromophores of molecules that do not possess any chiral center but that coaggregate with [147] (or bind to aggregates of [5, 148]) chiral molecules. Such induced CD signals may emerge in a chiral assembly even though chirality may not be observed at the level of the fiber.

One of the main uses of CD is to monitor self-assembly. For a small molecule gelator, a typical experiment consists in following the CD spectrum from a temperature above T_{gel} to a temperature below T_{gel}. A sharp change in the CD spectrum is often observed at the onset of aggregation which may not coincide exactly with gelation. Indeed, the latter occurs only when assemblies become long enough to entrap the solvent and is associated with an increase of viscosity. Using this method, transition temperatures [68, 149, 150], critical aggregation concentrations [150], and the influence of added ions [150] or of the solvent [82] have been determined. Photochemical reactions within gel fibers and the subsequent changes in molecular organization have also been monitored by CD [151].

As mentioned above, the onset of CD signals does not necessarily imply that a structure exhibiting chirality at a mesoscopic scale has emerged. However, when such a structure exists, CD spectroscopy may be useful to monitor changes in aggregate morphology. This is the case, for example, for the transformation from nanotubes to twisted ribbons observed in mixtures of diacetylenic phospholipid **22** and of a short-chain lipid, 1,2-bis(dinonanoyl)-*sn*-glycero-3-phosphocholine (see Sect. 3.1) [89]. Slight changes in the CD spectrum of tubules of the diacetylenic phospholipid **22** have also been used to determine the number of lipid bilayers in the tubule walls [152].

CD spectroscopy may give insights into whether some chiral molecules mix in the same fibers or, on the contrary, phase separate into different aggregates. In particular, when the two enantiomers of a given gelator are mixed, various behaviors may be expected (see Sect. 2.3). If a complete phase separation of the enantiomers into aggregates of opposite handedness occurs, the CD intensity is expected to vary linearly with the enantiomeric excess (ee) of the mixture. Similarly, if the two enantiomers cocrystallize to form a racemic (nonchiral) crystal, CD will arise from the remaining single enantiomer and a linear dependence of CD intensity on ee is expected. However, if the two enantiomers mix, at least partially, regardless of their proportions, coaggregation may occur with negative or positive cooperativity and it is likely that the CD intensity would deviate from a linear dependence on the ee. Thus, a linear dependence of CD intensity on ee in tubules of chi-

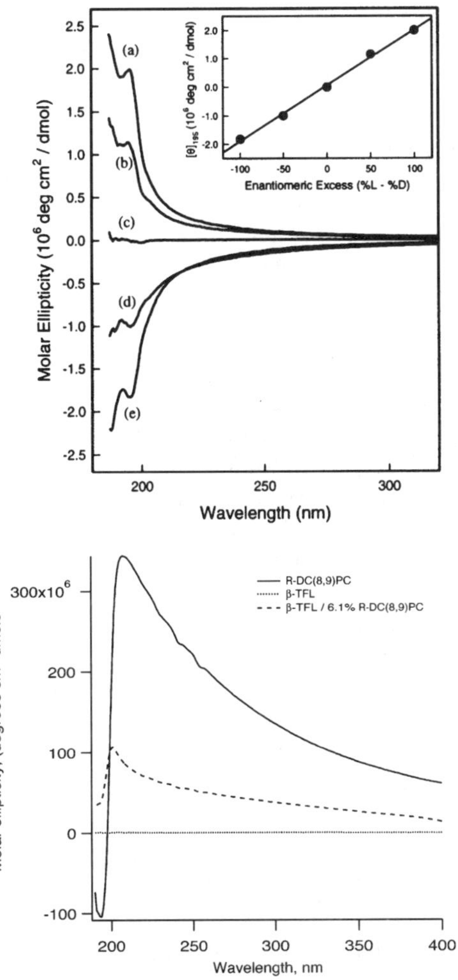

Fig. 13 *Top*: CD spectra of tubules of various mixtures of the two enantiomers of diacetylenic phospholipid **22** (inset: variation of the molar ellipticity as a function of enantiomeric excess) prepared from methanol/water. *Bottom*: CD spectra of tubules of diacetylenic phospholipid **22** (*solid line*), of an achiral analogue (*dotted line*), and of a mixture of the achiral analogue with 6.1% of **22** (*dashed line*). Reprinted with permission from [34] and [102]. Copyrights 1998 American Chemical Society and 2003 National Academy of Sciences (USA)

ral diacetylenic lipids was interpreted as a phase separation (Fig. 13) [34]. This behavior contrasts with that of an achiral diacetylenic lipid that forms equivalent amounts of right- and left-handed tubules. Upon adding only 6% of a similar chiral phospholipid, the intensity of the CD signal increases from zero to approximately 30% of the signal obtained for the enantiopure phospholipids, thus suggesting that the nonchiral lipids may express chirality induced by the added chiral lipid (Fig. 13) [102]. This amplification of chirality by achiral lipids may be compared to the sergeant-and-soldiers effects observed in helical polymers [153]. Other examples of the use of CD spectroscopy to evidence chiral interactions and recognition include the gels of compounds **19a** and **19b** (Scheme 3) [29], and the spontaneous formation of complementary hydrogen-bond pairs of mixtures of melamine and cyanuric acid derivatives and their hierarchical self-assembly in chiral supramolecular membranes [147]. Similarly, CD has been used to study a double-helical arrangement of A–T base pairs in an oligonucleotide-templated self-assembly of nucleotide bolaamphiphiles [154].

It is difficult to draw definite conclusions from a CD spectrum about the conformations of molecules and their relative positions in a gel fiber. The assignment of CD bands to the transitions of the chromophores to which they are allied is not always straightforward because significant bathochromic or hypsochromic shifts may occur upon aggregation; assignments performed from the absorption spectrum in solution may not provide reliable references. However, when this assignment can be made and when artifacts due to potential linear dichroism can be eliminated, the appearance of exciton coupling in the CD spectra may be interpreted to determine the sign (clockwise or counterclockwise) of a chiral twist of the molecular arrangement. In molecules that possess a single chromophore, exciton coupling occurs intermolecularly and arises from close chiral packing of the chromophores in the fibers. The sign of the exciton coupling indicates the relative orientation of the transition moments allied to the CD band. For example, such assignments have been performed for nitro-substituted derivatives of the 4,6-*O*-benzylidene monosaccharides shown in Scheme 2 [82], and for azobenzene-based sugar derivatives [155]. In the latter case, the sense of helicity at the molecular scale determined by CD spectroscopy corresponds to the handedness observed by SEM at the fiber level. However, to fully validate this correspondence, the orientation of the molecules with respect to the fibers should be determined.

Vibrational circular dichroism (VCD) may provide extra information compared to CD in the UV/Vis range. As this technique measures CD in the absorption region of vibrators, it may be helpful to study molecules lacking chromophores. Its use in the study of chiral aggregates is not yet widespread. But since VCD apparatus are now commercially available, there is no doubt that this method will develop progressively. Vibrational absorption and circular dichroism spectra provide richer information and can, in principle, be more easily predicted and assigned than in the electronic excitation range.

An example of the use of VCD was provided in a study of ammonium gemini surfactants bearing chiral tartrate counterions such as **23** (Scheme 4) [112]. These compounds aggregate into chirally twisted ribbons that express the chirality of the counterions despite the fact that the surfactants are not chiral. However, VCD measurements show induced CD bands in the symmetric and antisymmetric stretching modes of the CH_2 groups of the alkyl chains of the achiral cations. This unambiguously demonstrates that the cations adopt chiral conformations in the chiral ribbons, induced by tartrate anions.

4.2
Methods for Probing Chirality at the Fiber Level

Microscopy is by far the most commonly used technique for structural studies of chiral fiber morphologies. Small-angle neutron scattering and small-angle X-ray scattering have occasionally been used to determine some parameters of helical fibers such as the pitch and the width [76]. But microscopy is usually preferred as it provides direct images of the chiral structures. This preference partly contributes to the aesthetic appeal of some micrographs. However, the preparation of samples and the microscopic observations themselves can induce artifacts, leading to false interpretations such as misassignment of helix handedness. In the following, we intend to briefly compare of some of the methods often used to probe chirality at the fiber level, with their advantages and limitations. These methods are divided into three groups: optical microscopy, electron microscopy and derived techniques, and scanning probe microscopy.

Optical microscopy can provide information about the morphology of chiral fibers if their dimensions exceed a few hundred nanometers (0.2 μm is the lower theoretical limit in the wavelength range of visible light). Numerous examples of chiral fibers observed through this technique can be found, as illustrated in Figs. 9 and 14, which show optical micrographs of both helical and twisted ribbons [88, 92, 98, 99, 101, 109].

Despite the relatively limited accessible magnification, optical microscopy offers the possibility of direct observation of the native structures, since it requires *neither* drying *nor* staining, thus reducing the risk of artifacts. It has been demonstrated that sometimes the same solution displays much richer morphism of chiral structures when probed with optical microscopy than with electron microscopy [88]. However, optical microscopy and, more generally, techniques based on transmission observations of three-dimensional structures may be limited when it comes to an unambiguous determination of chiral aggregate handedness. This problem is particularly well described by Thomas et al. [98]. For instance, in the case of helical ribbons, it may be difficult to distinguish the two sides of the ribbon closer to and farther from the microscope objective, making it difficult to distinguish the apparent handedness. This problem may be partially overcome by using Nomarski differential

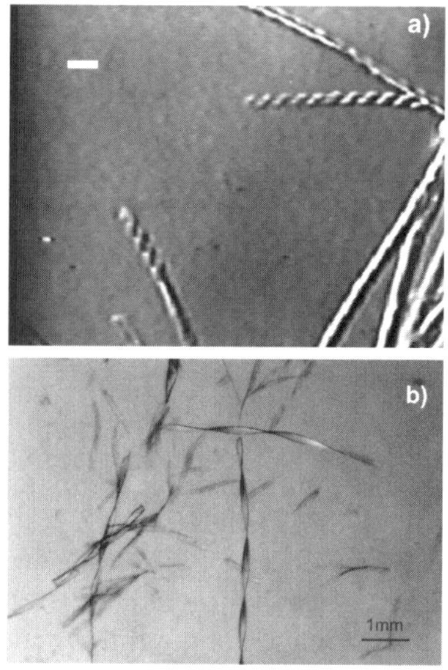

Fig. 14 Optical micrographs of left- and right-handed helical ribbons (**a**) and twisted ribbons (**b**). Reprinted with permission from [99] and [109]. Copyrights 1999 and 2000 American Physical Society and American Chemical Society

interference contrast (DIC) microscopy. This technique has a high sensitivity to the sample thickness, leading to a small focal distance that permits unambiguous placement of the microscope focal plane on either side of the chiral object, provided that the dimensions of the helices allow focalization solely to one side of the ribbons. In this case, one should be very careful about the position of the focus plane to avoid potential confusion (Fig. 15).

As can be noticed from the various figures of Sect. 3, electron microscopy is a very common tool for studying the morphology of chiral fibers. Both scanning (SEM) and transmission (TEM) electron microscopies are used to observe molecular assemblies at much higher resolutions compared to optical microscopy, and allow determination of the morphology and dimensions of chiral structures. However, one must underline the limits of these techniques. First, observations by electron microscopy operated under high vacuum require desolvation or thermal fixation (cryogenic techniques) of the samples. Desolvation of molecular assemblies in solution may cause a collapse or shrinkage of the structures. Moreover, organic molecules in the fibers and the carbon films that support them have a similar electron density. The samples often require staining or metal shadowing to improve contrast. These contrasting methods may cause artifacts for estimating the dimensions of chiral fibers as they result in larger apparent diameters [114]. Additionally, staining with heavy metal salts can modify gelation behaviors [71].

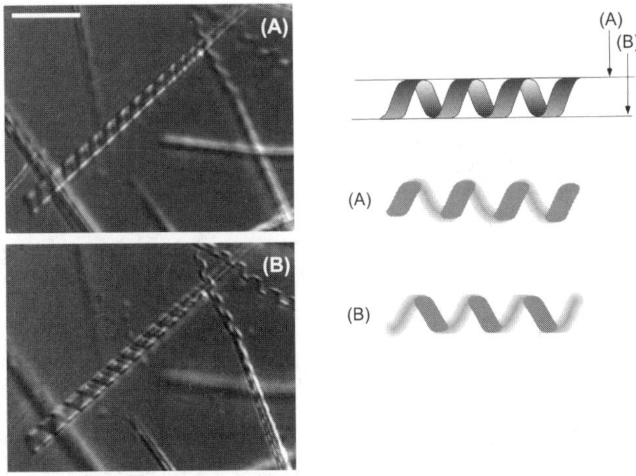

Fig. 15 Optical micrographs through a focus sequence of a large hydrated phosphonate tubule. Illustration of apparent shift in handedness upon changing the height of the focal plane. The scale bar is 5 μm. Photographs reprinted with permission from [98]. Copyright 1998 American Physical Society

Provided these potential artifacts do not affect the chiral morphologies, unambiguous assignment of their right- or left-handedness may still not be straightforward [33, 34, 155] especially when observing negatively stained images by TEM. In the case of negative staining, the total projections do not provide clear information about the orientation of fiber curvature in the plane perpendicular to the sample (Fig. 16). Using appropriate staining methods may help overcome this problem. When carefully contrasting chiral objects with a correct angle, for example using platinum, preferential shadowing can be obtained, thus helping the correct determination of handedness [28].

Finally, the preparation of grids for TEM requires successive steps which should all be performed very carefully to avoid confusion in handedness assignment: (1) deposition of the sample; (2) staining or shadowing; (3) introduction of the grid into the microscope; (4) production of images or of negatives; and (5) scanning or development on paper. At many of these steps, a lack of care in the manipulation will produce the mirror image of the desired micrograph and thus a wrong assignment of handedness (Fig. 17).

Cryogenic techniques have also been successfully used for the study of chiral effects in fibrillar networks. The great difference when using these techniques is that samples undergo an ultrafast freezing to avoid crystallization of the solvent before observation, therefore upholding the native morphologies. This is hardly practical for gels produced in solvents which cannot be frozen in an amorphous state and is most commonly performed

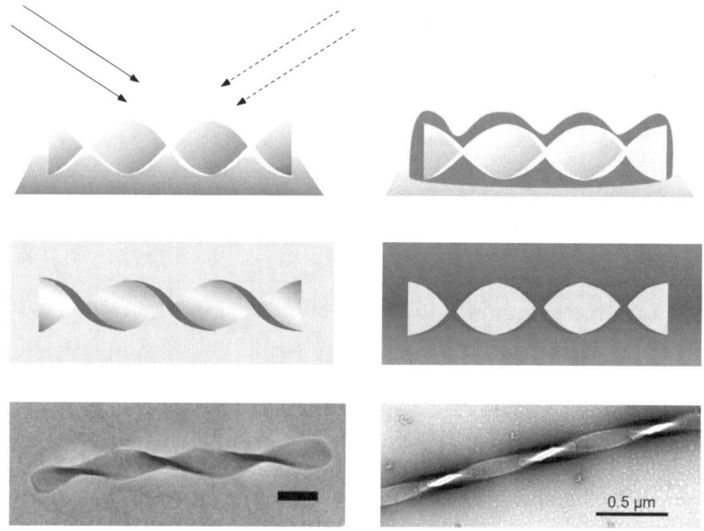

Fig. 16 Metal shadowing (*left*) versus negative staining (*right*). Metal shadowing is deposited to one side of the object along a preferential incidence (*arrow*), allowing easier handedness evaluation. The *bottom left* picture is a Pt shadowed TEM image of a twisted ribbon formed by a mixture of **23** (Scheme 4, $n = 16$) and its analogue bearing bromide counterions. The *bottom right* picture is a TEM image of a twisted ribbon of **23** stained with uranyl acetate with no metal shadowing

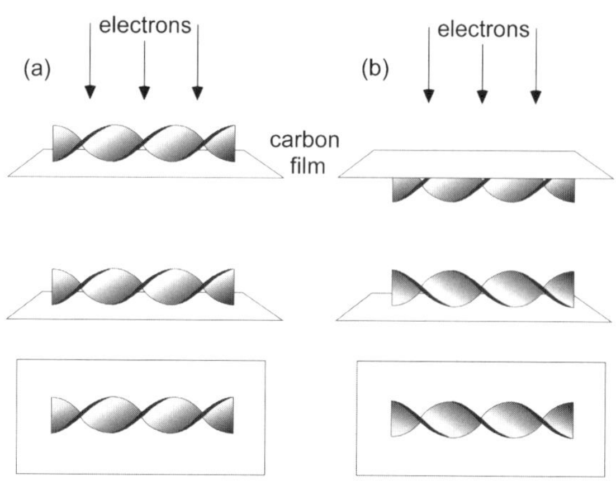

Fig. 17 Effect of sample position on handedness assignment: **a** when electrons hit the helix before the support film (*top*), the original handedness is observed on the micrograph (*bottom*), whereas **b** if the grid is put upside down, the images show an inverted handedness

with aqueous samples. Thus, cryo-TEM has been used to observe twisted fibers [105], as well as twisted [128] or helical [23] ribbons. This technique involves the formation of a thin vitrified film of the solution and the direct observation of this film, i.e., of the solvated objects, with TEM. An advantage of cryo-TEM is that it does not require staining because of the intrinsic contrast between vitrified ice and organic materials. But this particular feature also constitutes a drawback when studying chiral objects, since the absence of shadowing makes the determination of handedness uncertain (Fig. 18). It should also be mentioned that cryo-TEM is not always well adapted to the study of highly viscous systems such as gels, because of the difficulty of forming a sufficiently thin film with well-defined structures. These are probably the reasons why cryo-TEM observations are often complemented by other measurements, such as TEM on carbon–platinum shadowed samples.

Freeze fracture is another technique associated with TEM. Here, frozen samples are fractured and replicated by carbon–platinum spread at a specific angle under vacuum. Replicas are then observed under standard TEM operating conditions. Due to the shadowing procedure, the images obtained through this technique can provide more accurate information about chiral fiber handedness [56, 118, 149, 156] (Fig. 19). Upon replication, the "topography" of the native sample is conserved leading to 3D images. The handedness of a chiral fiber can then be assigned unambiguously, provided that it can be determined whether a concave or a convex replica is observed. If this latter aspect is not obvious at first sight, the assignment may be facilitated upon slight tilting of the sample grid. Freeze fracture also allows the analysis of viscous samples.

Analogous to freeze fracture, cryo-HRSEM (high-resolution scanning electron microscopy) also permits observation of the sample without prior removal of the liquid phase. Although this technique has been used for the observation of fibrillar structures in gels [157, 158], to the best of our knowledge it has not yet been applied to the study of chiral fibers. However, given the quality of the images that may be obtained, it represents a very promising method for probing the chirality of fibers in solution. Finally, scanning probe microscopies such as AFM or STM also have the potential to image hydrated samples in situ, without desolvation of the fibers, under highly humid conditions (Fig. 20). Yet, examples in the literature generally show dried samples

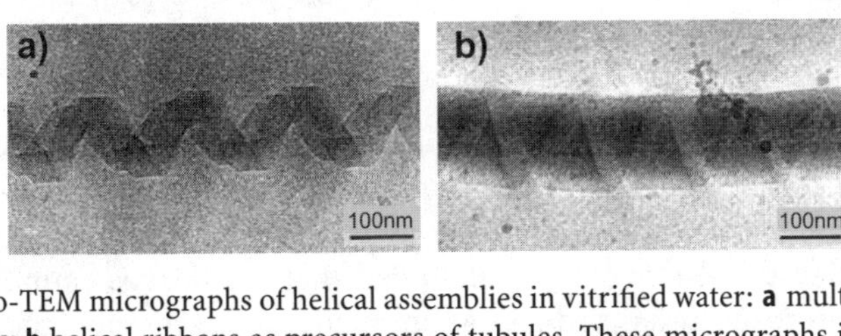

Fig. 18 Cryo-TEM micrographs of helical assemblies in vitrified water: **a** multilamellar helical ribbons; **b** helical ribbons as precursors of tubules. These micrographs illustrate how difficult handedness assignment can be when using cryo-TEM. Reprinted with permission from [23]. Copyright 2001 American Chemical Society

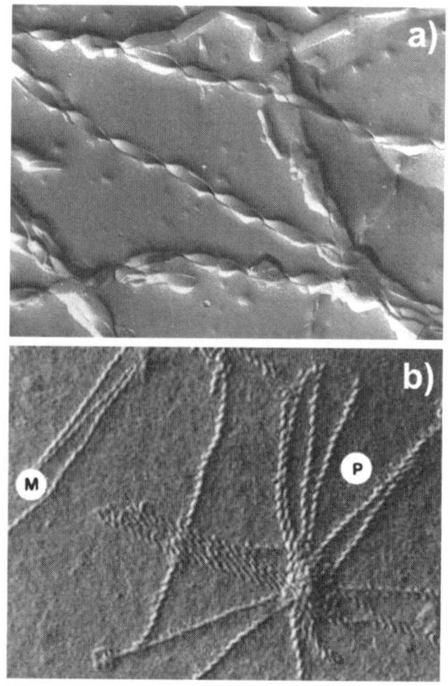

Fig. 19 Freeze-fracture images of **a** twisted ribbons and **b** right- and left-handed helical ribbons. Reprinted with permission from [149] **a** and [156] **b**. Copyrights 2001 and 1990 American Chemical Society

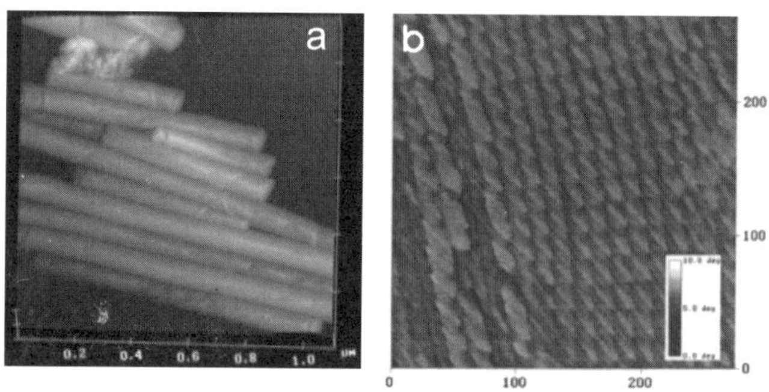

Fig. 20 a AFM image of tubular J-aggregates characterized by a "cigar-like" morphology kindly provided by Prof. H. Von Berlepsch [104]. b High-resolution AFM phase image of quadruple helices. Reprinted with permission from [160]. Copyright 2000 American Chemical Society

that are easier to observe and less mobile. Using these techniques, 3D maps of the samples can be generated that permit unambiguous handedness assignment [99, 159, 160]. However, artifacts may be introduced upon drying. For instance, dried tubules may be so flattened that the images cannot be interpreted with respect to handedness assignment, as for the phosphonate lipid tubules described by Thomas et al. [98] (Fig. 8).

5 Applications and Perspectives

The numerous examples cited in the previous sections illustrate the considerable efforts that have been devoted to the design and characterization of chiral self-assembled fibrillar networks over the last two decades. Supramolecular chemistry has been successful at creating a great diversity of chiral structures with twisted, helical, or cylindrical tubular morphologies that often express the chirality of their molecular constituents at a mesoscopic scale. These chiral structures represent excellent models for studying the emergence of specific shapes at a macroscopic scale through cooperative interactions between a large number of very small building blocks. In addition to this fundamental aspect, they possess a great potential for applications in the development of new functional supramolecular devices, taking advantage of the chirality of the molecular constituents organized in a hierarchical manner or (and) of the supramolecular chirality of the fibers that can be generated.

5.1 Applications Based on Chiral Recognition

Applications of fibers of low molecular weight gelators may emerge directly from the chirality of their molecular constituents. For instance, chiral recognition may occur at the molecular level between various guest molecules and chiral gelators, but only when the latter are engaged in a fiber-like organization. A first example has already been given in Sect. 2 of this chapter: the enantioselective incorporation of (*S*, *S*)-**19b** in the gels of (*R*, *R*)-**19a** (Scheme 3) [29]. Another example is the aqueous gelation of tripodal cholic acid derivative **29** (Scheme 5) reported by Maitra et al. [148]. The gelation of **29** creates highly hydrophobic chiral pockets that recognize the sodium salt of achiral bromophenol blue (BPB; **30**). When the latter is entrapped in the cavities of the gel fibers, circular dichroism bands are observed in the absorption region of BPB, suggesting that chiral conformations are induced in BPB upon recognition. When associating chiral gelators with liquid crystals (LC), original behaviors may also be observed [161]. For instance, cholesteric LC can be induced in the presence of a chiral gel, leading to unique liquid-crystalline physical gels. This topic is presented in detail in the following chapter of this book, and will therefore not be examined further here.

Selective chiral recognition by self-assembled fibrillar networks paves the way to important applications of gels. A first type of application is chiral discrimination/separation: enantioselective elution phenomena using a chiral organic gel as the stationary phase [162]. Chiral stationary phases are normally produced upon grafting chiral groups onto polymer beads. Alternatively, the solid-like fibers of a chiral gelator may also be used in chiral separation. This is the case for the benzene gel of a mix of chiral amino acid derived lipids **31a** and **31b** (Scheme 5). Thus, when eluting through this gel an aqueous solution of L- or D-*N*-dansyl-phenylalanine (**32**), a selectivity E_L/*E*_D (*E* = elution rate) of 1.5 was observed. Selective interaction of L-**32** and D-**32** with the gel is supported by the appearance of induced circular dichroism signals of opposite sign belonging to the dansyl chromophore. The intensity of the induced signal is twice as large for the D than for the L enantiomer. Moreover, when a gel of **31a** alone is used as the stationary phase, no enantioselectivity is observed, suggesting that enantioselectivity arises from specific interactions between the positively charged head groups of **31b** with the carboxylate of **32**.

A second potential application of chiral recognition in gel fibers is asymmetric synthesis. Shinkai et al. recently published pioneering work along this line using a derivative of compound **17** [27] as a chiral gelator (Scheme 3) [163]. In fact, they do not use the organic fibers directly. Instead, they use silica inorganic helices produced by replication of the organic fibers (see Sect. 5.3) [164]. Specifically, when applying either left- or right-handed helical silica to the addition of diisopropylzinc to aldehyde **33** (Scheme 6), product **34** is obtained mainly as a single enantiomer (96–98% optical purity). This result is remarkable in that the enantioselectivity of the reaction

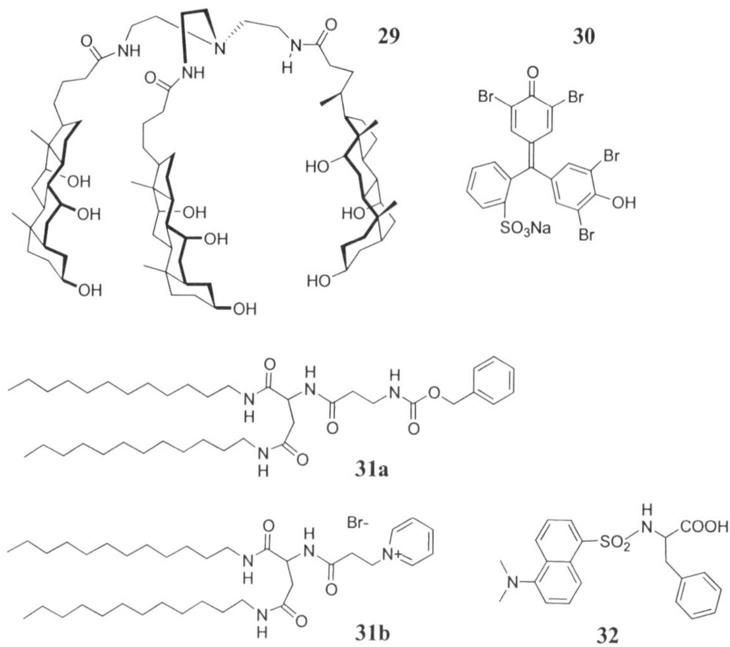

Scheme 5 Molecular components of gels involved in chiral recognition

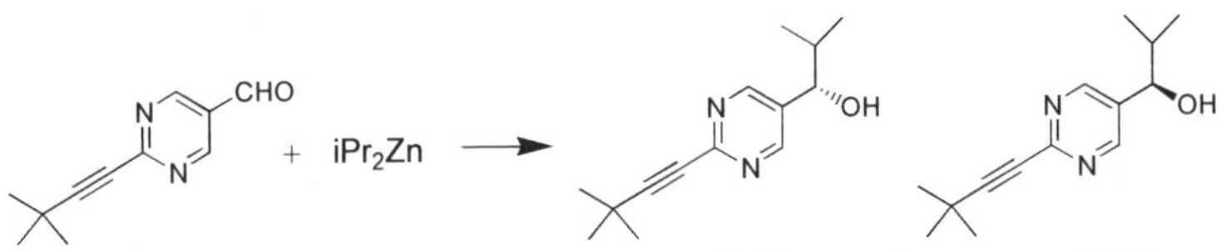

33 S-34 R-34

Scheme 6 Enantioselective reaction induced by chiral fibers

is not guided by any chiral organic substances, which have all been removed from the silica.

5.2 Chiral Fibers as Templates for Helical Crystallization of Proteins

Chiral fibers can prove potentially useful in biology as templates for the helical crystallization of biological macromolecules. This application is based on the idea that helical crystallization, e.g., of a protein, may be promoted at the surface of tubules formed by self-assembled chiral lipids bearing specific ligands. The helical morphology of the lipid bilayer induces the formation of the tubules and the ligands provide specific binding sites for the protein of interest (Fig. 21). A proof of concept was first proposed by Brisson et al. [165] who showed that biotinylated lipids such as **35** can form unilamellar tubular structures under specific conditions. The biotin residues of the lipids confer the tubules with the ability to recognize the protein streptavidin, which contains four biotin binding sites. Transmission electron micrographs revealed that streptavidin spontaneously assembles into ordered helical arrays on the tubule surface. Importantly, this helical array of streptavidin is induced by the preexisting chiral morphology of the biotinylated lipid tubules. For instance, no organized array of streptavidin was obtained using biotinylated lipids assembled as spheroidal liposomes. Another interesting feature of this system is that these helical crystals of streptavidin now possess biotin binding sites at their periphery. Thus, they may in turn serve as templates to bind a variety of biotinylated molecules. For example, secondary binding of biotinylated ferritin on the streptavidin tubules has been achieved by the authors [165].

Wilson-Kubalek et al. also produced specifically and nonspecifically functionalized unilamellar lipid tubules by using mixtures of a tubule-forming galactosylceramide and various charged or derivatized lipids [166]. Thus, nickel-doped lipids allowed the helical crystallization of histidine-tagged proteins. The authors also reproduced the helical crystallization of streptavidin. They even obtained helical arrays of relatively small proteins, such as actin and annexin, as well as large macromolecules, such as RNA polymerase (Fig. 21).

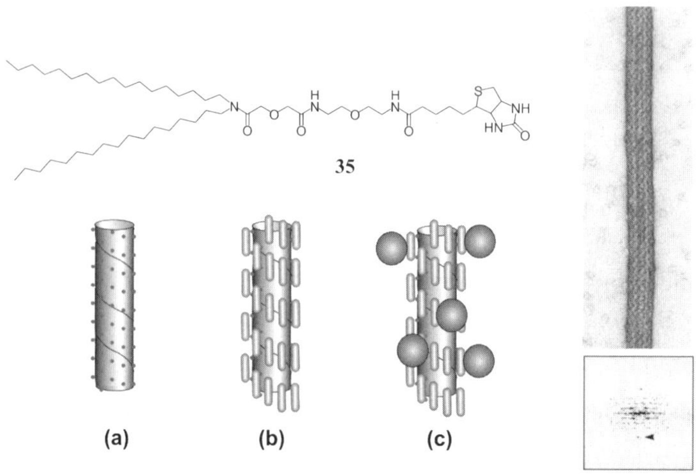

Fig. 21 Structure of biotinylated lipid **35** [165]; schematic representation of the helical crystallization of streptavidin on a chiral tubular structure. **a** Formation of a chiral tubule functionalized with biotin; **b** helical crystallization of streptavidin on the preexisting chiral tubes; **c** secondary binding of biotinylated macromolecules on the remaining binding sites of streptavidin. *Right*: TEM image of a helical array of RNA polymerase on a functionalized ceramide tubule (positive surface charge). The diffraction pattern below, with visible peaks to $1/38\,\text{Å}^{-1}$, illustrates the crystalline nature of the helices. Photographs reprinted with permission from [166]. Copyright 1998 National Academy of Sciences USA

The main advantage of these crystalline helical arrays of proteins is that they represent ideal specimens for 3D electron microscopy. The two-dimensional crystals of proteins at air/water interfaces usually prepared for electron microscopy studies provide a single view of the protein. To allow a 3D structure to be calculated from the images by reconstruction methods, the plane of the crystal can be slightly tilted so as to offer views at different angles. However, this operation requires the recording and combination of many images of tilted specimens and has intrinsic limitations, resulting in the loss of resolution along the direction perpendicular to the crystal plane. On the other hand, the cylindrical symmetry of the helical crystals readily offers all orientations of the repeat motifs. The analysis of helical specimens can be performed using computational tools that have been developed for proteins which spontaneously form helical assemblies, i.e., in the absence of any template. In addition, helical crystals form in solution and can be easily transferred onto electron microscopy grids. In contrast, the transfer of two-dimensional crystals from air/water interfaces is extremely tricky, inefficient, and prone to result in structural damage. Helical crystals may thus be very useful to structure determination.

5.3 Chiral Fibers as Templates for the Growth of Inorganic Replicas

Chiral fibers are representative examples of the variety of shapes, sizes, chemical compositions, and functions that can be reached through the self-assembly of small organic molecules. In contrast, inorganic objects with the same type of morphologies are not readily prepared. One approach to develop inorganic materials closely corresponding to organic assemblies is to transcribe them to produce inorganic replicas by sol–gel methods. Transcription occurs upon adding inorganic derivatives, often metal oxide precursors, to the organic assemblies which may or may not be preformed. Polymerization of the inorganic precursors at the surface of the organic template gives rise to a new robust inorganic material. If the template expresses chirality at a mesoscopic scale, the inorganic replica often does as well. The final product of the transcription process can be an organic–inorganic hybrid or purely inorganic depending on whether or not the organic template has been removed. This approach has developed very rapidly in the last few years. It constitutes a major advance in the design of functionalized materials with applications in the fields of electronics, optics, chromatography, or asymmetric synthesis with, for example, the tailoring of advanced chiral catalysts [163]. Self-assembled organic fibers are very attractive objects for transcription into inorganic materials because their high aspect ratio is expected to produce much sought for rodlike inorganic structures [167]. In the following, we focus mostly on examples where chiral fibers were used as templates. Even so, the literature is already far too rich to make an exhaustive presentation in the context of this chapter, and we have limited the scope of this section to a few examples. For more details, we refer the reader to several review articles [168–170].

Whilst the largest number of transcription protocols of organic templates into inorganic materials have been developed for metal oxides, and in particular silica, some reports deal with metals or semiconductors. Matsui et al. described the electroless metallization of bolaamphiphile nanotubes with nickel and copper baths containing reducing agents, leading to Ni and Cu nanowires that might find applications in nanoelectric circuits (Fig. 22a) [108]. In this case, the bolaamphiphile does not possess any stereogenic center and chirality is not apparent in the inorganic replica. On the other hand, when an achiral precursor gives rise to (both right- and left-handed) twisted ribbons, the twist of the fibers is transcribed into the inorganic replica. For example, a coiled fiber of CdS was obtained in this way (Fig. 22b) [171]. The absence of chirality in the system results in a mixture of right- and left-handed CdS. A remarkable aspect of this process is that CdS grows along only one face of the twisted-ribbon template. The unique coiled morphology of the semiconductor may be suitable for photovoltaic applications [171]. An example of true chirality transfer from organic to in-

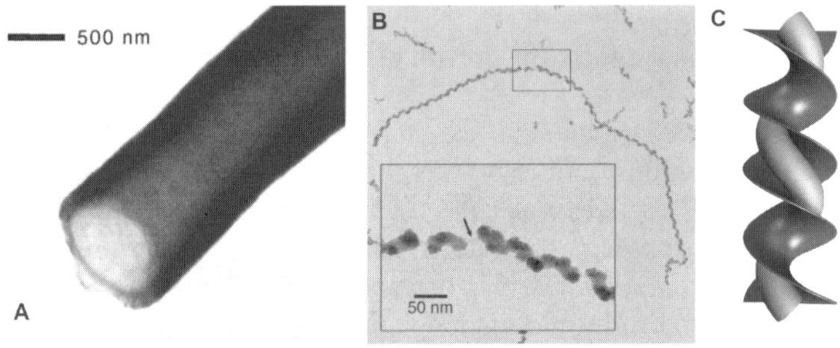

Fig. 22 **a** TEM micrograph of a Ni-coated bolaamphiphile nanotube; **b** TEM micrograph of a helical fiber of CdS with a pitch of 40–50 nm, produced on a twisted-ribbon template as shown in **c**. Reprinted with permission from [108] and [171]. Copyrights 2000 American Chemical Society and 2002 Wiley-VCH

organic structures that is also potentially applicable in the field of electronics and optics is the preparation of helical transition metal oxide tubes, achieved by Hanabusa et al. [172]. Gelator **36** (Fig. 23) is immediately derived from compound **17** [27] (Scheme 3), and possesses positively charged groups that promote the aggregation of the metal oxide precursors at the surface of the gel. Using the two *R*, *R* and *S*, *S* enantiomers of **36**, left-handed and right-handed inorganic helices are produced, respectively.

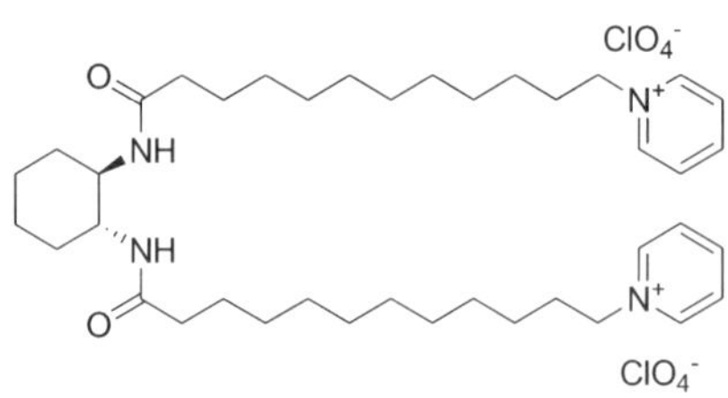

36

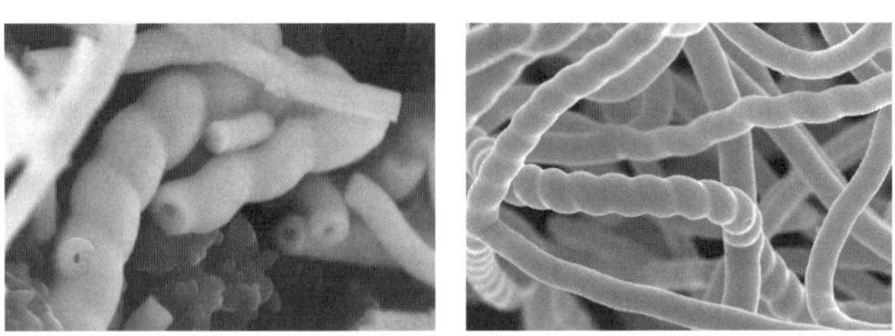

Fig. 23 Structure of **36** and SEM images of tantalum oxide fibers obtained from (*R*, *R*)-**36** (*right*) and (*S*, *S*)-**36** (*left*). Reprinted with permission from [172]. Copyright 2002 American Chemical Society

Tubules and helical ribbons formed by diacetylenic lipids such as **22** (Scheme 4) have been extensively used as templates for metallic coatings [173], and the deposition of gold nanoparticles [174] or polypyrrole threads [175]. Their coassembly with silica precursors gives rise to organic/inorganic hybrids having mesoscopic helical and tubular shapes and multilamellar walls [176].

The largest body of work on transcription of chiral fibers has been devoted to silica because of the ease with which it can be handled. The strong activity the Shinkai research group should be underlined in this respect. Gel fibers can be coated with silica, and yield hollow inorganic tubules after removal of the organic template [177]. If the organic template is hollow in the first place, as in the organic tubules of a crown-appended cholesterol gelator, both the interior and the exterior of the organic tube can be accurately coated with silica, which finally results in double-walled silica tubes (Fig. 24) [178].

Another example of very accurate transcription is the generation of both right- and left-handed silica fibers by sol–gel transcription of helical organogel fibers of a derivative of compound **17** [27] as a chiral gelator (Scheme 3) [164]. The authors clearly established that the final helicity of the inorganic fiber reflects the original chirality of the organic template (Fig. 25). A similar result was obtained with silyl derivatives of gelator **19b** (Scheme 3) [179]. A right-handed helix is produced by the R,R enantiomer, and a left-handed helix is produced by the S,S enantiomer. In most examples of helical fiber transcription into silica, the silica precursors are driven to the organic fiber surface by noncovalent interactions. In this case, however, the silica precursors are covalently attached to the gelator.

To illustrate the diversity of chiral inorganic objects that can be obtained by transcription of chiral self-assembled fibers as organic templates, even double-helical silica has been produced. Gels of a mix of sugar-based gelators produce double-helical silica nanotubes by transcription (Fig. 26) [180]. In addition, gels of gemini surfactant **23** (Scheme 4) produce double-helical fibrils of silica [181]. In the latter case, the continuous variation of the pitch

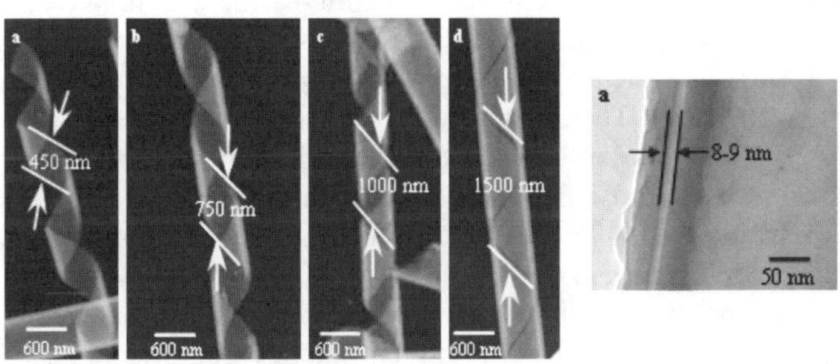

Fig. 24 TEM images of double-walled helical silica obtained after calcination of tubular ribbons in growth. Reprinted with permission from [178]. Copyright 2003 Wiley-VCH

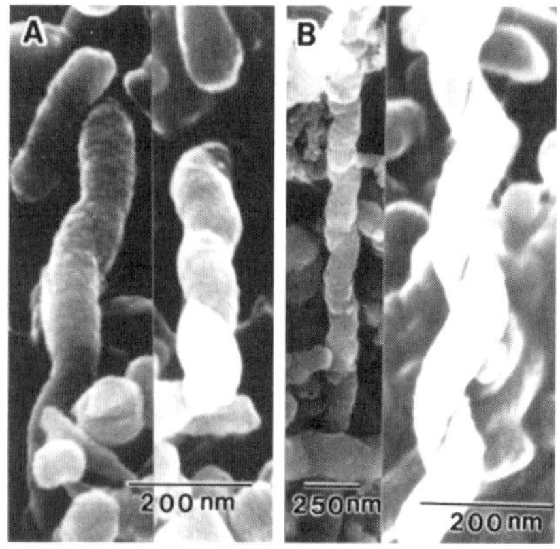

Fig. 25 SEM images of right-handed and left-handed helical silica fibers obtained after transcription of organic chiral templates. Reprinted with permission from [164]. Copyright 2000 American Chemical Society

of the organic template as function of the enantiomeric enrichment of the gelator [43] allows tuning of the pitch of the final double-helical silica.

5.4 Perspectives

The first four sections of this chapter illustrate not only the considerable efforts that have been devoted to the design and characterization of chiral self-assembled fibrillar networks, but also the difficulty in understanding at the molecular scale the mechanisms through which chirality is expressed in self-assembled fibers and how it affects the structure and properties of the gels. To progress along these lines, it will be necessary to gather accurate information about the molecular structure of gel fibers. However, this remains a very challenging task. As is often the case in the chemical sciences, progress may be expected from the improvement of instrumental techniques. The resolution of microscopic tools is still increasing and may reach a level of practical use to distinguish molecules within a fiber. The use of Rietvel methods for X-ray or neutron powder diffraction may lead to the resolution of some gel structures. The improvement of theoretical tools may allow more accurate assignment and interpretation of circular dichroism spectra, both in the electronic absorption range and the vibrational range. Molecular modeling tools have not yet been applied to the study of gel fibers and may also provide useful information.

Future developments will thus aim at understanding how specific shapes assembled from large numbers of much smaller building blocks emerge at a macroscopic scale through cooperative interactions. This fundamental question not only pertains to the science of fibrillar networks, but also to all

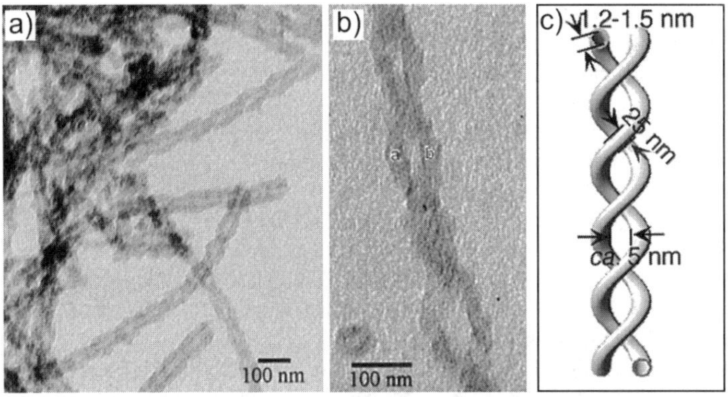

Fig. 26 TEM images **a, b** and schematic representation **c** of double-helical silica nanotubes. Reprinted with permission from [180]. Copyright 2002 American Chemical Society

hierarchically organized structures in biology and physics. Answers are not easily found. Yet, applications of these objects have already started to be developed. As seen in this last section, chiral fibrillar networks have proven to be useful in areas as diverse as the helical crystallization of biological macromolecules, chiral separation, and the production of chiral inorganic objects.

References

1. Fuhrhop J-H, Helfrich W (1993) Chem Rev 93:1565
2. Schnur JM (1993) Science 262:1669
3. Kunitake K (1992) Angew Chem Int Ed 31:709
4. Spector MS, Selinger JV, Schnur JM (2003) Chiral molecular self-assembly. In: Green MM, Nolte RJM, Meijer EW, Denmark SE, Siegel J (eds) Materials chirality—Topics in stereochemistry, vol 24. Wiley, Hoboken, NJ, chap 5
5. Trager O, Sowade S, Bottcher C, Fuhrhop J-H (1997) J Am Chem Soc 119:9120
6. Terech P, Weiss RG (1997) Chem Rev 97:3133
7. Estroff LA, Hamilton AD (2004) Chem Rev 104:1201
8. Luo X, Li C, Liang Y (2000) Chem Commun 2091
9. van Esch J, De Feyter S, Kellog RM, De Shryver F, Feringa BL (1997) Chem Eur J 3:1238
10. Tomioka K, Sumiyoshi T, Narui S, Nagaoka Y, Iida A, Miwa Y, Taga T, Nakano M, Handa T (2001) J Am Chem Soc 123:11817
11. Schoobeek FS, van Esch J, Hulst R, Kellog RM, Feringa BL (2000) Chem Eur J 6:2633
12. Hanabusa K, Koto C, Kimura M, Shirai H, Kakehi A (1997) Chem Lett 429
13. Kölbel M, Menger FM (2001) Chem Commun 275
14. Yasuda Y, Takebe Y, Fukumoto M, Inada H, Shirota Y (1996) Adv Mater 8:740
15. Brotin T, Utermöhlen R, Fages F, Bouas-Laurent H, Desvergnes J-P (1991) Chem Commun 416
16. Babu P, Sangeetha NM, Vijaykumar P, Maitra U, Rissanen K, Raju AR (2003) Chem Eur J 9:1922

17. Lu L, Weiss RG (1996) Chem Commun 2029
18. Terech P, Allegraud JJ, Garner CM (1998) Langmuir 14:3991
19. Saikaigudin Y, Shikata T, Urakami H, Tamura A, Hirata HJ (1987) Electron Microsc 36:168
20. Zana R, Xia J (eds) (2004) Gemini surfactants. Marcel Dekker, New York
21. Yoza K, Amanokura N, Ono Y, Akao T, Shinmori H, Takeuchi M, Shinkai S, Reinhoudt DN (1999) Chem Eur J 5:2722
22. Gronwald O, Shinkai S (2001) Chem Eur J 7:4329
23. Boettcher C, Schade B, Fuhrhop J-H (2001) Langmuir 17:873
24. Bhattacharya S, Ghanashyam Acharya SN, Raju AR (1996) Chem Commun 2101
25. Fuhrhop J-H, Bedurke T, Hahn A, Grund S, Gatzmann J, Riederer M (1994) Angew Chem Int Ed Engl 33:350
26. Mamiya JI, Kanie K, Hiyama T, Ikeda T, Kato T (2002) Chem Commun 1870
27. Hanabusa K, Yamada M, Kimura M, Shirai H (1996) Angew Chem Int Ed Engl 35:1949
28. Fuhrhop J-H, Schnieder P, Rosenberg J, Boekema E (1987) J Am Chem Soc 109:3387
29. de Loos M, van Esch J, Kellog RM, Feringa BL (2001) Angew Chem Int Ed 40:613
30. Jokić M, Makarević J, Žinić M (1995) J Chem Soc Chem Commun 1723
31. Becceril J, Escuder B, Miravet JF, Gavara R, Luis SV (2005) Eur J Org Chem 481
32. Fuhrhop J-H, Boettcher C (1990) J Am Chem Soc 112:1768
33. Singh A, Burke TG, Calvert JM, Georger JH, Herendeen B, Price RR, Schoen PE, Yager P (1988) Chem Phys Lipids 47:135
34. Spector MS, Selinger JV, Singh A, Rodriguez JM, Price RR, Schnur JM (1998) Langmuir 14:3493
35. Gulik-Krzywicki T, Fouquey C, Lehn J-M (1993) Proc Natl Acad Sci USA 90:163
36. Yamada N, Sasaki T, Murata H, Kunitake T (1989) Chem Lett 205
37. Terech P, Rodriguez V, Barnes JD, McKenna GB (1994) Langmuir 10:3406
38. Makarević J, Jokić M, Raza Z, Štefanić Z, Kojić-Prodić B, Žinić M (2003) Chem Eur J 9:5567
39. Watanabe Y, Miyasou T, Hayashi M (2004) Org Lett 6:1547
40. Čaplar V, Žinić M, Pozzo J-L, Fages F, Mieden-Gundert G, Vögtle F (2004) Eur J Org Chem 4048
41. Oda R, Huc I, Candau SJ (1998) Angew Chem Int Ed 37:2689
42. Oda R, Huc I, Candau SJ, MacKintosh FC (1999) Nature 399:566
43. Jacques J, Collet A, Wilen SH (1994) Enantiomers, racemates and resolutions, 3rd edn. Krieger, Malabar, FL
44. Wallach O (1885) Liebigs Ann Chem 286:90
45. Pratt Brock C, Schweizer WB, Dunitz JD (1991) J Am Chem Soc 113:9811
46. Kitaigorodskii AI (1961) Organic chemical crystallography. Consultants Bureau, New York, pp 65–112
47. Makarević J, Jokić M, Perić B, Tomišić V, Kojić-Prodić B, Žinić M (2001) Chem Eur J 7:3328
48. Menger FM, Yamasaki Y, Catlin KK, Nishimi T (1995) Angew Chem Int Ed Engl 34:585
49. Masuda M, Shimizu T (1996) Chem Commun 1057
50. Snijder CS, de Jong JC, Meetsma A, van Bolhuis F, Feringa BL (1995) Chem Eur J 1:549
51. Kinbara K, Hashimoto Y, Sukegawa M, Nohira H, Saigo K (1996) J Am Chem Soc 118:3441

52. Becceril J, Burguete MI, Escuder B, Galindo F, Gavara R, Miravet JF, Luis SV, Peris G (2004) Chem Eur J 10:3879
53. Yager P, Schoen PE (1984) Mol Cryst Liq Cryst 106:371
54. Georger JH, Singh A, Price RR, Schnur JM, Yager P, Schoen PE (1987) J Am Chem Soc 109:6175
55. Yamada K, Ihara H, Ide T, Fukumoto T, Hirayama C (1984) Chem Lett 1713
56. Kulkarni VS, Boggs JM, Brown RE (1999) Biophys J 77:319
57. Sommerdijk NAJM, Lambermon MHL, Feiters MC, Nolte RJM, Zwanenburg B (1997) Chem Commun 1423
58. Tachibana T, Kambara H (1965) J Am Chem Soc 87:3015
59. Tachibana T, Kambara H (1969) Bull Chem Soc Jpn 42:3422
60. Schnur JM, Ratna BR, Selinger JV, Singh A, Jyothi G, Easwaran KRK (1994) Science 264:945
61. McCrea JF, Angerer S (1960) Biochim Biophys Acta 42:357
62. Ramanathan N, Currie AL, Colvin JR (1960) Nature 190:779
63. Nakashima N, Asakuma S, Kim J-M, Kunitake T (1984) Chem Lett 1709
64. Fuhrhop J-H, Schnieder P, Boekema E, Helfrich W (1988) J Am Chem Soc 110:2861
65. Yanagawa H, Ogawa Y, Furuta H, Tsuno K (1989) J Am Chem Soc 111:4567
66. Kunitake T, Yamada N (1986) Chem Commun 655
67. Selinger JV, Spector MS, Schnur JM (2001) J Phys Chem B 105:7158
68. John G, Jung JH, Minamikawa H, Yoshida K, Shimizu T (2002) Chem Eur J 8:5494
69. Jung JH, John G, Masuda M, Yoshida K, Shinkai S, Shimizu T (2001) Langmuir 17:7229
70. Sommerdijk NAJM, Buynsters PJJA, Akdemir H, Geurts DG, Pistorius AMA, Feiters MC, Nolte RJM, Zwanenburg B (1998) Chem Eur J 4:127
71. Fuhrhop J-H, Köning J (1994) Membranes and molecular assemblies: the synkinetic approach. Royal Society of Chemistry, Cambridge
72. Boettcher C, Stark H, van Heel M (1996) Ultramicroscopy 62:133
73. Aggeli A, Nyrkova IA, Bell M, Harding R, Carrick L, Mcleish TCB, Semenov AN, Boden N (2001) Proc Natl Acad Sci USA 98:11857
74. Fuhrhop J-H, Demoulin C, Böttcher C, Köning J, Siggel U (1992) J Am Chem Soc 114:4159
75. Chiruvolu S, Warriner HE, Haranjo E, Idziak SHJ, Rädler JO, Plano RJ, Zasadzinski JA, Safinya CR (1994) Science 266:1222
76. Imae T, Hayashi N, Matsumoto T, Tada T, Furusaka M (2002) J Colloid Interface Sci 225:285
77. Engelkamp H, Middelbeek S, Nolte RJM (1999) Science 284:785
78. Takafuji M, Ihara H, Hirayama C, Hachisako H, Yamada K (1995) Liq Cryst 18:97
79. Lu L, Cocker TM, Bachman TE, Weiss RG (2000) Langmuir 16:20
80. Wang R, Geiger C, Chen L, Swanson B, Whitten DG (2002) J Am Chem Soc 122:2399
81. Lyon RP, Atkins WM (2001) J Am Chem Soc 123:4408
82. Gronwald O, Shinkai S (2001) J Chem Soc Perkin Trans 2 1933
83. Lin Y-C, Kachar B, Weiss RG (1989) J Am Chem Soc 111:5542
84. Hafkamp RJ, Feiters M, Nolte RJM (1999) J Org Chem 64:412
85. Koumoto K, Yamashita T, Kimura T, Luboradzki T, Shinkai S (2001) Nanotechnology 12:25
86. Thomas B, Safinya CR, Plano RJ, Clark NA (1995) Science 267:1635
87. Murata K, Aoki M, Suzuki T, Harada T, Kawabata H, Komori T, Ohseto F, Ueda K, Shinkai S (1994) J Am Chem Soc 116:6664
88. Nakashima N, Asakuma S, Kunitake T (1985) J Am Chem Soc 107:509

89. Svenson S, Messersmith PB (1999) Langmuir 15:4464
90. Hanabusa K, Maesaka Y, Kimura M, Shirai H (1999) Tetrahedron Lett 40:2385
91. Becerril J, Burguete MI, Escuder B, Luis SV, Miravet JF, Querl M (2002) Chem Commun 738
92. Shimizu T, Masuda M (1997) J Am Chem Soc 119:2812
93. Nakazawa I, Masuda M, Okada Y, Hanada T, Yase K, Asai M, Shimizu T (1999) Langmuir 15:4757
94. Schnieder J, Messerschmidt C, Schulz A, Gnade M, Schade B, Luger P, Bombicz P, Hubert V, Fuhrhop J-H (2002) Langmuir 16:8575
95. Tachibana T, Kayama S, Takeno H (1972) Bull Chem Soc Jpn 45:415
96. Shimizu T, Hato M (1993) Biochim Biophys Acta 1147:50
97. Hafkamp RJ, Kokke BPA, Danke IM, Geurts HPM, Rowan AE, Feiters M, Nolte RJM (1997) Chem Commun 545
98. Thomas BN, Corcoran RC, Cotant CL, Lindemann CM, Kirsch JE, Persichini PJ (1998) J Am Chem Soc 120:12178
99. Thomas BN, Lindemann CM, Clark NA (1999) Phys Rev E 59:3040
100. Thomas BN, Lindemann CM, Corcoran RC, Cotant CL, Kirsch JE, Persichini PJ (2002) J Am Chem Soc 124:1227
101. Zastavker YV, Asherie N, Lomakin A, Pande J, Donovan JM, Schnur JM, Benedek GB (1999) Proc Natl Acad Sci USA 96:7883
102. Pakhomov S, Hammer RP, Mishra BK, Thomas BN (2003) Proc Natl Acad Sci USA 100:3040
103. Yang W, Chai X, Chi L, Liu X, Cao Y, Lu R, Jiang Y, Tang X, Fuchs H, Li T (1999) Chem Eur J 5:1144
104. von Berlepsch H, Kirstein S, Böttcher C (2003) J Phys Chem B 107:9646
105. von Berlepsch H, Böttcher C, Ouart A, Regenbrecht M, Akari S, Keiderling U, Schnablegger H, Dähne S, Kirstein S (2000) Langmuir 16:5908
106. Giuleri F, Krafft M-P, Riess JG (1994) Angew Chem Int Ed Engl 33:1514
107. Giuleri F, Guillod F, Greiner J, Krafft M-P, Riess JG (1995) Chem Eur J 2:1335
108. Matsui H, Pan S, Gologan B, Jonas SH (2000) J Phys Chem B 104:9576
109. Lindsell WE, Preston PN, Seddon JM, Rosair GM, Woodman TA (2000) Chem Mater 12:1572
110. Shimizu T, Iwaura R, Masuda M, Hanada T, Yase K (2001) J Am Chem Soc 123:5947
111. Lvov YM, Price RR, Selinger JV, Singh A, Spector MS, Schnur JM (2000) Langmuir 16:5932
112. Berthier D, Buffeteau T, Léger J-M, Oda R, Huc I (2002) J Am Chem Soc 124:13486
113. Selinger JV, MacKintosh FC, Schnur JM (1996) Phys Rev E 53:3804
114. Imae T, Takahashi Y, Muramatsu H (1992) J Am Chem Soc 114:3414
115. Tachibana T, Mori T, Hori K (1980) Bull Chem Soc Jpn 53:1714
116. Shirakawa M, Kawano S, Fujita N, Sada K, Shinkai S (2003) J Org Chem 68:5037
117. Svenson S, Köning J, Fuhrhop J-H (1994) J Phys Chem 98:1022
118. König J, Boettcher C, Winkler H, Zeitler E, Talmon Y, Fuhrhop J-H (1993) J Am Chem Soc 115:693
119. van Esch J, Schoonbeek F, de Loos M, Kooijman HL, Spek A, Kellog RM, Feringa BL (1999) Chem Eur J 5:937
120. Estroff LA, Hamilton AD (2000) Angew Chem Int Ed 39:3447
121. Wang G, Hamilton AD (2003) Chem Commun 310
122. Ihara H, Yamagishi M, Takafuji M, Hachisato H, Hirayama C, Yamada K (1990) Nippon Kagaku Kaishi 1047
123. Nakashima T, Kimizuka N (2002) Adv Mater 14:1113

124. Sagawa T, Fukugawa S, Yamada T, Ihara H (2002) Langmuir 18:2002
125. Yun YJ, Park SM, Kim BH (2003) Chem Commun 254
126. Itojima Y, Ogawa Y, Tsuno K, Handa N, Yanagawa H (1992) Biochemistry 31:4757
127. Abdallah DJ, Sirchio SA, Weiss RG (2000) Langmuir 16:7558
128. Estroff LA, Leiserowits L, Addadi L, Weiner S, Hamilton AD (2003) Adv Mater 15:39
129. Svenson S, Kirste B, Fuhrhop J-H (1994) J Am Chem Soc 116:11969
130. Placin F, Desvergne J-P, Belin C, Buffeteau T, Desbat B, Ducasse L, Lassègues J-C (2003) Langmuir 19:4563
131. Hoffmann H, Krämer U, Thurn H (1990) J Phys Chem 92:2027
132. van der Schoot P, Cates ME (1994) J Chem Phys 101:5040
133. Lachenmayer K, Oppermann W (2002) J Chem Phys 116:392
134. Hong MK, Narayan O, Goldstein R, Shyamsunder E, Austin RH, Fisher DS, Hogan M (1992) Phys Rev Lett 68:1430
135. Oizumi J, Kimura Y, Ito K, Hayakawa R (1996) J Chem Phys 104:9137
136. Oda R, Lequeux F, Mendes E (1996) J Phys II (France) 6:1429
137. Prosser SR, Hunt SA, DiNatale JA, Vold RR (1996) J Am Chem Soc 118:269
138. Katsaras J, Donaberger RL, Swainson IP, Tennant DC, Tun Z, Vold RR, Prosser RS (1997) Phys Rev Lett 78:899
139. Rosenblatt C, Yager P, Schoen P (1987) Biophys J 52:295
140. Terech P, de Geyer A, Struth B, Talmon Y (2002) Adv Mater 14:495
141. Shklyarevskiy IO, Jonkheijm P, Christianen PCM, Schenning APHJ, Del Guerzo A, Desvergne J-P, Meijer EW, Maan JC (2005) Langmuir (in press)
142. Lescanne M, Colin A, Mondain-Monval O, Heuzé K, Fages F, Pozzo J-L (2002) Langmuir 18:7151
143. Xing B, Yu C-W, Chow K-H, Ho P-L, Fu D, Xu B (2002) J Am Chem Soc 124:14846
144. Mukhopadhyay S, Maitra U, Ira Krishnamoorthy G, Schmidt J, Talmon Y (2004) J Am Chem Soc 126:15905
145. Glattli A, Daura X, Seebach D, van Gunsteren WF (2002) J Am Chem Soc 124:12972
146. Craig DP, Power EA, Thirunamachandran T (1974) Chem Phys 27:149
147. Kawasaki T, Tokuhiro M, Kimizuka N, Kunitake T (2001) J Am Chem Soc 123:6792
148. Maitra U, Mukhopadhyay S, Sarkar A, Rao P, Indi SS (2001) Angew Chem Int Ed 40:2281
149. Spector MS, Singh A, Messersmith PB, Schnur JM (2001) Nano Lett 1:375
150. Ihara H, Sakurai T, Yamada T, Hashimoto T, Takafuji M, Sagawa T, Hachisako H (2002) Langmuir 18:7120
151. de Jong JJD, Lucas LN, Kellog RM, van Esch JH, Feringa BL (2004) Science 304:278
152. Spector MS, Easwaran KRK, Jyothi GJ, Selinger JV, Singh A, Schnur JM (1996) Proc Natl Acad Sci USA 93:12943
153. Green MM, Reidy MP, Johnson RJ, Darling G, O'Leary DJ, Willson G (1989) J Am Chem Soc 111:6452
154. Iwaura R, Yoshida K, Masuda M, Oshini-Kameyama M, Yoshida M, Shimizu T (2003) Angew Chem Int Ed 42:1009
155. Kobayashi H, Friggeri A, Koumoto K, Amaike M, Shinkai S, Reinhoudt DN (2002) Org Lett 4:1423
156. Fuhrhop J-H, Svenson S, Boettcher C, Rossler E, Vieth H-M (1990) J Am Chem Soc 112:4307
157. Menger FM, Caran KL (2000) J Am Chem Soc 122:11679
158. Menger FM, Peresypkin AV (2003) J Am Chem Soc 125:5340
159. Marini DM, Hwang W, Lauffenburger DA, Zhang S, Kamm RD (2002) Nano Lett 2:295

160. Messerschmidt C, Svenson S, Stocker W, Fuhrhop J-H (2000) Langmuir 16:7445
161. Moriyama M, Mizoshita N, Yokota T, Yokota T, Kishimoto K, Kato T (2003) Adv Mater 15:1335
162. Ihara H, Shudo K, Hirayama C, Hachisako H, Yamada K (1996) Liq Cryst 20:807
163. Sato I, Kadowaki K, Urabe H, Jung JH, Ono Y, Shinkai S, Soai K (2003) Tetrahedron Lett 44:721
164. Jung JH, Ono Y, Hanabusa K, Shinkai S (2000) J Am Chem Soc 122:5008
165. Ringler P, Müller W, Ringsdorf H, Brisson A (1997) Chem Eur J 3:620
166. Wilson-Kubalek EM, Brown RE, Celia H, Milligan RA (1998) Proc Natl Acad Sci USA 95:8040
167. Xia Y, Yang P, Sun Y, Wu Y, Mayers B, Gates B, Yin Y, Kim F, Yan H (2003) Adv Mater 15:353
168. Van Bommel KJC, Friggeri A, Shinkai S (2003) Angew Chem Int Ed 42:980
169. Mann S, Burkett SL, Davis SA, Fowler CE, Mendelson NH, Sims SD, Walsh D, Whilton NT (1997) Chem Mater 9:2300
170. Jung JH, Shinkai S, Shimizu T (2003) Chem Rec 3:212
171. Sone ED, Zubarev ER, Stupp SI (2002) Angew Chem Int Ed 41:1706
172. Kobayashi S, Hamasaki N, Suzuki M, Kimura M, Shirai H, Hanabusa K (2002) J Am Chem Soc 124:6550
173. Schnur JM, Price R, Schoen P, Yager P, Calvert JM, Georger J (1987) Thin Solid Films 152:181
174. Burkett SL, Mann S (1996) Chem Commun 321
175. Goren M, Qi Z, Lennox RB (2000) Chem Mater 12:1222
176. Seddon AM, Patel HM, Burkett SL, Mann S (2002) Angew Chem Int Ed 41:2988
177. Ono Y, Nakashima K, Sano M, Kanekiyo Y, Inoue K, Hojo J, Shinkai S (1998) Chem Commun 1477
178. Jung JH, Lee S, Yoo JS, Yoshida K, Shimizu T, Shinkai S (2003) Chem Eur J 9:5307
179. Moreau JEJ, Vellutini L, Wong Chi Man M, Bied C (2001) J Am Chem Soc 123:1509
180. Jung JH, Yoshida K, Shimizu T (2002) Langmuir 18:8724
181. Sugiyasu K, Tamaru S-I, Takeuchi M, Berthier D, Huc I, Oda R, Shinkai S (2002) Chem Commun 1212

Top Curr Chem (2005) 256: 219–236
DOI 10.1007/b107176

Published online: 1 July 2005

Gelation of Liquid Crystals with Self-Assembled Fibers

Takashi Kato (✉) · Norihiro Mizoshita · Masaya Moriyama · Tetsu Kitamura

Department of Chemistry and Biotechnology, School of Engineering,
The University of Tokyo, Hongo, Bunkyo-ku, 113-8656 Tokyo, Japan
kato@chiral.t.u-tokyo.ac.jp

Abstract Physical gelation of liquid crystals with low molecular weight gelators leads to the formation of a new class of anisotropic gels that have great potentials for optical, electrical, and photofunctional materials. The liquid crystalline (LC) physical gels are microphase-separated anisotropic composites consisting of liquid crystals and self-assembled solid fibers. For these materials, the isotropic–anisotropic transitions due to liquid crystals and the sol–gel transitions due to gelators occur reversibly and independently. The thermal, optical, and electrical properties of the LC gels are tuned by the selection and combination of the components, which determine the microphase-separated structures. LC gels based on room temperature nematic liquid crystals show electro-optical switching on twisted nematic and light scattering modes. The electro-optical properties can be improved in the presence of fiber additives. Discotic liquid crystals that function as hole transport materials have been used as LC components of anisotropic gels. The discotic gels exhibit hole mobilities higher than those of the liquid crystal alone. Chemical modification of gelators with functional moieties is another versatile approach for functionalization of LC gels. Hydrogen-bonded gelators with photoswitchable azobenzene moieties have been developed and complexed with liquid crystals. The resultant photoresponsive LC gels show light-induced structural changes, which are applicable to rewritable information recording.

Keywords Liquid crystals · Gelation · Self-assembly · Hydrogen bonding · Electro-optical properties

Abbreviations

LC	Liquid crystalline
TN	Twisted nematic
Col_h	Hexagonal columnar

1 Introduction

Physical gelation of functional fluids such as liquid crystals [1–5] and electrolytes [6–8] with low molecular weight gelators is a new approach to the development of functional molecular materials. A large number of gelators for common organic solvents and aqueous solutions, which have isotropic structures, have been exploited [9–11]. Liquid crystal is one of the functional soft materials that form anisotropic structures with orientational order and positional disorder [12]. Thermotropic liquid crystalline (LC) materials have been widely used in the field of electro-optics [12–15]. The electronic properties of liquid crystals as organic semiconductors have also attracted attention from the viewpoint of the fabrication of organic devices [16–20].

LC physical gels are obtained by fibrous self-assembly of low molecular weight gelators in thermotropic liquid crystals [1–5, 21–37]. Three-dimensional dispersion of these self-assembled nanofibers in liquid crystals results in the physical gelation of liquid crystals, as shown in Fig. 1. In these composites, liquid crystals (LC phase) and self-assembled fibers of gelators (solid phase) form microphase-separated structures. Figures 2 and 3 show some examples of the components used for LC gels. The liquid crystals (Fig. 2) are effectively gelated by using 0.2–5.0 wt % of the gelators shown in Fig. 3. They have hydrogen bonding or π–π stacking moieties that promote the

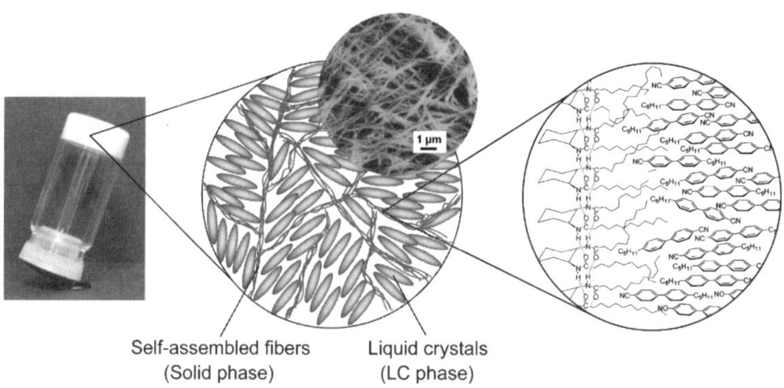

Fig. 1 Schematic illustration of a hierarchical structure of LC physical gels

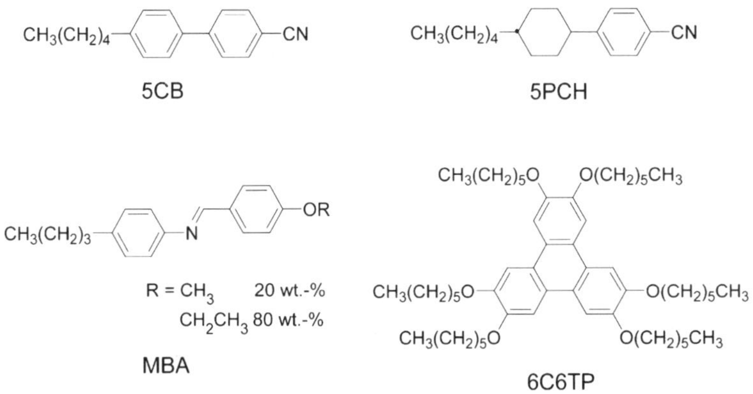

Fig. 2 Chemical structures of liquid crystals studied for the preparation of LC physical gels

formation of one-dimensional molecular assemblies. Self-aggregation of the gelators leads to the formation of solid fibers of 30–100 nm diameter.

Two transitions for the LC gels are observed independently: the sol–gel transition induced by the fibrous aggregation of gelators, and the isotropic–anisotropic transition of liquid crystals. Figure 4 shows the phase transition behavior of the mixtures of nematic liquid crystal 5CB and cyclohexane diamide gelator $Cy(11)_2$ [3]. The mixtures exhibit three states thermoreversibly: isotropic liquid (sol), isotropic gel, and LC gel states. The transition from isotropic liquid to isotropic gel states corresponds to the fibrous aggregation of $Cy(11)_2$. Variable-temperature infrared measurements for the mixtures show the formation of intermolecular hydrogen bonding in fibrous aggregates [3, 26]. The sol–gel transition temperatures increase as the concentration of $Cy(11)_2$ increases (Fig. 4). On the other hand, the isotropic gel–LC gel transition is due to the isotropic–nematic transition of 5CB in the fiber networks. The gel–gel transition temperatures are almost the same with the isotropic–nematic transition temperature of 5CB alone, which indicates the microphase separation of liquid crystals and fibrous aggregates of gelators in the mixtures.

In the LC gels, the transition temperatures are tunable by the choice of the components. The different order of sol–gel transition temperatures ($T_{\text{sol-gel}}$) and isotropic–anisotropic transition temperatures ($T_{\text{iso-lc}}$) gives two types of phase behavior. Figure 5 shows the two types of transitions for nematic LC gels. When $T_{\text{sol-gel}}$ is higher than $T_{\text{iso-lc}}$ (Type I, Fig. 5a), randomly dispersed networks of gelators are formed in the isotropic states of solvents on cooling [3–5, 25–35]. Then the LC gel states are induced by the isotropic–anisotropic transitions of liquid crystals. In contrast, when $T_{\text{iso-lc}}$ is higher than $T_{\text{sol-gel}}$ (Type II, Fig. 5b), aligned fibrous aggregates of gelators are formed by the template effects of the LC media, resulting in oriented LC gels [22–24]. Self-assembled fibers aligned

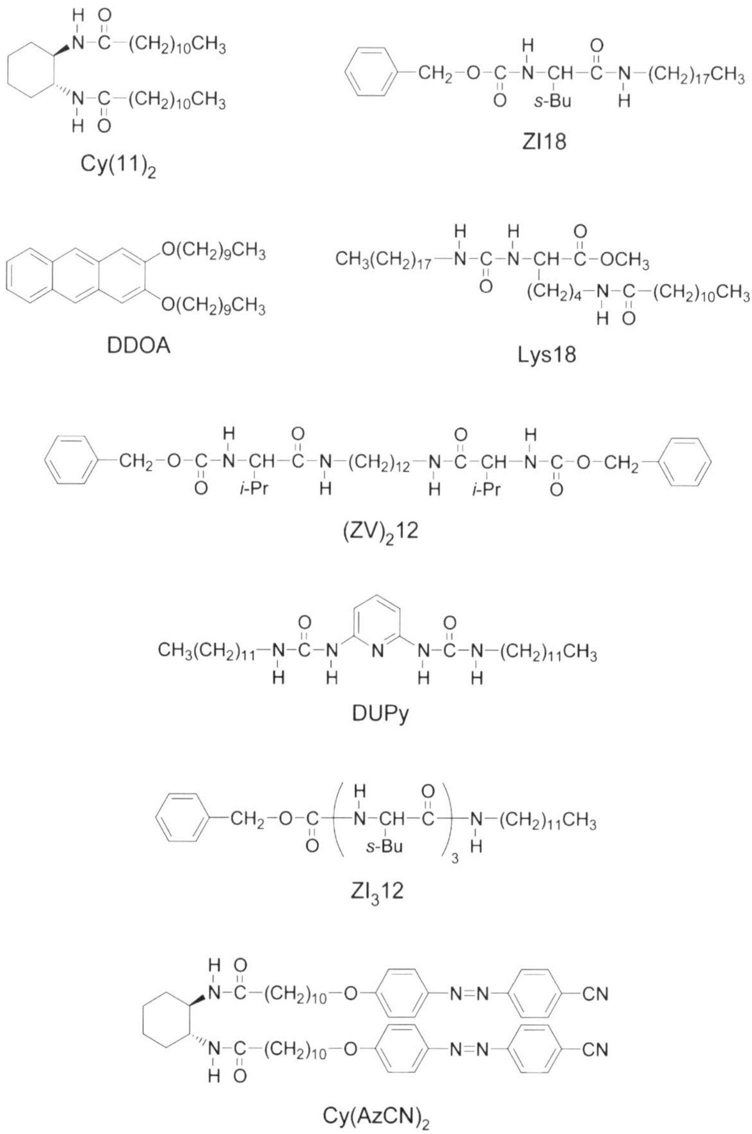

Fig. 3 Chemical structures of gelators used for the preparation of LC physical gels

parallel and/or perpendicular to the LC director can be obtained by the appropriate combination of the components. Fibrous assembly parallel to liquid crystal alignment (Fig. 5b) is observed for the nematic mixtures containing Lys18 [23, 24]. Fibrous aggregates of non-hydrogen-bonded anthracene derivative DDOA are also aligned in oriented cyanobiphenyl liquid crystals [22].

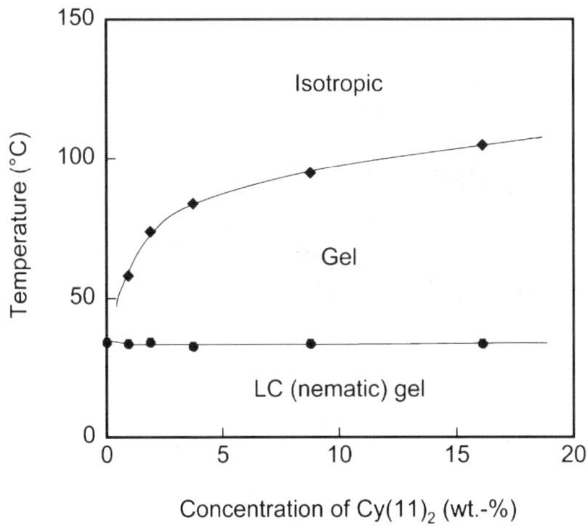

Fig. 4 Phase behavior of a mixture of 5CB and $Cy(11)_2$ on cooling

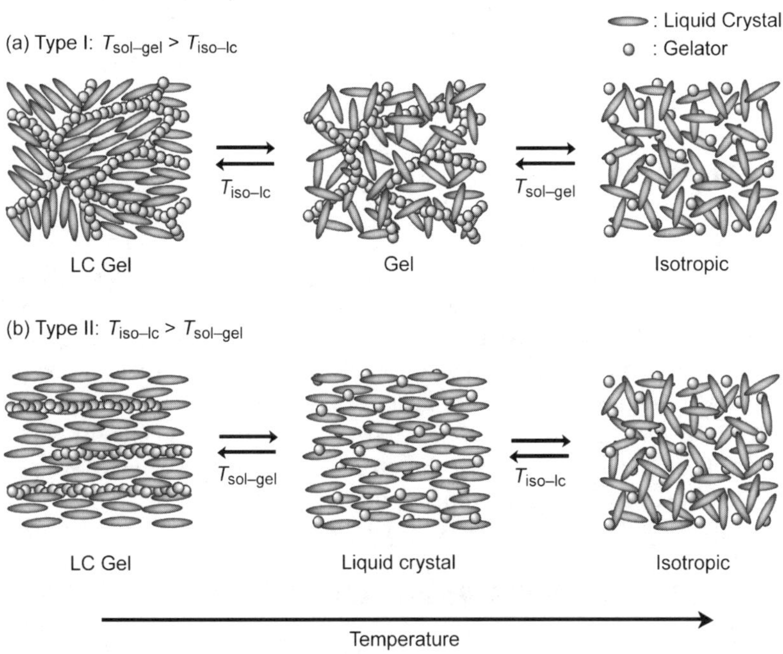

Fig. 5 Schematic illustration of two types of structural transitions of LC gels: (**a**) $T_{\text{sol-gel}} > T_{\text{iso-lc}}$ (Type I); (**b**) $T_{\text{sol-gel}} < T_{\text{iso-lc}}$ (Type II)

Smectic LC gels also exhibit two types of transitions [21]. For Type II, aligned fibrous aggregates of ZI18 are formed in the homogeneously oriented smectic A phase of the liquid crystal in a parallel rubbed cell (Fig. 6a) [21]. The direction of aligned fibers is perpendicular to the molecular orientation

(a) (b)

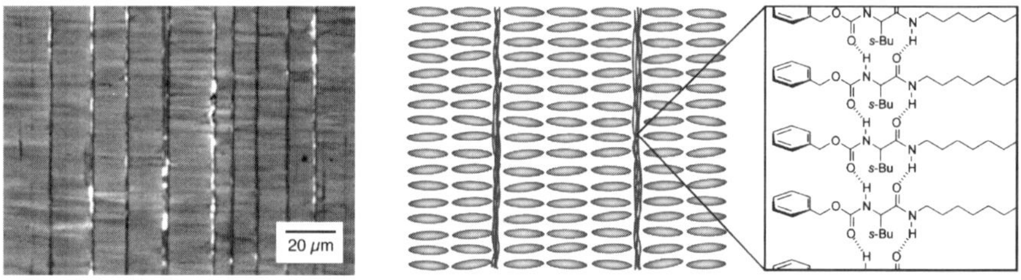

Fig. 6 (**a**) Polarized optical photomicrograph of a mixture of a smectic liquid crystal (SCE8) containing 1.0 wt % of ZI18 at 60 °C in a parallel rubbed cell. It is cooled from the isotropic state. (**b**) Schematic illustration of an anisotropically oriented structure of the mixture of SCE8 and ZI18

of the liquid crystal, that is, parallel to the layered structure of the smectic LC media. The fibers maintain a distance of ca. 10 μm. Polarized infrared measurements show that the hydrogen-bonded chains of ZI18 are oriented in the fiber direction as shown in Fig. 6b.

2 Electro-Optical Properties

In this part, we describe the electro-optical behavior of LC physical gels. LC components in the anisotropic gels can respond to external electric fields under the effect of submicrometer-scale phase separation of liquid crystals and self-assembled solid fibers. Electro-optical properties of LC gels have been tuned and improved by small amounts of the fiber additives with appropriate morphologies. As shown in Sect. 2.1, fast and high-contrast responses are observed for nematic LC gels in twisted nematic (TN) cells. In Sect. 2.2, nematic LC gels forming LC polydomain structures are shown to have potentials for light scattering electro-optical materials.

2.1 Electro-Optical Effects in Twisted Nematic Cells

Room temperature nematic liquid crystals have been developed for electro-optical applications [13–15]. In particular, twisted nematic (TN) liquid crystal displays have been widely used for practical display devices [13–15, 38]. In the TN cells, nematic liquid crystals form twisted alignment due to the influence of rubbed alignment polymer layers coated on the substrates (Fig. 7a). The TN cells are placed between two crossed polarizers. Without electric fields, the twisted LC alignment induces optical rotation of incident polarized

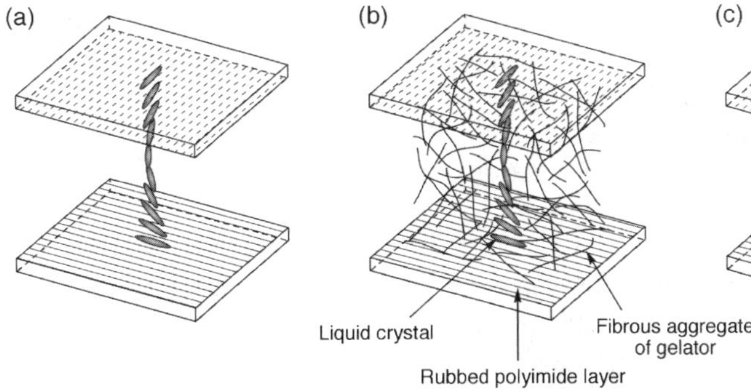

Fig. 7 Schematic illustration of TN cells incorporating nematic LC physical gels: (**a**) exhibiting TN alignment without aggregates of hydrogen-bonded molecules; (**b**) exhibiting TN alignment with a randomly dispersed fibrous network; (**c**) exhibiting TN alignment with an oriented fibrous network

light, resulting in light transmission states. The light transmittance for the cells can be electrically tuned by the reorientation of nematic liquid crystals from TN to homeotropic alignment.

Electro-optical properties of the nematic LC gels have been examined on the TN mode [3, 5, 23, 25–30]. The TN alignment behavior and the electro-optical responses of the nematic gels can be observed for the LC gels (Fig. 7b, Fig. 7c). The behavior is dependent on the chemical structures of the gelators and their concentrations. For example, the LC gels of 5CB formed by gelators $Cy(11)_2$, ZI18, and Lys18 exhibit TN alignment and response to electric fields when the concentrations of the gelators are less than 1.0 wt % [3, 23, 25]. In contrast, the LC gels based on $(ZV)_2 12$ show neither TN alignment nor electro-optical response even though the concentration of the gelator is 0.5 wt % [25]. Microscope observation shows that gelators $Cy(11)_2$, ZI18, and Lys18 form dispersed fibrous aggregates in the TN cells (Fig. 7b). Compound $(ZV)_2 12$ forms thick aggregates in the cell. The poor TN alignment and poor response of the gels formed by $(ZV)_2 12$ are due to the strong interactions between thick aggregates of the gelator and the liquid crystals.

Introduction of small amounts of self-assembled fibers into nematic liquid crystals in TN cells is found to accelerate electro-optical responses of the materials [25, 27–30]. In 16 μm-thick TN cells, the response time of the 5CB gel containing dispersed self-assembled fibers of ZI18 (0.5 wt %) is 6 ms at 10 V [25]. This response is twice as fast as that of 5CB alone (12 ms). Similar effects of the self-assembled fibers have been observed for several LC physical gels. For the mixture of 5PCH and ZI18 (0.5 wt %), the response time is 7 ms at 10 V, which is 10 ms shorter than that of 5PCH alone (17 ms) [27–29]. When gelator DUPy (1.0 wt %) is mixed with 5CB, the response time of the gel in the TN cell is 4 ms [30]. A slight lowering of threshold voltages has also been observed for these fast responsive nematic gels. For example, the threshold voltages of 5CB alone and the 5CB gel containing 0.5 wt % of ZI18 are 1.0 and 0.8 V, respectively [25]. These results indicate that the appropriate interactions between the liquid crystals and the fibrous aggregates of the gelators induce TN alignment that is metastable and responsive. The presence of the aggregates should weaken the interactions between the LC molecules and the rubbed polymer surfaces.

The faster responses described above have been observed for the gels containing randomly dispersed self-assembled fibers (Fig. 7b). Such random dispersion of the fibers in TN cells induces partial light scattering, leading to the decrease in display contrast [29].

In order to improve optical properties, nematic gels oriented along TN alignment (Fig. 7c) have been prepared by using the LC physical gels of Type II [23]. Such an oriented structure has been obtained for the 5PCH/Lys18 gels. For the mixture of 5PCH and Lys18 containing 0.4 wt % of Lys18, the $T_{\text{iso-lc}}$ and $T_{\text{sol-gel}}$ are 53 and 43 °C, respectively. On cooling from $T_{\text{iso-lc}}$ to $T_{\text{sol-gel}}$, the homogeneous LC mixture dissolving the gelator shows TN alignment. On the sol–gel transition, gelator Lys18 forms oriented fibrous aggregates in the liquid crystals exhibiting TN alignment, resulting in the anisotropic composite structures as shown in Fig. 7c. In the TN cell, the 5PCH gel responds in 7 ms, while the response time of 5PCH alone is 17 ms. The threshold voltage is also decreased from 1.7 to 0.7 V by introducing oriented fibers of Lys18. Moreover, the Type II gels forming the ordered aggregates show higher light transmission than the Type I gels. The efficiency of light use in the TN cell of the 5PCH/Lys18 gel reaches 88% for that in the TN cell of 5PCH alone. In contrast to nematic gels with randomly dispersed fibers, the oriented 5PCH/Lys18 gel exhibits a faster response, a lower threshold voltage, and a smaller decrease of contrast. The selection of the components and the morphology control of the aggregates are essential for the induction of significant electro-optical properties.

2.2
Light Scattering Electro-Optical Effects of Nematic Gels

LC physical gels have been applied to light scattering electro-optical materials (Fig. 8) [24, 31–33]. For LC composites containing non-LC solid components, random dispersion of the solid components in liquid crystals induces the formation of LC polydomain structures showing light scattering [39–41]. Such light scattering states can be electrically switched to transparent monodomain states. As no polarizer is needed in light scattering electro-optical materials, bright and high-contrast electro-optical switching is available with simple device structures.

LC physical gels containing random network aggregates of gelators, which are formed through Type I phase behavior, are suitable for the light scattering materials as shown in Fig. 8. This is because small amounts of nanofibers

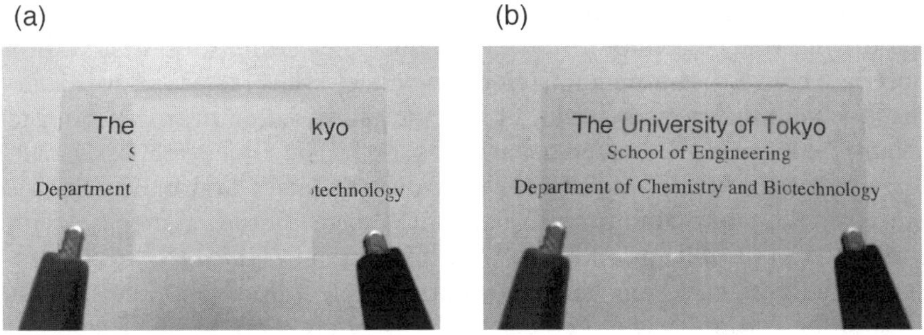

Fig. 8 Photographs of light scattering liquid crystal cells filled with a gel consisting of 5CB and ZI_312: (**a**) light scattering state (electric field off); (**b**) light transmission state (electric field on)

dispersed in liquid crystals effectively induce the formation of LC polydomain structures that can scatter light (Fig. 9a). Light scattering electro-optical properties have been reported for LC physical gels formed by amino acid and

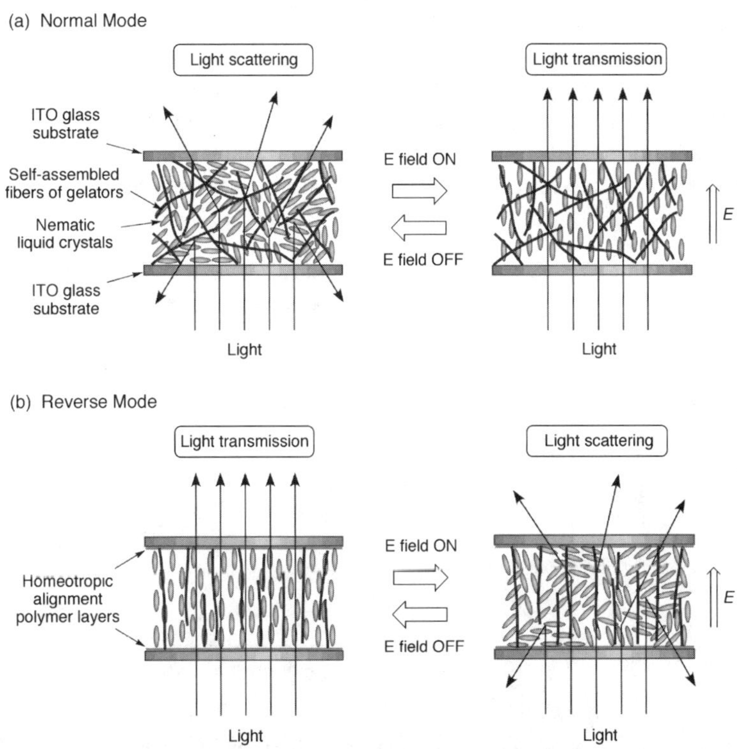

Fig. 9 Schematic illustration of light scattering electro-optical switching of nematic LC gels: (**a**) normal mode; (**b**) reverse mode

sugar derivatives [31–33, 42]. For the LC gels formed by a sorbitol derivative, the driving voltage is high (~100 V) and a large electro-optical hysteresis is observed between step-up and step-down voltage scans [42]. We have examined the light scattering behavior of nematic gels based on hydrogen-bonded gelators containing L-isoleucine moieties [31]. The isoleucine derivatives form hydrogen-bonded aggregates with different sizes and morphologies. For example, mono-amino acid gelator ZI18 forms fibrous aggregates with a diameter of ca. 100 nm, while the fiber thickness of $ZI_3 12$ with three amino acid moieties is ca. 30 nm. The increased number of hydrogen-bonded moieties in the gelators leads to efficient gel formation, thermal stabilization of the gels, and the induction of high light scattering states. The nematic gels formed by gelator $ZI_3 12$ show a significant light scattering electro-optical behavior (Fig. 10, curve A). The driving voltage for the 5CB gel containing 0.2 wt % of the gelator is lower than 40 V. The ratio of the light transmittance between light scattering and light transmission states is more than 60. The LC gels exhibit a stable orientation behavior of the LC molecules both in the step-up and step-down voltage scans. Electro-optical properties of nematic LC gels are also tunable by the choice of LC components as well as gelators. Liquid crystal mixture E63, whose refractive anisotropy is higher than that of 5CB, is suitable for preparing light scattering LC gels [32].

These normal light scattering electro-optical materials exhibit turbid (light scattering) states in electric field off states and transparent (light transmission) states in on states (Fig. 8, Fig. 9a). Light scattering switches on reverse mode (transparent in off states and turbid in on states) are useful for various applications. Reverse mode switching for LC physical gels has been attained

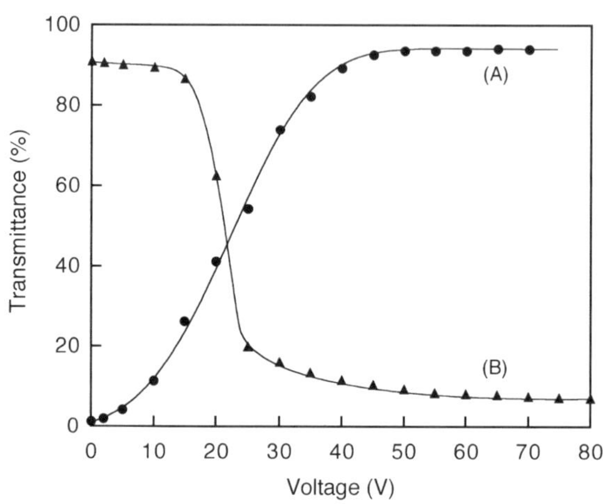

Fig. 10 Relationships between light transmittance and applied voltage in 16 μm-thick cells: *A* 5CB gel containing 0.2 wt % of $ZI_3 12$ (normal mode switching); *B* MBA gel containing 0.2 wt % of Lys18 (reverse mode switching)

by forming homeotropically oriented structures through Type II phase transition (Fig. 9b) [24]. To attain high transparency in off states, homeotropically oriented LC gels are prepared in liquid crystal cells coated with homeotropic alignment polymer films. As an LC component, nematic liquid crystal MBA with negative dielectric anisotropy is used because it aligns the LC molecules perpendicularly to the applied electric field. Due to the presence of the fibrous aggregates of gelators, LC polydomains are formed under the application of electric fields, leading to the induction of light scattering states (Fig. 9b). The reverse mode switching can be driven by low voltages around 30 V (Fig. 10, curve B).

3
Photoconductive Materials

3.1
Structures and Phase Behavior

Photoconductive properties are observed for some discotic molecules having abundant π electrons [16–20]. For example, triphenylene derivatives can show high charge carrier mobility in the hexagonal columnar (Col_h) LC phase due to their ordered molecular packing [16–20, 43–45].

Triphenylene derivative 6C6TP shown in Fig. 2 is one of the discotic LC materials having charge transporting properties. On cooling 6C6TP exhibits a Col_h phase from 98 to 54 °C. The discotic LC physical gels are formed when the mixtures of 6C6TP and gelator $(ZV)_2 12$ are cooled down from the isotropic liquid state (Fig. 11) [34]. For example, the mixture containing

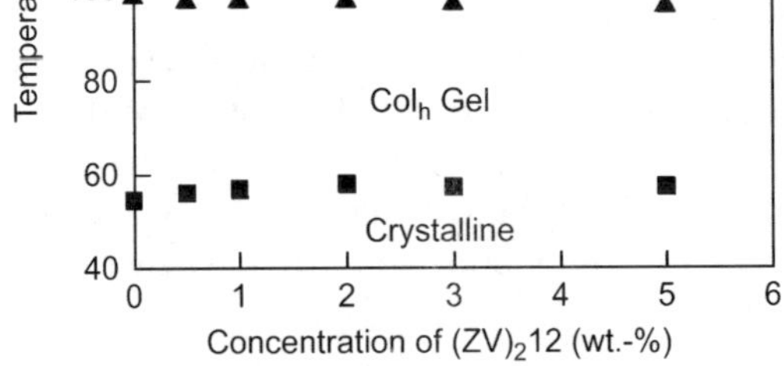

Fig. 11 Phase transition behavior of mixtures of 6C6TP and $(ZV)_2 12$

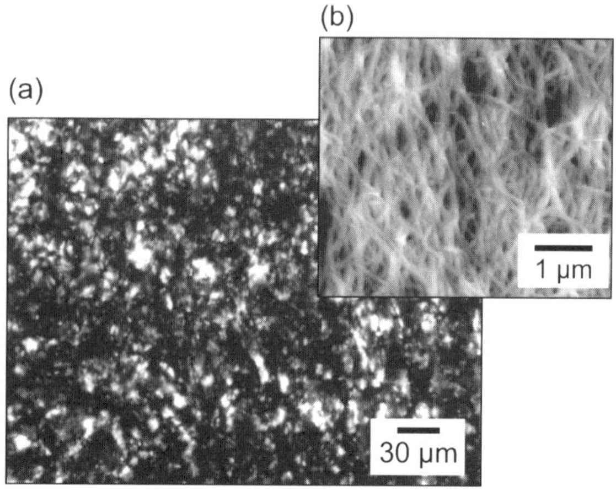

Fig. 12 (**a**) Polarized optical photomicrograph of the Col_h gel based on 6C6TP and $(ZV)_2 12$ at 76 °C; (**b**) AFM image of a xerogel obtained from 6C6TP/$(ZV)_2 12$

3.0 wt % of $(ZV)_2 12$ exhibits a transition from the isotropic liquid to the normal gel at 133 °C and then a transition to the Col_h gel state at 97 °C. The Col_h gel is exhibited until 57 °C at which temperature a Col_h–crystalline transition occurs.

Figure 12a shows a polarized optical photomicrograph of the Col_h gel based on $(ZV)_2 12$ and 6C6TP at 76 °C. In the dark area of the image, columns of 6C6TP align homeotropically to the substrates. Figure 12b shows an AFM image of the xerogel, which indicates the formation of a fine network of fibrous aggregates with a diameter of 50–100 nm.

3.2 Photoconductivities

We have found that the hole mobility of discotic liquid crystals is enhanced by the physical gelation with hydrogen-bonded fibrous aggregates [34]. The hole mobility of the Col_h gels can be measured by the time-of-flight method. For the Col_h phase of 6C6TP alone exhibiting homeotropic alignment, the hole mobility is 4.5×10^{-4} cm^2 V^{-1} s^{-1} (Fig. 13). Interestingly, the mobility of the Col_h gel states is 1.2×10^{-3} cm^2 V^{-1} s^{-1}, which is ca. three times higher than that of 6C6TP alone. The dispersed network based on the aggregates of $(ZV)_2 12$ may effectively suppress the fluctuation of the 6C6TP molecules and enhance the molecular order of the columnar structure, leading to the positive effect on the hole mobility. The combination of fibrous molecular network and discotic LC compounds will provide new electro-active organic materials.

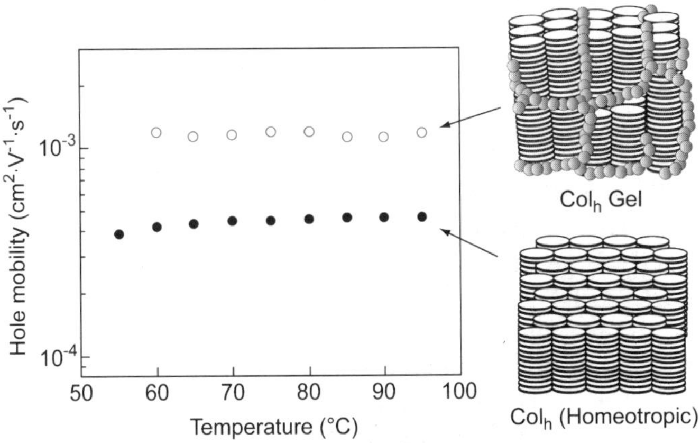

Fig. 13 Plot of the hole mobility versus temperature of 6C6TP/(ZV)$_2$12 3.0 wt% (*open circle*) and 6C6TP alone (*filled circle*) in an electric field of 4.0×10^4 V cm^{-1}

4
Photoresponsive Materials

4.1
Photoinduced Structural Changes

The composite structures of LC physical gels change in response to external stimuli such as heat and electric fields. In addition, if photoresponsibility is introduced into LC physical gels, the dynamic change in anisotropic composite structures can be induced by the light stimuli. Among the methods to afford the photoresponsive functions to the LC gels, the introduction of photochromic reactions is effective for the development of new photoresponsive materials. The photochromic reactions, in particular, the *trans–cis* photoisomerization of azobenzene derivatives are often used for induction of phase transition and control of the molecular alignment of liquid crystals [46–53]. Moreover, gel–sol transitions can be induced by the photochromic reactions in the organogel systems [54–56].

The photoresponsive nematic LC gels can be formed from mixtures consisting of nematic liquid crystal 5CB and gelator Cy(AzCN)$_2$ having photochromic azobenzene moieties [35]. When the mixtures are cooled down from an isotropic liquid (Fig. 14a), the nematic LC gel forms at room temperature (Fig. 14b) via an isotropic gel state. In the gel state, the gelator with *trans*-azobenzene moieties (*trans*-Cy(AzCN)$_2$) aggregates through hydrogen bonds of their amide units. The nematic gel has fine micrometer-scale phase-separated structures due to the finely and randomly dispersed aggregates of *trans*-Cy(AzCN)$_2$.

The photostimulated structural changes of the LC gels can be induced by *trans–cis* photoisomerization of azobenzene units in $Cy(AzCN)_2$. After UV irradiation, the structural transition from the nematic gels to chiral nematic LC phases occurs, which is revealed by the observation of fingerprint textures (Fig. 14c). In these chiral nematic phases, chiral cyclohexane compound $Cy(AzCN)_2$ with *cis*-azobenzene moieties (*cis*-$Cy(AzCN)_2$) acts as a chiral dopant by dissolving in 5CB. In the photoinduced phase transition from nematic gels to chiral nematic LC states, dissociation of hydrogen bonds of $Cy(AzCN)_2$ is observed by infrared spectroscopy [37]. Upon UV irradiation, the hydrogen-bonded N–H (3280 cm^{-1}) and C=O (1637 cm^{-1}) stretching bands of amide units in $Cy(AzCN)_2$ observed before UV irradiation shift to free N–H (3400 (br) cm^{-1}) and C=O (1662 cm^{-1}) bands. The molecular polarity of azobenzene dramatically changes between *trans*- and

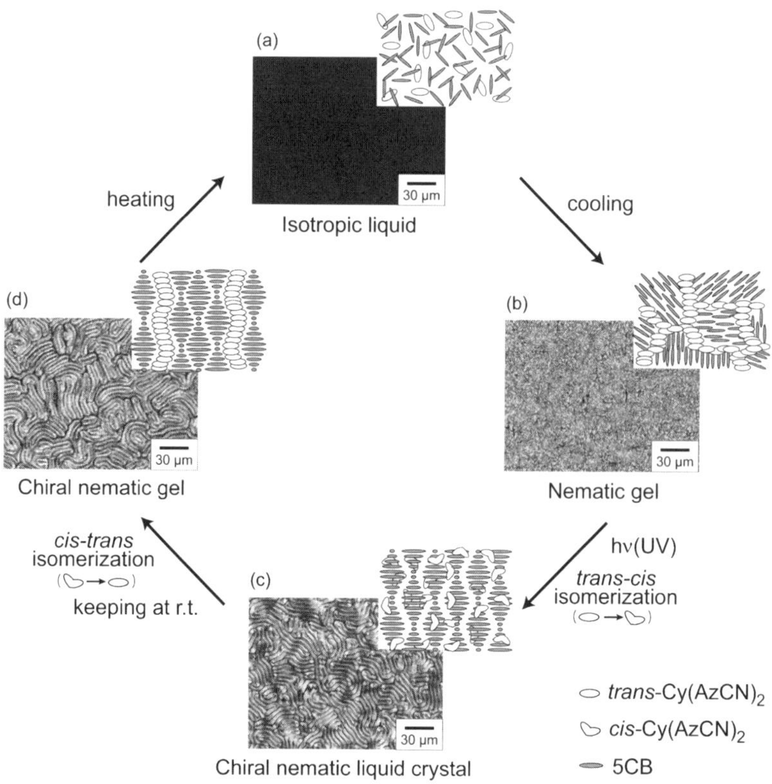

Fig. 14 Polarized optical photomicrographs of a mixture of $Cy(AzCN)_2$ (3 wt %) and 5CB and schematic illustrations of their structures: (**a**) isotropic liquid state at 120 °C; (**b**) nematic gel state at room temperature before UV irradiation; (**c**) chiral nematic LC phase (LC sol state) at room temperature after UV irradiation of the nematic gel; (**d**) chiral nematic gel state at room temperature after maintaining a chiral nematic phase

cis-isomers. The polarity change of azobenzene moieties induced by *trans–cis* photoisomerization increases the solubility of $Cy(AzCN)_2$ in 5CB, leading to the dissociation of hydrogen bonds and the transition to LC states.

The chiral nematic LC states (Fig. 14c) return to gel states by the *trans–cis* back-isomerization. In this stage, *trans*-$Cy(AzCN)_2$ reaggregates along the fingerprint structures, resulting in the formation of chiral nematic gels (Fig. 14d). The fingerprint structures are maintained even after most of the *cis*-$Cy(AzCN)_2$ are isomerized to *trans*-$Cy(AzCN)_2$. The aggregation of *trans*-$Cy(AzCN)_2$ with fingerprint structures has been confirmed by the observation of the xerogels [35]. Although the fingerprint structures of chiral nematic LC states should behave as templates for self-aggregation of *trans*-$Cy(AzCN)_2$ in the initial stage, these fingerprint structures are eventually fixed by the aggregates of *trans*-$Cy(AzCN)_2$. The chiral nematic gels (Fig. 14d) as well as nematic gels (Fig. 14b) are stable at room temperature. However, once the chiral nematic gels are heated to an isotropic state and cooled to room temperature, the nematic gels form again (Fig. 14, d→a→b). This result indicates that bistable LC gel states consisting of the same components can be repetitively produced by providing light stimuli and thermal treatments as shown in Fig. 14.

The reversible structural changes of discotic LC gels consisting of $Cy(AzCN)_2$ and 6C6TP can also be induced by light stimuli [36]. In this case, the *trans–cis* photoisomerization of azobenzene moieties induces the change in aggregation behavior of the gelator and consequently causes the change in columnar domain size and orientation.

4.2
Structural Pattern Formation

The photocontrol of reversible structural changes of LC gels leads to the fabrication of micropatterns [35–37]. Figure 15 shows the polarized photomicrographs of the patterned gels based on nematic liquid crystal 5CB (Fig. 15a) and discotic Col_h liquid crystal 6C6TP (Fig. 15b), prepared by UV irradiation through a photomask. In Fig. 15a, the regions of the nematic gels with fine domains and the chiral nematic gels with fingerprint structures are arranged. Because of stability of both the gels, these patterns are stable at room temperature for more than two years. Figure 15b exhibits the photopatterns of the Col_h gels. Although the gels in the irradiated region have larger domains of the Col_h phase and partially homeotropic alignment (dark area), the gels in the non-irradiated region have fine domains. As these photopatterns can be erased by heating to isotropic phases and subsequent cooling, we can repetitively fabricate photopatterns. The repetitive formations of photopatterns based on the bistable gel structures can be applied to photon-mode rewritable information recordings.

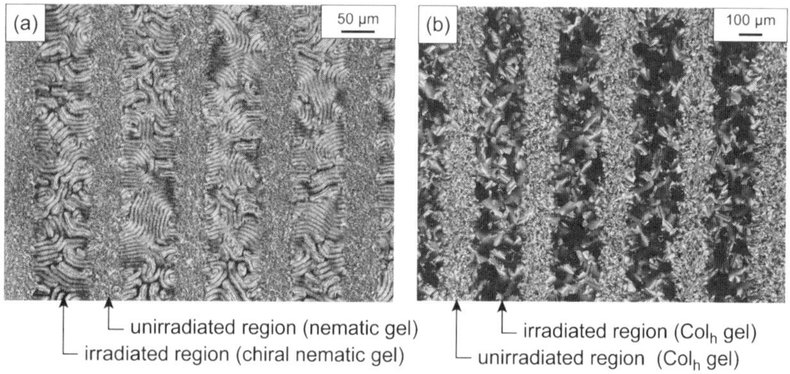

Fig. 15 Polarized optical photomicrographs of the patterned samples prepared by UV irradiation through a photomask: (**a**) 5CB/Cy(AzCN)$_2$ (3 wt%, r.t.); (**b**) 6C6TP/Cy(AzCN)$_2$ (3 wt%, 65 °C)

5 Conclusion

Liquid crystalline physical gels are new functional soft materials. They are obtained by orthogonal self-assembly of liquid crystals and gelators. The electro-optical or electric properties can be tuned by the morphology control of the phase-separated structures. The interactions between liquid crystals and fibrous aggregates of gelators improve or enhance the properties of the anisotropic gels. These materials may become interesting dynamically functional systems in advanced technologies and biorelated fields [1, 2, 57–59].

References

1. Kato T (2002) Science 295:2414
2. Kato T, Mizoshita N, Kanie K (2001) Macromol Rapid Commun 22:797
3. Kato T, Kutsuna T, Hanabusa K, Ukon M (1998) Adv Mater 10:606
4. Kato T, Kondo G, Hanabusa K (1998) Chem Lett 193
5. Kato T, Kutsuna T, Hanabusa K (1999) Mol Cryst Liq Cryst 332:377
6. Placin F, Desvergne JP, Lassègues JC (2001) Chem Mater 13:117
7. Ikeda A, Sonoda K, Ayabe M, Tamaru S, Nakashima T, Kimizuka N, Shinkai S (2001) Chem Lett 1154
8. Kubo W, Kitamura T, Hanabusa K, Wada Y, Yanagida S (2002) Chem Commun 374
9. Abdallah DJ, Weiss RG (2000) Adv Mater 12:1237
10. van Esch JH, Feringa BL (2000) Angew Chem Int Ed 39:2263
11. van Bommel KJC, Friggeri A, Shinkai S (2003) Angew Chem Int Ed 42:980
12. Demus D, Goodby JW, Gray GW, Spiess HW, Vill V (eds) (1998) Handbook of liquid crystals. Wiley-VCH, Weinheim
13. van Haaren J, Broer D (1998) Chem Ind 1017

14. Gray GW, Kelly SM (1999) J Mater Chem 9:2037
15. Pauluth D, Tarumi K (2004) J Mater Chem 14:1219
16. Boden N, Movaghar B (1998) In: Demus D, Goodby JW, Gray GW, Spiess HW, Vill V (eds) Handbook of liquid crystals, vol 2B. Wiley-VCH, Weinheim, p 781
17. Adam D, Schuhmacher P, Simmerer J, Häussling L, Siemensmeyer K, Etzbach KH, Ringsdorf H, Haarer D (1994) Nature 371:141
18. van de Craats AM, Warman JM, Fechtenkötter A, Brand JD, Harbison MA, Müllen K (1999) Adv Mater 11:1469
19. Schmidt-Mende L, Fechtenkötter A, Müllen K, Moons E, Friend RH, MacKenzie JD (2001) Science 293:1119
20. O'Neill M, Kelly SM (2003) Adv Mater 15:1135
21. Mizoshita N, Kutsuna T, Hanabusa K, Kato T (1999) Chem Commun 781
22. Kato T, Kutsuna T, Yabuuchi K, Mizoshita N (2002) Langmuir 18:7086
23. Mizoshita N, Hanabusa K, Kato T (2003) Adv Funct Mater 13:313
24. Suzuki Y, Mizoshita N, Hanabusa K, Kato T (2003) J Mater Chem 13:2870
25. Mizoshita N, Hanabusa K, Kato T (1999) Adv Mater 11:392
26. Yabuuchi K, Rowan AE, Nolte RJM, Kato T (2000) Chem Mater 12:440
27. Mizoshita N, Kutsuna T, Hanabusa K, Kato T (2000) J Photopolym Sci Technol 13:307
28. Mizoshita N, Kutsuna T, Hanabusa K, Kato T (2000) Proc SPIE Int Soc Opt Eng 4107:108
29. Mizoshita N, Hanabusa K, Kato T (2001) Displays 22:33
30. Yabuuchi K, Marfo-Owusu E, Kato T (2003) Org Biomol Chem 1:3464
31. Mizoshita N, Suzuki Y, Kishimoto K, Hanabusa K, Kato T (2002) J Mater Chem 12:2197
32. Mizoshita N, Suzuki Y, Hanabusa K, Kato T (2003) Proc SPIE Int Soc Opt Eng 5003:159
33. Mizoshita N, Suzuki Y, Kishimoto K, Kato T, Hanabusa K (2004) Mol Cryst Liq Cryst 409:175
34. Mizoshita N, Monobe H, Inoue M, Ukon M, Watanabe T, Shimizu Y, Hanabusa K, Kato T (2002) Chem Commun 428
35. Moriyama M, Mizoshita N, Yokota T, Kishimoto K, Kato T (2003) Adv Mater 15:1335
36. Moriyama M, Mizoshita N, Kato T (2004) Polym J 36:661
37. Moriyama M, Mizoshita N, Kato T (2004) Proc SPIE Int Soc Opt Eng 5518:160
38. Schadt M, Helfrich W (1971) Appl Phys Lett 18:127
39. Crawford GP, Zumer S (eds) (1996) Liqiud crystals in complex geometries formed by polymer and porous networks. Taylor & Francis, London
40. Doane JW, Vaz NA, Wu BG, Zumer S (1986) Appl Phys Lett 48:269
41. Rajaram CV, Hudson SD, Chien LC (1996) Chem Mater 8:2451
42. Janssen RHC, Stümpflen V, Broer DJ, Bastiaansen CWM, Tervoort TA, Smith P (2000) J Appl Phys 88:161
43. Bushby RJ, Lozman OR (2002) Curr Opin Colloid Interface Sci 7:343
44. Boden N, Bushby RJ, Clements J, Movaghar B (1995) Phys Rev B 52:13274
45. Inoue M, Ukon M, Monobe H, Sugino T, Shimizu Y (2001) Mol Cryst Liq Cryst 365:439
46. Ichimura K (2000) Chem Rev 100:1847
47. Ikeda T (2003) J Mater Chem 13:2037
48. Seki T (2004) Polym J 36:435
49. Mallia VA, Tamaoki N (2004) Chem Soc Rev 33:76
50. Gibbons WM, Shannon PJ, Sun ST, Swetlin BJ (1991) Nature 351:49

51. Stumpe J, Läsker L, Fischer Th, Rutloh M, Kostromin S, Ruhmann R (1996) Thin Solid Films 284:252
52. Moriyama M, Tamaoki N, Song S, Matsuda H (2001) J Mater Chem 11:1003
53. Kato T, Hirota N, Fujishima A, Fréchet JMJ (1996) J Polym Sci Part A: Polym Chem 34:57
54. Murata K, Aoki M, Suzuki T, Harada T, Kawabata H, Komori T, Ohseto F, Ueda K, Shinkai S (1994) J Am Chem Soc 116:6664
55. Geiger C, Stanescu M, Chen L, Whitten DG (1999) Langmuir 15:2241
56. Ayabe M, Kishida T, Fujita N, Sada K, Shinkai S (2003) Org Biomol Chem 1:2744
57. Kato T, Mizoshita N (2002) Curr Opin Solid State Mater Sci 6:579
58. Kato T (2000) Struct Bonding 96:95
59. Goodby JW (1999) Curr Opin Solid State Mater Sci 4:361

Top Curr Chem (2005) 256: 237–273
DOI 10.1007/b107178

Published online: 1 July 2005

Dendritic Gelators

Andrew R. Hirst · David K. Smith (✉)

Department of Chemistry, University of York, York YO10 5DD, UK
dks3@york.ac.uk

Abstract Dendritic molecules fall somewhere between small-molecule organic systems and polymers. Like polymers, they are constructed from a repeating motif, often have nanoscopic dimensions, and are capable of forming multiple non-covalent interactions. However, they are synthesized using organic chemistry methods and, unlike polymers, have well-defined, discrete structures which can be precisely controlled. This combination of properties makes dendritic molecules of particular interest for application in the assembly of gel-phase materials. In particular, this review focusses on the way in which molecular-scale information, put into place using organic synthesis, is transcribed up to the nanoscale, as visualised by electron microscopy techniques. Furthermore, it is illustrated that the molecular and nanoscale structures have a direct impact on the macroscopic materials properties of the gel-phase network. We discuss the structural effects on macroscopic gelation in terms of molecular size, shape and chirality, and clearly outline the specific advantages of using dendritic structures for this type of soft materials application.

Keywords Dendrimer · Gel materials · Nanotechnology · Self-assembly · Supramolecular chemistry

1 General Introduction to Dendritic Gels

The first attempts to investigate and define the gel state were made well over 100 years ago [1]. However, as pointed out by Dorothy Jordan-Lloyd in 1926 and often acknowledged since, a gel is easier to recognise than define [2]. Originally, gelators were generally polymeric systems, and it was Flory who first developed a detailed understanding of the ways in which covalently crosslinking polymeric architectures could give rise to an extended sample-spanning network, capable of behaving macroscopically as a gel [3]. Interestingly, it is worth noting that Flory appreciated that branched building blocks could give rise to hyperbranched polymeric networks with gel-phase properties. Crosslinked silicates (sol–gel matrices) also constitute a major class of inorganic gel-phase material [4]. However, it is not only permanent, covalently crosslinked polymeric architectures which give rise to gels—associative polymers are also extensively used for the formation of gel-phase materials [5]. In associative polymers, the building blocks are held together by multiple non-covalent interactions to form an extended sample-spanning network. In many cases, gelation occurs as a consequence of phase separation, e.g. self-assembly of block copolymers, with "similar" monomeric repeat units or blocks forming complementary interactions with one another.

In recent years, as the remainder of this volume illustrates [6], significant attention has begun to focus on the fact that small molecules can assemble into fibrous architectures. These assembled fibres can be considered to be supramolecular (non-covalent) polymers [7] which, in certain cases, are capable of aggregating further to yield gel-phase materials. One of the major attractions of using self-assembly methods [8] to generate gel-phase materials is the fact that the chemical architecture of low molecular weight gelators (LMWGs) determines the nanostructure and morphology of the self-assembled state. In this way, subtle control of the macroscopic gel-phase properties is ultimately achieved by manipulation of the individual molecular recognition events. Applying a supramolecular approach to the assembly of fibrillar aggregates therefore offers a controlled synthetic route to develop functional nanoscale materials. Indeed, the emerging fields of supramolecular hydrogel [9] and organogel [10] formation are of considerable importance in understanding self-assembly (or bottom-up) fabrication processes for the development of nanochemistry. Great emphasis is currently being placed on understanding the complex relationship between molecular recognition and self-assembly, which ultimately is expressed macroscopically as gelation. As described elsewhere in this volume [6], a wide range of different small molecules have been used for the formation of gel-phase materials with a variety of different nanoscale features. Due to their potential applications in templated materials synthesis, drug de-

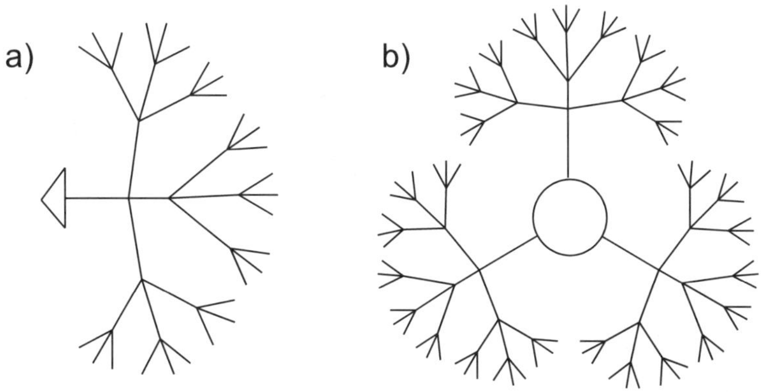

Fig. 1 Dendritic architectures: **a** dendron, **b** dendrimer

livery, chiral separations, environmental technology and biomimetics, these supramolecular soft materials constitute an important class of functional systems.

One class of molecular structure (or building block) which has been of particular recent interest in a wide range of different research arenas is the dendritic macromolecule (Fig. 1) [11]. Dendrimers are oligomers constructed in a controlled manner using branched repeat units. "Bottom-up" fabrication using simple dendritic building blocks has recently been widely exploited in the self-assembly of nanoscale architectures [12].

The specific aspect of dendrimer self-assembly of interest in this article is their ability to form nanoscale assemblies with gel-phase materials properties. The principal mechanism by which dendrimers form gels is supramolecular, and relies on complementary non-covalent interactions between individual dendritic building blocks. This chapter will demonstrate that dendrimers are of particular interest as gelators because their molecular size lies somewhere between that of small molecules frequently used in LMWGs and polymers traditionally used for making gels (Fig. 2). As such, dendritic molecules possess some of the advantages of both low molecular weight and polymeric gelators, which makes them a unique class of molecule for application in gel technology.

Part of the appeal of dendrimers lies in their precisely defined oligomeric molecular structures—in stark contrast to polymers traditionally used for gel formation. By controlling precisely the three-dimensional branched architecture, explicit manipulation of molecular interactions (hydrogen bonding, donor–acceptor effects etc.) between dendritic building blocks becomes possible. In analogy with traditional LMWGs, it therefore becomes possible to "program" structural information into dendritic building blocks in order that they can effectively self-assemble. This provides a range of materials with

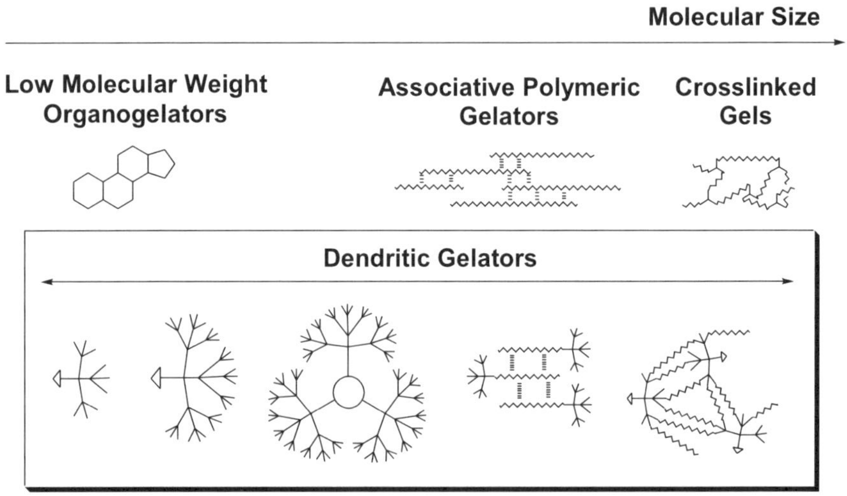

Fig. 2 Dendritic molecules can act as gelators using the full range of gelation mechanisms—from low molecular weight gelators through associative polymer-type gels to covalently crosslinked gels

regular structures that would otherwise be inaccessible via conventional synthetic routes.

In contrast to simple LMWGs, however, the dendritic architecture can effectively multiply up the number of recognition units present in the branching units. This is analogous to multiple non-covalent interactions, which are possible between polymeric gelator building blocks. It is well-known that such multivalency [13] can strengthen molecular recognition processes on dendritic surfaces [14], and this review will show that it can also give rise to stable gel-phase assemblies which, in addition, show remarkable degrees of tunability. Dendritic branches can also, as a consequence of their well-defined molecular sizes and shapes, play significant steric roles [15] in controlling the self-assembly of gel-phase materials. Simple LMWGs usually do not exert such significant steric impacts, as a consequence of their smaller sizes.

For the purpose of this article we have sub-divided dendritic gelators into two distinct types: one-component dendritic gelators (based on the self-assembly of a single dendritic building block) and two-component dendritic gelators (which require the addition of a second component in order to induce gelation). We will see how dendritic gelators can effectively "fill the space" between LMWGs and polymeric gelators—many dendritic gelators behave like LMWGs, while some can be considered as more traditional polymeric gelators. We will discuss the structure–property relationships and the underlying principles which drive the gelation process, particularly in terms of the relationship between the spatial organisation of the building blocks, the designed manipulation of molecular interactions via the synthesis (hydrogen bonding, donor–acceptor effects etc.) and chirality. In particular, we will focus on the roles played by the dendritic branching and the extent to which it is proactive in encouraging (or discouraging) the assembly of gel-phase fibrillar aggregates. In this way, we hope to illustrate the prospects and potential of gels constructed using branched building blocks.

2 Introduction to Dendrimers

2.1 Dendritic Structures: Dendrons and Spherical Dendrimers

Before considering dendritic gel-phase materials, it is first worthwhile to briefly consider some key aspects of dendrimer chemistry. Dendritic structures, as stated above, are monodisperse, well-defined oligomers constructed using a branched repeat unit [16]. In general terms, dendritic structures can be classified as either dendrons or dendrimers. A dendron is a single branched unit emanating from a focal point (Fig. 1a). Spherical dendrimers (Fig. 1b), on the other hand, possess a number of dendrons attached to a core unit. In both cases, a dendritic molecule consists of three essentially different architectural regions: a core (or focal point), layers or "generations" of repeat units emanating from this core, and terminal functional groups on the outer layer of repeat units. It is generally considered that the multiple surface groups of a dendritic structure control its interactions with the surrounding solvent environment and neighbouring molecules. The core (or focal point) is effectively buried within the branched shell and experiences a unique encapsulated microenvironment, which is conceptually similar to the microenvironments found within biological macromolecules [17]. For their applications as building blocks capable of self-assembly, the ability to vary precisely the size and functionality of dendritic structures is a crucial advantage that provides this class of molecule with immense synthetic tunability [18], which can in principle enable access to a vast array of different nanoscale assemblies.

2.2 Synthetic Approaches to Dendritic Architectures: Divergent, Convergent and Hyperbranched

Given the emphasis of this review on the synthetic tunability of dendritic systems, it is worth reflecting briefly on the different synthetic approaches used to build dendritic macromolecules. These syntheses have been the subject of extensive and excellent reviews [11, 16] and the basic strategies are only covered in outline here. Dendrimers are synthesised by stepwise iter-

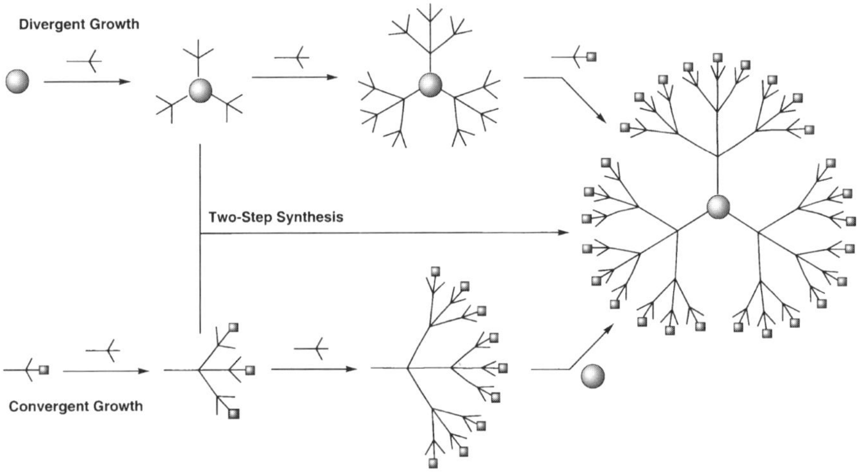

Scheme 1 Synthesis of dendritic architectures. Adapted from reference [16b] with the kind permission of Elsevier

ative synthetic procedures. Such methods of synthesis allow the branched architectures to be constructed in a controlled way, one layer of branching at a time. Two distinct stepwise synthetic methodologies have achieved positions of prominence for the preparation of dendrimers (Scheme 1): the divergent (inside-out) and convergent (outside-in) approaches.

The divergent strategy, which was pioneered by Vögtle [19], Denkewalter [20], Tomalia [21], Newkome [22], Meijer [23] and Mülhaupt [24], builds the dendritic structures from the central core or focal point out to the periphery. Using this approach, reaction of the peripheral functionalities of the core with a complementary reactive group on the branched monomer introduces a new layer of branching at each coupling point. The branched monomer must have its own peripheral functionalities present in a latent (or protected) form. After a layer of branching has been introduced, the new surface can then be activated and another layer of branching introduced. The cycle is then repeated. This divergent approach results in an increase in the number of peripheral functionalities on the dendritic structure with each synthetic step. Thus, the number of coupling reactions required increases exponentially with each successive generation of growth. This can make it increasingly difficult to get complete reaction of all the surface functionalities as the dendritic architecture increases in size.

In contrast, the convergent approach initiates the growth of the dendritic structure from what will eventually become the exterior of the molecule, and progresses inwards. Hawker and Fréchet [25], Miller and Neenan [26], and Moore and Xu [27] pioneered convergent approaches to different dendritic architectures. A recent review by Grayson and Fréchet [28] provides an excellent overview of this synthetic approach. Convergent growth is achieved by coupling the focal point of the growing structure to the peripheral groups of the branched monomer. In this case, the focal point of the branched monomer must be present in latent (or protected) form, and after completion of the coupling, the single functional group located at the focal point of the dendritic fragment or dendron can be activated. This approach avoids an exponential increase in the number of reaction sites, which is inherent in the divergent approach. However, the focal point functional group often decreases in reactivity as it becomes increasingly sterically hindered.

It is also possible to combine the most attractive features of convergent and divergent methodologies in a semi-convergent (or double-exponential) growth method [29]. This approach provides more rapid access to higher-generation dendrimers. Using this approach (instead of building up the branching a single layer at a time) larger branched fragments are coupled together. For example, coupling two second-generation fragments can yield a fourth-generation system without the need to synthesise the intermediate-stage third-generation molecule.

Regardless of the approach employed, the choice of synthetic route is justified by the features desired for the target molecule, the chemistry available for growth, and the specific building blocks used for the construction of the dendritic framework. Generally, the convergent approach provides better overall structural control, in part as a result of its enhanced potential for purification at intermediate stages of growth, and in part as a result of its innate ability to introduce differentiated functionalities at the focal point and the periphery of the dendrimer. Irrespective of the approach employed, however, the dendritic molecule must be purified, sometimes by simple crystallisation or precipitation or more often employing time-consuming column chromatography.

The explosion of interest in dendrimer chemistry over recent years has generated a huge range of dendritic architectures, making them ripe for exploitation in the areas of nanochemistry and self-assembly. The most widely used dendritic systems rely on amide, ester and ether connectivities, primarily as a consequence of the synthetic simplicity of introducing branched building blocks via the formation of these types of bonds. (It is important that reactions used in dendrimer synthesis are particularly robust.)

It is also worth pointing out that in order to expedite the availability of branched macromolecules, chemists have also attempted to mimic properties of dendrimers through the use of hyperbranched polymers [30]. Whereas dendritic structures have to be prepared in a stepwise iterative fashion as described above, hyperbranched polymers are usually conveniently obtained in a non-iterative one-pot polymerisation of multifunctional monomers of the AB_n type. However, this simple procedure yields imperfect structures that often exhibit irregular architectures with incomplete branching and broad molecular weight distributions. Considerable effort has been devoted to controlling the degree of branching and polydispersity, with Frey and co-workers, for example, reporting considerable success [31]. Nonetheless, hyperbranched systems will always, by definition, be imperfectly branched and polydisperse. They are, therefore, less appealing than dendritic structures for precise studies of self-assembly in which the impact of individual chemical modifications at the molecular level has a profound impact on material properties. It is often stated that hyperbranched systems have greater potential for commercial exploitation than structurally perfect dendrimers as a consequence of their lower production cost. However, it should be noted that many of the well-defined dendritic gelators discussed in this article are readily synthetically accessible. Indeed, the self-assembly step effectively multiplies up relatively small dendritic structures in a non-covalent manner to generate assembled architectures which have high levels of inherent branching. As such, the self-assembly of small dendritic building blocks is one of the most exciting areas of dendrimer chemistry for potential commercial exploitation, as it does not necessarily rely on the time-consuming covalent synthesis of expensive high-generation branched architectures.

2.3 Dendritic Structures in Materials and Supramolecular Chemistry

Dendritic structures have been widely exploited in materials and supramolecular chemistry—a full evaluation is certainly beyond the scope of this article [12]. It is simply worth noting here that in recent years, attention has turned to the assembly of a wide variety of nanoscale materials using dendritic building blocks. To provide some insight into this explosion of interest in bottom-up fabrication with dendritic building blocks, the following selected examples are instructive:

- Nanoparticles have been synthesised using dendritic stabilisers as ligands [32]. The dendritic architecture and size control the diameter and stability of the spherical nanoparticles formed.
- Cylindrical and tubular architectures have been assembled [33].
- Dendritic surfactants which exhibit superamphiphilic behaviour and generation-dependent self-assembly have been reported [34].
- Polycationic dendrimers and dendrons have also been assembled with DNA in order to generate complexes which have well-defined nanoscale sizes (ca. 50–200 nm) that are capable of transfecting DNA into cells [35].
- A wide range of dendritic systems capable of self-ordering on the nano- and mesoscales in order to form liquid crystalline domains have also been reported [36].

It is therefore of little surprise that attention has recently begun to focus on the potential of dendritic systems to assemble into nanoscale arrays which exhibit gel-phase materials behaviour. As outlined above, in order to generate tunable gel-phase materials, it is necessary to understand and modulate the molecular recognition event, i.e. the specific non-covalent interactions between individual molecular building blocks. This allows the transcription of information from the molecular to the nanoscopic level to be modulated, ultimately controlling the macroscopic material properties. Given the opportunities for exercising exquisite control over dendritic composition, the intrinsic potential of this class of molecule for molecular recognition, and the inherent nanoscale size of well-defined dendritic structures, this class of macromolecule provides a fascinating opportunity for exploring the phenomena of supramolecular gelation.

3
One-Component Dendritic Gelators

3.1
Dendritic Hydrogels

3.1.1
Low Molecular Weight Dendritic Hydrogelators

As early as 1986, Newkome and co-workers initiated the field of dendritic gelators by reporting dendritic bola-amphiphiles or "arborols" with hydrophilic surface head groups connected by a flexible lipophilic hydrophobic spacer (Fig. 3) [37]. Unlike an amphiphile, which has an apolar chain functionalised on one end with a polar head group, bola-amphiphiles have an apolar chain which is functionalised on both ends with polar head groups.

In their preliminary communication, Newkome and co-workers reported that their arborols formed a gel in aqueous solution. They commented in particular on the "low molecular weight (1052) compared to a typical gel". Using transmission electron microscopy (TEM) they illustrated that a fibrous assembly was formed, and noted that a 1 wt/vol% solution exhibited no steady-state flow— analogous to a traditional crosslinked polymer gel. The authors suggested that the polar head groups became "crosslinked via hydro-

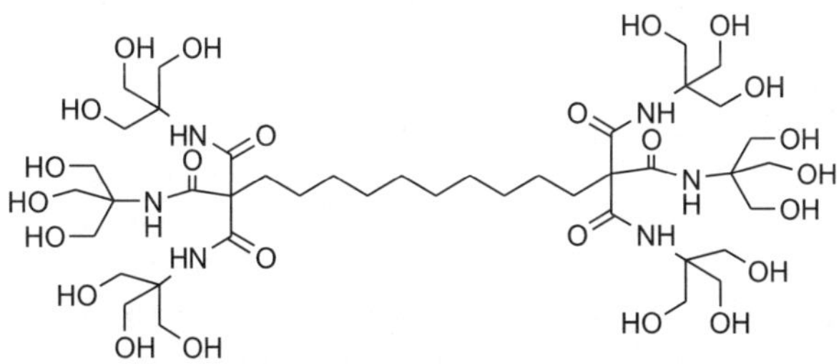

Fig. 3 Dendritic bola-amphiphiles which act as effective hydrogelators

gen bonding through water molecules". This work was significantly ahead of its time. Not only was this an early report of a well-defined small-molecule gelator, it was also some 10 years ahead of most reports of dendritic compounds which possess materials properties.

It is worth noting that Newkome's structures have multiple surface hydroxyl groups. As outlined in the Introduction, it is these surface groups that determine the compatibility of dendritic architectures with bulk environments such as the surrounding solvent. These hydroxyl groups therefore ensure the compatibility of these systems with water. Furthermore, the relatively apolar hydrocarbon chain is shielded within the dendritic superstructure, and it may be expected that the structure will attempt to phase separate in some way in order to yield stabilising hydrophobic–hydrophobic and hydrophilic–hydrophilic interactions.

In a series of interesting studies [38], Newkome and co-workers modulated the chemical architecture of these gelators (Fig. 3). The structures can be abbreviated as $[m]$-n-$[m]$ arborols, where m and n designate the number of surface hydroxyl groups and the number of carbon atoms in the connecting hydrocarbon spacer chain, respectively. Structural changes were made using simple synthesis in order to assess the impact of chemical modification at the molecular level on the macroscopic ability of these dendrimers to induce physical gelation. Attachment of the hydrophilic head groups to the hydrophobic core was achieved via a relatively straightforward two-step procedure: reaction of an α,ω-dibromoalkane (which becomes the spacer chain) with either ethyl sodiomethane tricarboxylate or methyl sodiomalonate, followed by amidation with tris(hydroxymethyl)aminomethane (to generate the dendritic head groups) [22]. This procedure gave high overall yields and was sufficiently flexible to enable a variety of different dendritic structures to be generated.

Thermally reversible aqueous gels were formed from the [6]-n-[6] series where the number of carbon atoms in the spacer was $8 \leq n \leq 13$. In this case, gels were formed at concentrations of dendrimer as low as 1.0 wt/vol %. In the [9]-n-[9] series gels were only formed where $10 \leq n \leq 13$. It therefore became clear that the presence of larger polar head groups ($m = 9$) required a longer apolar spacer chain in order for gelation to be observed. These experiments thus indicated that an "optimum" relationship existed between the length of the hydrocarbon spacer and the "size" of the hydrophilic head group [38a]. By using experimental techniques (e.g. electron microscopy and light scattering) as well as computer simulations, Newkome and co-workers postulated that self-assembly was driven by uni-directional orthogonal stacking of building blocks that maximised both lipophilic–lipophilic and hydrophilic–hydrophilic interactions. In this way, the dumb-bell building blocks hierarchically self-assembled to form extended fibrous rod structures.

Newkome and co-workers also modified the properties of the spacer unit in a different way. Specifically, they introduced central acetylenic (a) [38b],

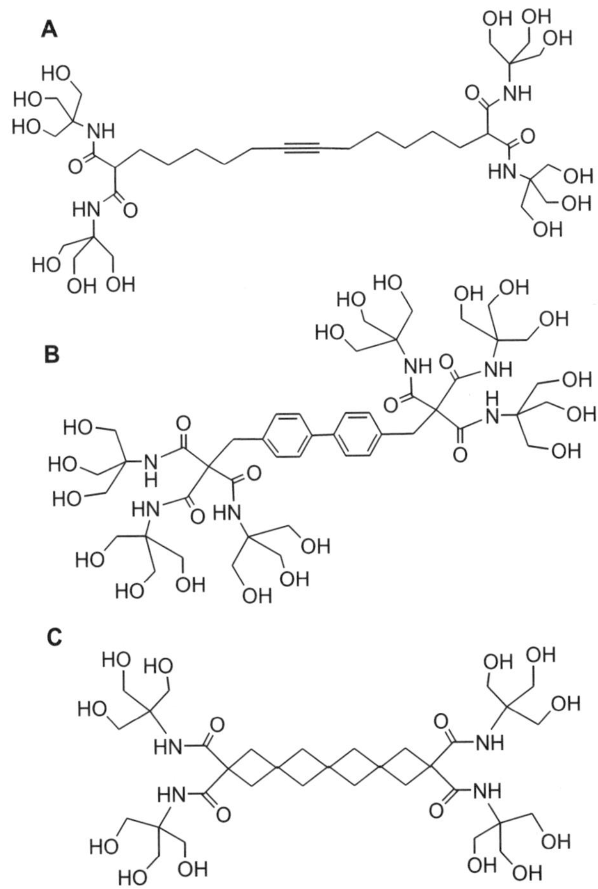

Fig. 4 Structures of dendritic bola-amphiphiles with varied spacer chains: alkyne (**a**), biphenyl (**b**), spirane (**c**)

biphenyl (b) and spirane moities (c) (Fig. 4) [38c]. It was noted that when the spacer chain contained an alkyne group, the self-assembled state was significantly altered. Electron microscopy revealed the presence of single-stranded self-assemblies that hierarchically self-assembled to form helical supramolecular structures. This aggregated state had a regular repeating twisted structure and a diameter of approximately 600 Å. Furthermore, these helical superstructures were not observed for the alkane analogues. Therefore, it was argued that the less flexible alkyne spacer unit induces a non-orthognal stacking and a favoured helical arrangement (Fig. 5).

Interestingly, the arborols possessing spirane or biphenyl moieties did not induce gelation in an aqueous environment. Newkome and co-workers concluded that the biphenyl and spirane spacer units were more sterically demanding than the related alkane series, which disrupted the molecular in-

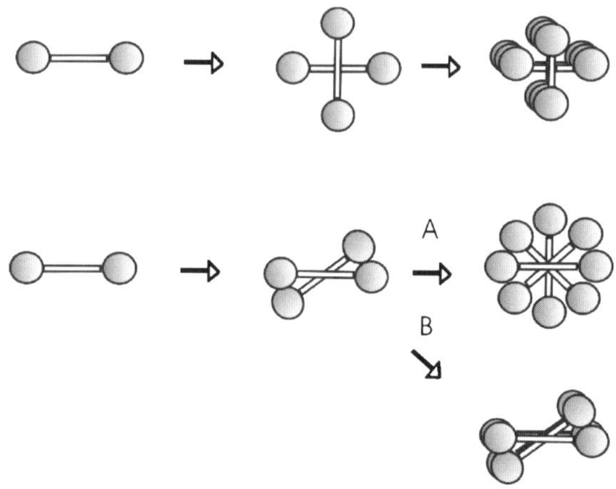

Fig. 5 Postulated stacking arrangements for dumb-bell-shaped arborols. *Top*: orthogonal stacking of the alkane arborols. *Bottom*: non-orthogonal relationships between adjacent molecules of alkyne arborols leads to A—helical structures, B—ribbon structures. Reproduced from reference [38b] with the kind permission of Wiley-VCH

teractions between building blocks. In summary, it was concluded that for this class of hydrogelator, there appeared to be three parameters that control gelation: the length of the spacer, the diameter/rigidity of the spacer and the size of the head group.

Another interesting approach, which used the molecular requirements of arborol self-assembly, was developed by Jørgensen et al. [39]. In this case, the self-assembling units of Newkome's arborols were covalently attached to a tetrathiafulvalene (TTF) core (Fig. 6) which has the potential for electrical conductivity in the presence of an appropriate dopant. Unfortunately, the synthetic route used to generate this molecule gave rise to a mixture of *cis* and *trans* isomers. It was anticipated that uni-directional stacking of the gelator building block may yield molecular wires with the TTF moieties forming an internally conducting stack, partitioned from the solvent by the arborol

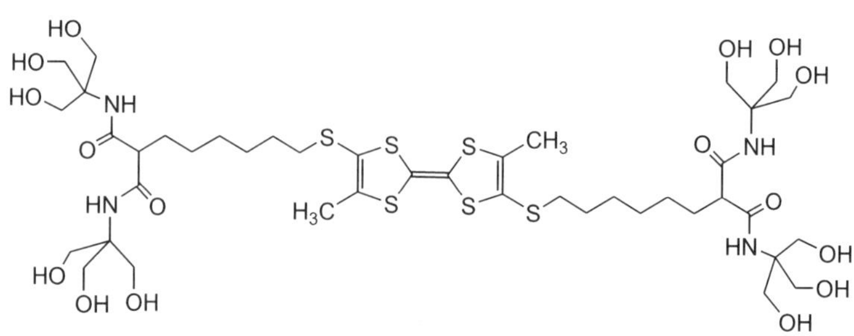

Fig. 6 Molecular structure of a bis-arborol-type molecule

head groups. The TTF gel was prepared by dissolving a known amount of material in hot solvent mixtures of 25% (v/v) ethanol/water or dimethylformamide/water. On cooling, these solutions became opaque, resulting in a gel-phase material. In this case, string-like superstructures were observed using both phase contrast optical microscopy and atomic force microscopy (AFM). These self-assemblies had lengths of the order of microns and diameters ranging from about 30 to several hundred nanometres. It was found that on oxidation of the gel with iodine, UV/Vis bands corresponding to oxidised TTF oligomers ($\lambda = 874$ nm) were observed in addition to a band corresponding to oxidised monomer ($\lambda = 673$ nm). In free solution, only the band corresponding to the oxidised monomer is usually seen for this type of compound. This observation suggested that the TTF units were brought into close proximity within the gel, perhaps in a stacked arrangement. Cyclic voltammetry in acetonitrile indicated that the redox potentials of the TTF core were not altered significantly by the arborol substituents. The redox properties within the gel state would also have been of interest; however, gel-phase electrochemical investigations were not performed.

As indicated above, the investigation of electroactive self-assemblies is an area of considerable current interest in terms of the development of insu-

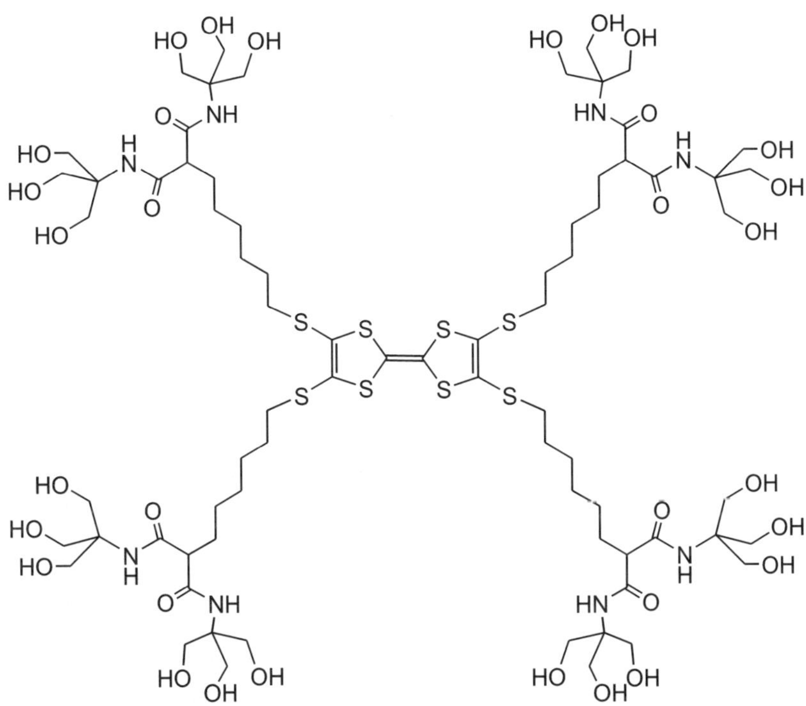

Fig. 7 A bis-arborol tetrathiafulvalene derivative

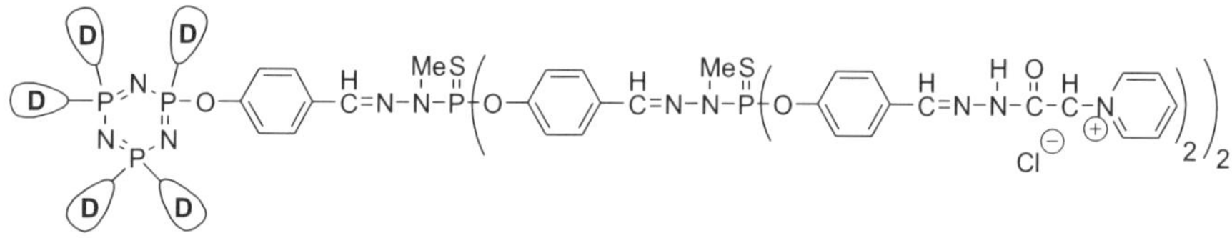

Fig. 8 Hydrogelators based on ammonium-functionalised phosphorus-based dendrimers

lated nanoscale wires [40]. With such goals in mind, Bryce and co-workers have recently reported a modified TTF arborol (Fig. 7) which, owing to the symmetry of the molecule, could not exist as a mixture of geometric isomers [41]. Interestingly, they formed thin films of the dendrimer using a spin-coating method, and noted that the arborol played a key role in enabling film formation, presumably due to its inherent potential for self-assembly. The thin films exhibited low-energy UV/Vis bands after iodine doping, and furthermore showed modest conductivities, which suggests that supramolecular ordering is present in the film. This indicates how the order inherent in a gel-phase material may be harnessed, even at concentrations below the gelation threshold, to yield advanced functional nanostructured materials.

Majoral et al. reported a different class of dendritic hydrogelator. This system was based on a symmetric (spherical) phosphorus-containing dendrimer (Fig. 8) [42]. These dendrimers were relatively straightforward to synthesise using divergent methodology well-established by Majoral and co-workers [43]. Given that the synthetic approach is a divergent one, it was possible to generate a range of dendrimers with different surface functionalities by simply varying the capping reagent used in the final step of the synthesis. This enabled these researchers to investigate dendrimers with a range of different surfaces. Given that it is the surface of the dendrimer which is exposed to the bulk solvent, it was to be expected that simple modulation of these groups should have a significant impact on the materials behaviour of the dendrimers; in fact, all the gel-forming dendrimers reported by this group have polycationic surfaces.

Majoral and co-workers also reported that dendrimers of generations 1 to 4 formed reversible rigid gels when dissolved in water, with gelation being observed at concentrations of 1.5–1.8 wt/vol %. As stated above, the dendrimers investigated had cationic surfaces: either ammonium or pyridinium chloride. The nature of the cation appeared to modify the kinetics of gelation, which was, in addition, strongly dependent on the experimental conditions and extremely slow. The best results were obtained when an aqueous solution of dendrimer was heated at 60–65 °C for 11–13 days. Ammonium-functionalised dendrimers formed gels more quickly than the pyridinium-functionalised system. Remarkably, gelation times were shortened by 10–30% in the presence of water-soluble components such as metal salts (Ni, Y, Er acetates) or a buffer (tris(hydroxymethyl)aminomethane (Tris)). It is not clear, however, why this is the case, although the authors speculated that acetate anions could act to enhance the process of gelation by bridging the non-covalent network of dendrimers. It is also somewhat unclear from simple inspection of the structures why they should give rise to the one-dimensional fibrillar assemblies observed by scanning electron microscopy (SEM) of the freeze-dried aerogels. However, the authors have recently published a paper in which the aggregation of this type of cationic dendrimer has been optimised using high salt conditions, and the process rationalised in terms of colloidal flocculation [44]. In addition, it was reported that macroscopic fibres could be produced under gentle flow conditions. These fibres were mechanically strong and, as the authors point out, the high inherent functionality of dendritic molecules means that these fibres have potential for a wide range of applications.

In the examples above, the dendrimers which form hydrogels are analogous to small molecules. They have well-defined structures and relatively low molecular masses. However, as indicated in the Introduction, dendritic molecules are able to span the range between small molecules and polymer systems. In the following section, the dendritic systems described benefit from the presence of a significant "polymeric" component, which plays a proactive role in the formation of the hydrogel.

3.1.2
Dendritic Hydrogelators with a Polymeric Component

Self-assembling biomaterials have been extensively developed by Stupp et al. [45] using a block copolymer strategy. The long-term goal of this work is to develop functional materials that act as a temporary molecular scaffold for cell growth and tissue engineering applications. The building block con-

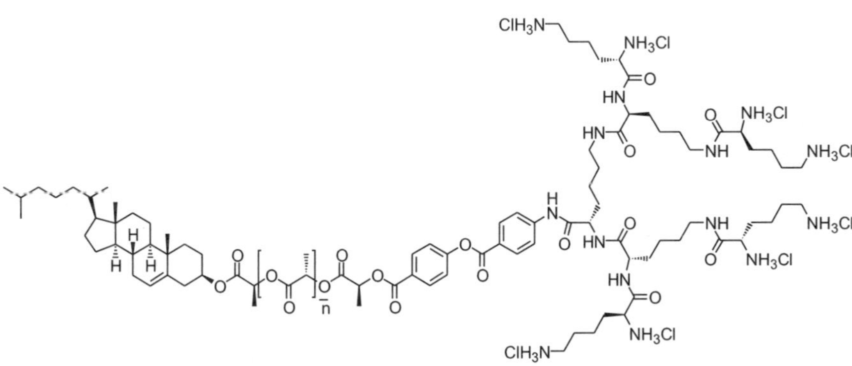

Fig. 9 Block copolymer based on an L-lysine dendron, a biodegradable L-lactic acid polymer chain and a cholesterol unit

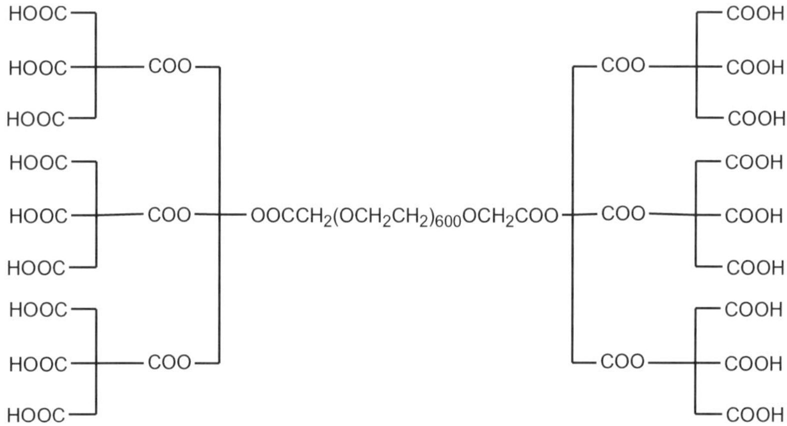

Fig. 10 ABA block copolymer, with dendritic A blocks based on citric acid, which exhibits hydrogelation properties

tains a cholesteryl moiety that is linked to a short chain of approximately 20 L-lactic acid residues (Fig. 9). The hydroxy terminus of this L-lactic acid chain was substituted with L-lysine dendrons of different generations. In this system, the cholesteryl unit assists the self-assembly process and the lactic acid unit is potentially biodegradable.

It was proposed that the steric effects of the dendritic branching would help to direct the self-assembly process. Indeed, it was observed that modulating the "size" of the dendron head group (from generation 1 to generation 3) controlled the mode of self-assembly and the aggregate morphology. It should be noted that Stupp and co-workers do not explicitly state that these materials form hydrogels; however, it is clear that this strategy to self-assemble nanoscale soft biomaterials, which exhibit a significant degree of hydration, is in some ways analogous to the process of gelation. In a recent paper, these authors reported the application of this type of system to form biological scaffolds [46]. A poly(L-lactic acid) fibrous network was modified with the second-generation dendron. It was demonstrated that this network allowed more effective cell growth—a feature which may, in the future, be useful in tissue engineering and cell transplantation.

In a similar general approach, Namazi and Adeli reported thermoreversible hydrogels formed from ABA-type copolymers. The A blocks were constituted from dendritic citric acid and the B block was a polyethylene glycol unit (Fig. 10) [47]. This dendrimer has a dumb-bell-type structure, somewhat reminiscent of Newkome's bola-amphiphiles, only with a polymeric spacer chain. First- and second-generation dendritic systems were reported and it was found that the gels could also be formed in the presence

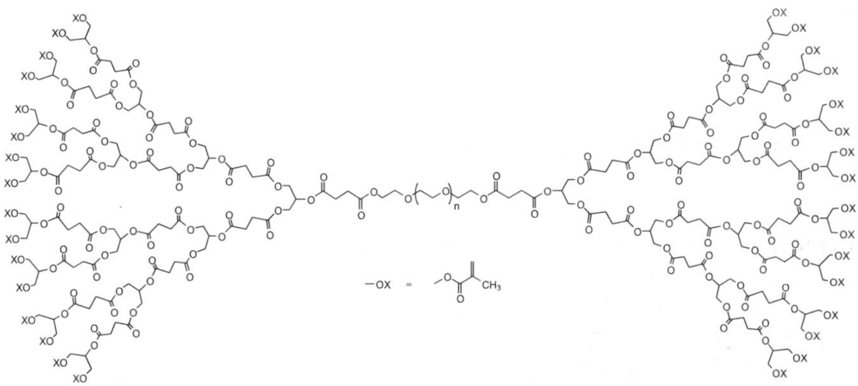

Fig. 11 Photocrosslinkable dendrimer with polymerisable methacrylate peripheral groups can be used as a macromer for the construction of crosslinked gels

of small molecules. The authors indicated that such systems could therefore have applications in drug delivery.

Grinstaff and co-workers utilised dendritic architectures for the synthesis of hydrogels using a crosslinked polymer chemistry approach. Firstly, they synthesised a series of well-defined polyester dendritic macromolecules composed of succinic acid, glycerol and polyethylene glycol [48] (Fig. 11). These branched structures are composed of building blocks that are known to be biocompatible or degradable in vivo to give natural metabolites. Derivatisation of the dendritic periphery with a methacrylate unit rendered the system photocrosslinkable. Photopolymerisation resulted in a gel network composed of permanent covalent contact points or linkages resulting in a hydrogel material. This illustrates how dendritic molecules can act, not only in a similar way to LMWGs (as described previously), but also as fascinating building blocks for the formation of more traditional crosslinked polymer gels. In situ photopolymerisation is currently being explored for dental, drug delivery, biological adhesive and ophthalmology applications. Indeed, in one of their papers, Grinstaff and co-workers illustrated that their biodegradable dendritic gel performed effectively as an adhesive for repairing corneal lacerations, and was more effective than conventional stitching [48b].

In the example above, the key role of the dendritic architecture is to act as a multivalent crosslinking unit in the formation of the gel-phase materials. Gitsov and Zhu have also utilised a similar approach to the application of dendritic architectures. They reacted Fréchet-type dendritic benzyl bromides and polyethylene glycol to generate hybrid covalently crosslinked dendritic–linear copolymers, which acted as hydrogels [49].

3.2 Organogels

3.2.1 Low Molecular Weight Dendritic Organogelators

In 2000, Aida and co-workers were one of the first groups to report dendritic molecules capable of forming physical gels with organic solvents at low concentration [50]. They made use of Fréchet-type dendrons functionalised at the focal point with a dipeptide (Tyr-Ala) building block (Fig. 12, category 1). They found that second- or third-generation dendritic branching was required for effective gelation, which occurred at very low concentrations (ca. 1.0 mM). Indeed, they calculated that one molecule of gelator could effectively

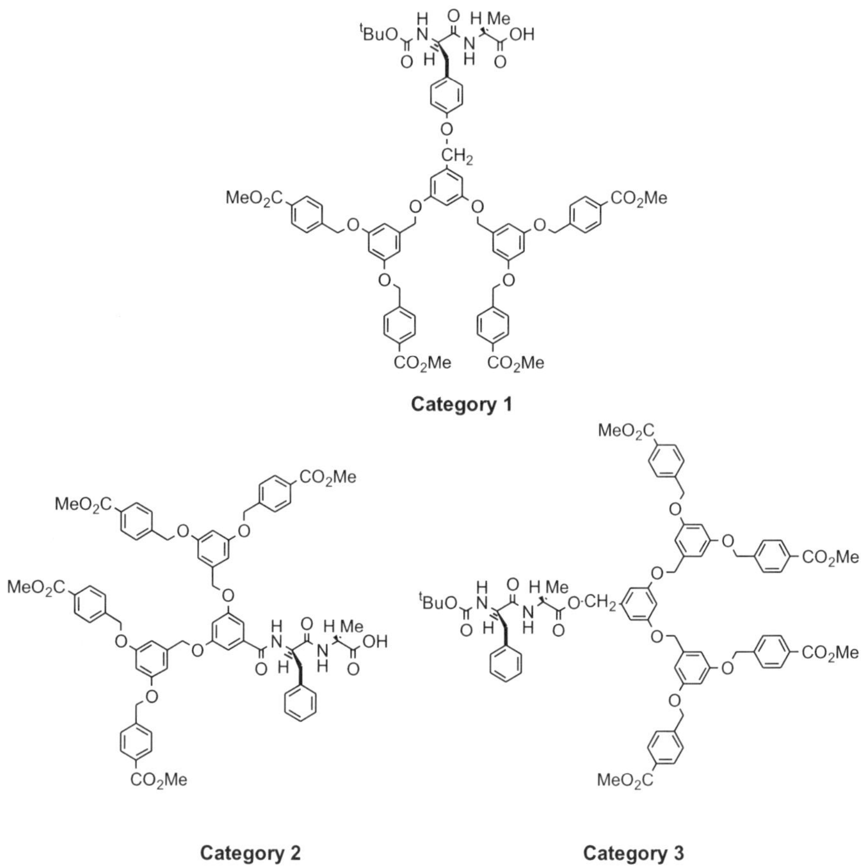

Fig. 12 Three categories of molecules, some of which are capable of forming gel-phase materials in organic solvents (see discussion)

"immobilise" almost 20 000 solvent molecules. (It should, however, be pointed out that in gel-phase materials, individual solvent molecules are not strictly immobilised but actually retain their molecular mobility, trapped within the interstices of the gel microstructure, with the overall flow of the fluid being inhibited as a consequence of capillary forces.) In contrast, lower-generation analogues only gave a crystalline solid, rather than a gel-phase assembly.

In a subsequent excellent paper, Jang and Aida uncovered the detailed structural parameters required for effective gelation by synthesising a series of different dendrons (Fig. 12) [51]. Three different categories of potential gelator were reported, in which the Fréchet-type dendron was attached either to the tyrosine side chain of the dipeptide (category 1), the *N*-terminus of the dipeptide (category 2) or the *C*-terminus of the dipeptide (category 3). The key conclusions of this study were as follows:

1. Dipeptides were required for effective gelation (rather than monopeptides).
2. Ester functionalities (rather than methoxy groups) on the surface of the dendron assisted the gelation process.
3. Peptides in categories 1 or 2 were required for gel formation.
4. Higher-generation dendrons gave more effective gelation.

The dendritic branching therefore plays a proactive role in encouraging gelation (perhaps steric). However, on applying differential scanning calorimetry (DSC) to the dried gels, only a small dendritic effect was observed. In this case, the effect of dendritic branching was not readily quantified.

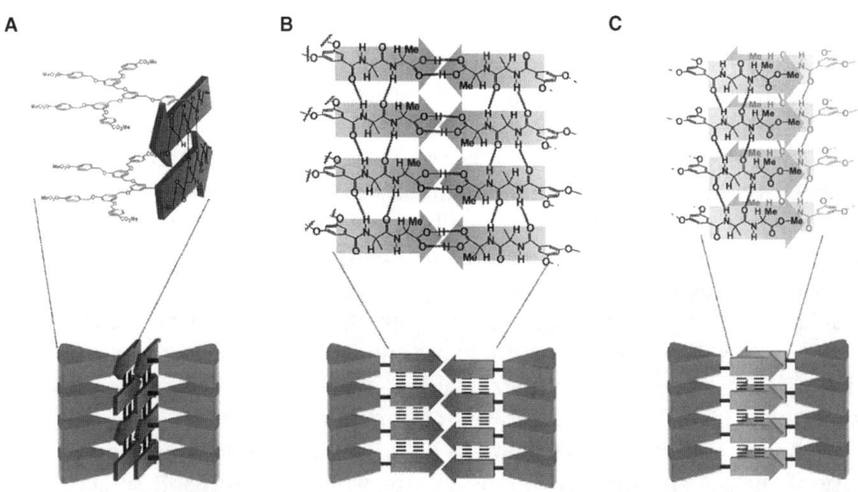

Fig. 13 Proposed assembly modes for the organogelators depicted in Fig. 12. This image is reproduced from reference [51] with kind permission of the American Chemical Society

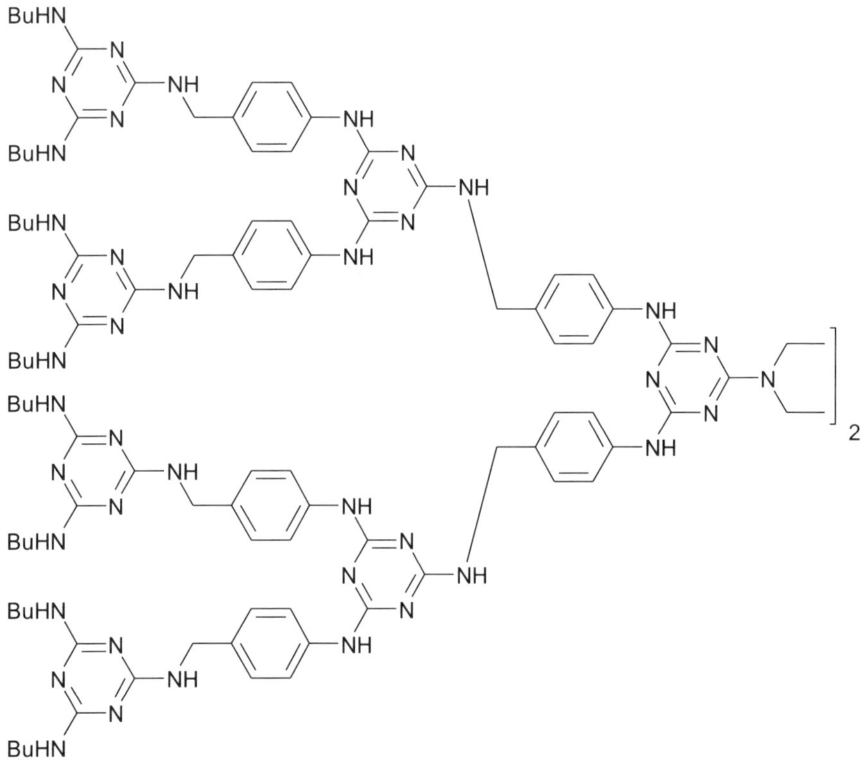

Fig. 14 Dendritic organogelator based on triazine building blocks

Furthermore, the xerogels were investigated using SEM. In certain cases, the self-assembled state exhibited a fibrous nanostructure, whilst others showed relatively wide (30–60 nm) self-assembled nanoribbons. X-ray diffraction (XRD) measurements on the dried xerogels combined with other results helped the authors to propose different modes of self-assembly for the different categories of dendritic gelators. These are illustrated in Fig. 13.

Recently, Jang and Aida functionalised a macrocyclic core with dendritic substituents [52]. The resultant dendrimer self-assembled into a solution-phase fibrous nanostructure with the dendritic branching assisting the solubilisation of the macrocycle and playing a proactive role in enabling effective stacking of the units into fibres. However, the authors did not report gel-phase materials properties for these assemblies, and it is worth noting here that gelation requires the hierarchical self-assembly of individual nanoscale supramolecular fibres into a network of fibre bundles.

Simanek and co-workers reported a small library of triazines with a variety of surface and interior linking groups which were capable of acting as organogelators (Fig. 14) [53]. They ascertained that dendrimers which incorporated interior and surface groups capable of donating hydrogen bonds (e.g.

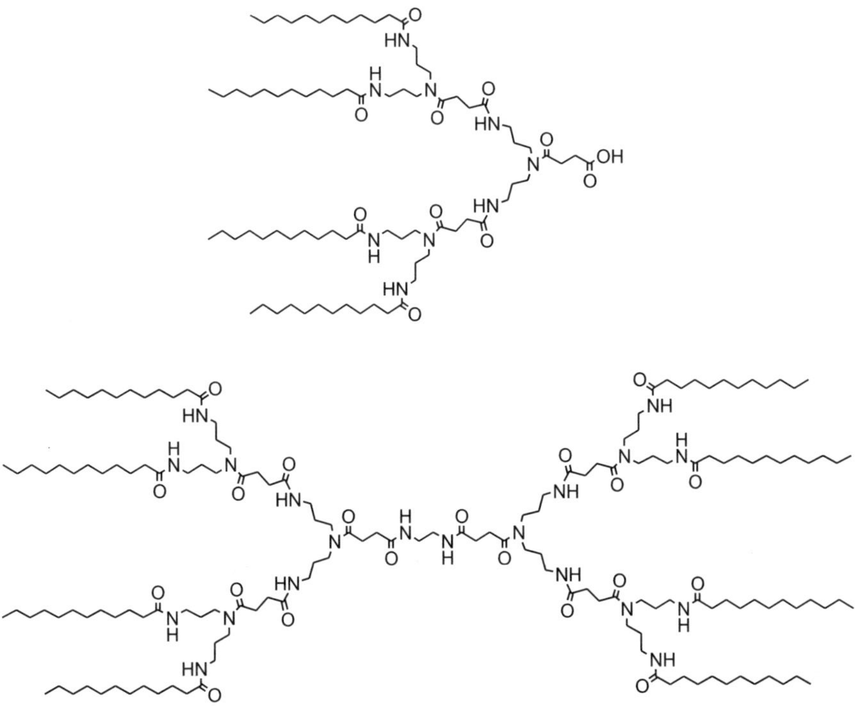

Fig. 15 Dendrons and symmetric dendrimers capable of gelating organic solvents

using an N – H group) were more effective gelators than those that did not. The authors also reported that halogenated solvents only formed gels when acidified with HCl. Furthermore, they observed that for gel-forming systems, the gel permeation chromatography (GPC) retention times were significantly different in acidified and neutral chloroform, which indicates that GPC was a good method for probing gelation behaviour. Transmission electron microscopy (TEM) confirmed the presence of fibres in the assembled structure. In this case, however, no clear dendritic effect on material properties was reported.

Kim and co-workers used peptide dendrons to form gel-phase assembled materials at relatively high concentrations (8 wt/vol %) [54]. They employed both dendrons and symmetric dendrimers constituted from repeating amide units with peripheral hydrophobic chains (Fig. 15). They observed a negative dendritic effect on the thermal stability of the gel as determined from the gel–sol transition temperature (T_{gel}) measurements.

In this example, the thermal stability decreased with increasing dendritic generation, for both the dendrons and the symmetric dendrimers. TEM and XRD studies indicated a lamellar structure. Interestingly, some of these asymmetric phase-separated dendrons also possess amphiphilic properties and

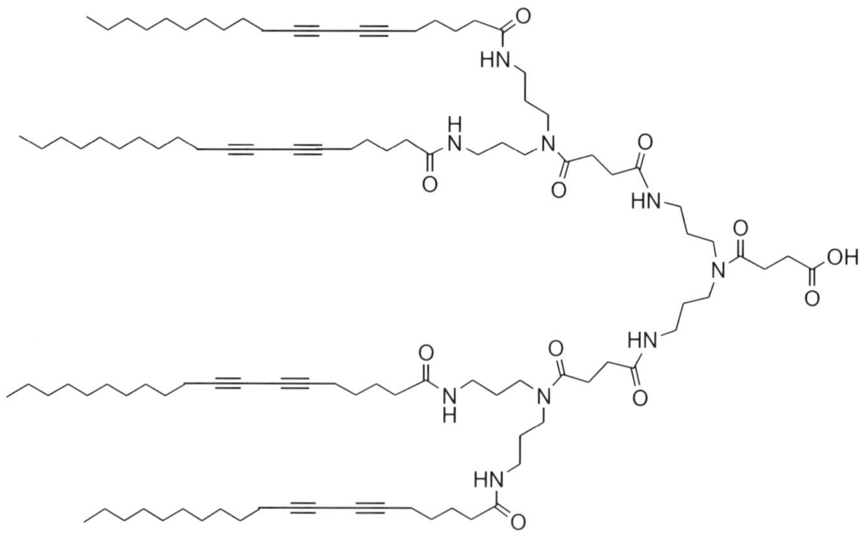

Fig. 16 Dendron used for assembly of gel-phase materials which can subsequently be crosslinked by in situ photopolymerisation of the triple bonds

were observed to form discrete vesicles (diameter ca. 160 nm) in an aqueous phase.

In an extension of this work, alkyne functionalities were incorporated into the long hydrophobic tails on the periphery of the dendritic structure (Fig. 16) [55]. TEM images of the second-generation dendron indicated the presence of fibrous bundles, and XRD of the dried gels indicated that these fibres had a hexagonal columnar structure. An attempt was made to covalently "capture" the assembled superstructure using in situ photopolymerisation of the triple bonds. UV irradiation of the assembled structures in toluene led to a colour change, with the samples becoming deep red. After irradiation for 2 h, part of the sample became insoluble in organic solvents, with this product maintaining its columnar hexagonal packed structure as indicated by XRD. It was argued that this material was the covalently fixed analogue of the gel-phase self-assembled state.

Smith and co-workers reported a one-component gelator based on L-lysine dendritic branches (dendrons) (Fig. 17) [56]. These symmetric dendrimers with a core disulfide moiety induced gelation of a number of organic solvents including cyclohexane, acetonitrile and dichloromethane. In particular, the T_{gel} value increased dramatically with increased dendritic branching. This was the first example of a positive dendritic effect on gelation which could be directly quantified. It was argued that this positive dendritic effect is a consequence of additional dendrimer–dendrimer hydrogen-bonding interactions, which are possible between amide groups in the more extensive dendritic branching. SEM indicated that a generation-dependent fibrous morphology

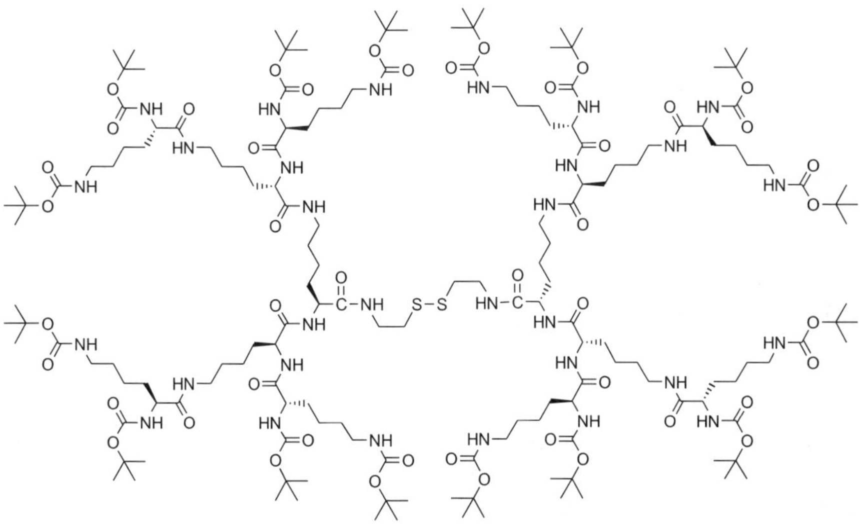

Fig. 17 Third-generation dendritic disulphides constructed using biocompatible L-lysine building blocks

was generated, with circular dichroism (CD) spectroscopy studies showing that the nanoscale self-assembly had a helical bias. This investigation indicated that peptidic dendritic branching can play a proactive role in the formation of highly functional organic materials.

It is worth noting that several other groups have reported peptidic dendrimers capable of non-specific self-assembly in organic solvents. Kraft and Osterod investigated dendritic aromatic amides [57], whilst Chow and co-workers investigated a series of dendrimers based on branching constructed from β-alanine and 3,5-diaminobenzoic acid [58]. In both cases, ill-defined nanoscale aggregates assembled as a consequence of hydrogen-bonding interactions. However, in neither case was gel-phase materials behaviour reported. The degree of aggregation was relatively low in both cases, and hence the dendrimer network was not sufficiently developed to induce macroscopic gelation.

3.2.2
Dendritic Organogelators with a Polymeric Component

Perhaps some of the most exciting work using one-component dendritic gelators has been performed by Stupp and co-workers. They have investigated the assembly properties of so-called dendron rod–coil molecules (Fig. 18) [59]. These dendron rod–coils are effectively tri-block copolymers, which have a flexible (coil) polymer unit attached to a well-defined linear rigid rod unit, which is, in turn, grafted to a dendritic head unit. The synthesis of

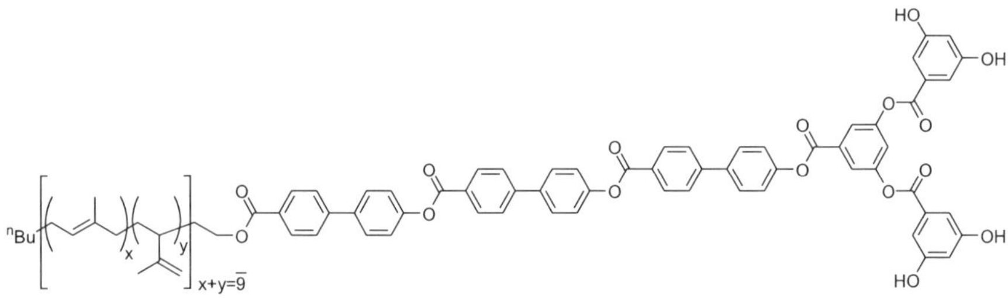

Fig. 18 Dendron rod–coil molecule used for self-assembly of a one-dimensional nanoscale architecture, hence leading to the formation of gel-phase materials

a whole range of different dendritic generations of dendron rod–coils has been reported [60]. Interestingly, even in the extremely dilute state (as low as 0.2 wt/vol %), these systems underwent self-assembly in various organic solvents to produce birefringent soft solids. TEM imaging indicated the formation of one-dimensional fibrous objects with a uniform width of 10 nm, which could be consistent with bimolecular packing. Strands as long as 10 μm could be isolated, which indicates the high aspect ratio of the assembled superstructure. In order to probe the structural requirements for assembly, a range of analogues have been prepared and their assembly properties investigated. It was found that:

1. At least four hydrogen-bonding OH groups were required on the surface of the dendron block for effective gelation, indicating the importance of hydrogen-bonding interactions.
2. The rigid-rod segment required a sufficient number of biphenyl ester units for assembly to occur. Indeed, the more biphenyl ester units, the greater the mechanical strength of the gel, suggesting a role for π–π interactions in the assembly.
3. A sufficiently long coil was found to be essential for gelation.
4. Dendrons of higher generation were not effective in forming gels, possibly a steric effect of the bulky head group.

AFM imaging indicated a uniform thickness of 2 nm and this led the authors to propose a ribbon-like structure, a proposal supported by light-scattering studies [61]. The crystal structure of an analogous small organic molecule was determined, and this allowed the authors to propose a structural model for the assembled ribbons.

Most interestingly, Stupp and co-workers have begun to develop applications for the self-assembled nanostructures. They exploited the self-assembly of molecular building blocks as scaffolding for toughening polymeric materials [62]. Judicious solvent selection provided a solvent capable of promoting self-assembly and providing a polymerisable medium. In this exciting development, the polymerisation reaction essentially takes place around the

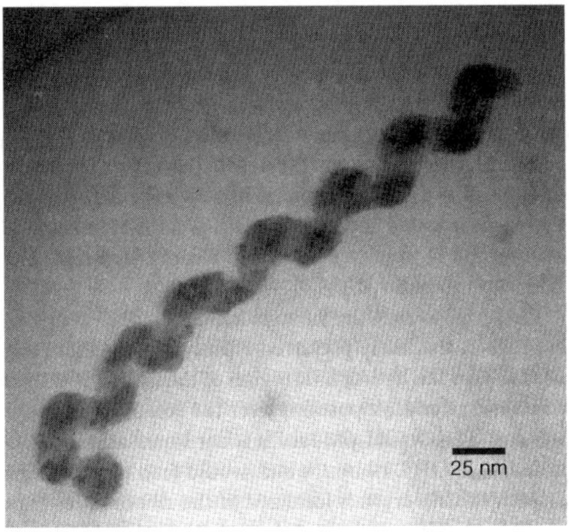

Fig. 19 TEM image of the CdS nanohelices generated using assemblies of dendron rod–coils as a templating matrix. This image is reproduced from reference [63] with kind permission of Wiley-VCH

nanoscale assembled structure generated by the dendron rod–coil. Small-angle X-ray scattering (SAXS) studies indicated that the polymer retained reflections corresponding to the fibrous scaffold and, furthermore, the polystyrene generated in this way had a significantly higher impact strength. It was argued that the supramolecular structure within the polymer may act as an impact-absorbing skeleton to dissipate strain energy. Dendron rod–coils have also been used for the preparation of CdS nanohelices (Fig. 19) [63]. Cadmium nitrate was added to THF and then mixed with a dendron rod–coil gel. The gel was aged for one week and then exposed to $H_2S(g)$. The sample was aged for a further week, and the solid inorganic product was isolated.

TEM analysis indicated the formation of right- and left-handed helices of CdS, which had been templated on the twisted nanoribbon architecture assembled by the dendron rod–coil. These workers have also utilised their dendron rod–coils for the assembly of ZnO nanocrystals [64]. When these organic–inorganic assemblies were placed in an electric field, a degree of alignment occurred (as determined using UV/Vis spectroscopy and second harmonic generation measurements). Most notably, these aligned nanocomposites had a lower threshold for lasing behaviour than pure ZnO nanocrystals. These results indicate the way in which self-assembly using dendritic building blocks can be used to generate hybrid nanomaterials with potential photonic applications.

Recently, Stupp and co-workers demonstrated that dendron rod–coil systems can have conjugated systems such as oligothiophene incorporated into their rod segments [65]. These molecules formed gel-phase materials in organic solvents and self-assembly of the systems gave rise to a blue-shifted absorption spectrum, and a red-shifted, quenched fluorescence spectrum. In the case of the oligothiophene system, self-assembly led to a three orders of magnitude enhancement in the conductivity of iodine-doped films. In addition, the supramolecular assemblies could be aligned in an electric field; this macroscopic orientation may well be useful in the fabrication of devices for use in nanotechnology.

4 Two-Component Dendritic Gelators

4.1 Introduction to Two-Component Gels

Two-component gelators are significantly rarer than single-component systems. In a two-component gel, the process of gelation is intrinsically dependent on two different molecular-scale components interacting to generate a complex. This complex subsequently undergoes hierarchical self-assembly to yield nanoscale fibres that interact with one another to form a sample-spanning entangled network. Macroscopically, this hierarchical process generates a gel-phase material.

A range of different systems have been used to generate two-component gels. For example, a number of systems are formed by two components which interact through complementary hydrogen-bonding interactions [66]. Other approaches to the formation of the intial complex have relied on π–π acceptor–donor interactions [67] and metal–ligand bonds [68]. A final method for generating two-component gel-phase systems involves the formation of a latent covalent bond between two molecules which in their own right are not gelators, but which form a gelator on reaction with one another; in some cases, this bond-forming step can itself be reversible (and therefore somewhat like a non-covalent interaction) [69].

In addition to exemplifying the way in which supramolecular chemistry can operate on multiple different levels, two-component systems have a vast degree of tunability. Either one of the two molecular components can be subtly varied in order to ascertain the effects on nanoscale architecture and macroscopic behaviour. In this way, there is the potential to access novel forms of functional gel-phase materials.

4.2 Two-Component Dendritic Organogels

In early 2001, Smith and co-workers communicated the basic design principles of the first dendritic two-component gelator [70]. This system was based

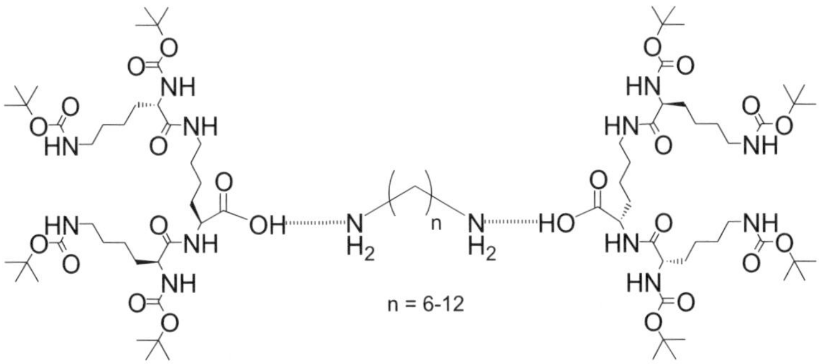

Fig. 20 Structure of dendritic two-component gelator unit

on the interaction of dendritic building blocks based on L-lysine repeat units with an aliphatic diamine (Fig. 20).

This system has been investigated in detail, and the tunability of this type of two-component gelation system has been exemplified, as explained below. The acid–base interaction between the two components plays a primary role in the formation of the gel—if the acid is protected as an ester, no gel formation takes place. It is proposed that this interaction (Fig. 20) generates the gelator complex, which is the species that hierarchically self-assembles to form fibrous gel-phase aggregates. Interestingly, it was reported that macroscopic gelation still occurred when the acid–base interaction was swapped for an interaction between a crown ether and a protonated amine (Fig. 21) [71]. This indicates that different supramolecular interactions between the components can easily be employed.

Smith, Hirst and co-workers reported the effect of the aliphatic diamine spacer chain length on the supramolecular chiral assembly [72]. Interestingly, the degree of structuring observed in these gel-phase materials was directly controlled by the length of the diamine spacer unit, as illustrated by the thermally reversible gel–sol phase transition (T_{gel}). Remarkably, as the length of

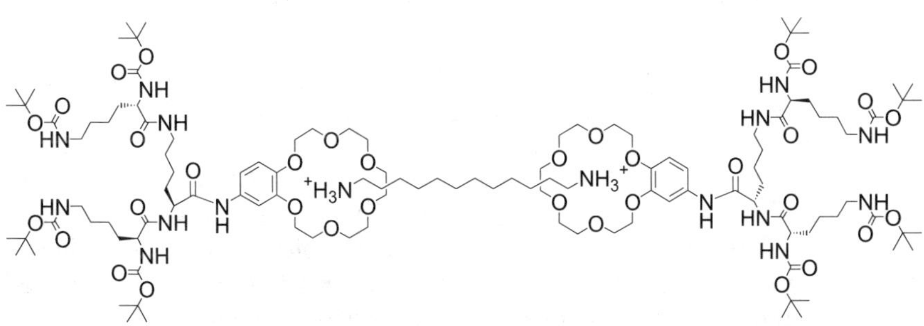

Fig. 21 Two-component gel based on interactions between a crown ether and a bis-protonated amine

the spacer unit was incrementally increased from six to twelve carbon atoms, T_{gel} increased profoundly from 4 to 105 °C. This major change in macroscopic behaviour results from a relatively small modulation in molecular-level structure. Scanning electron microscopy (SEM) showed that the length of the spacer unit dictated the aggregate morphology and that the formation of long, intertwined fibres with widths of ca. 20 nm is a prerequisite for a high degree of structuring. This situation was only achieved with aliphatic diamines that possess relatively longer spacer chains. Intriguingly, CD spectroscopy indicated that the spacer unit also controlled the level of helicity of the self-assembled state. Furthermore, ^{1}H NMR spectroscopy of the self-assembled state indicated that the spacer unit dictates the mobility of the peptide units via the formation of intermolecular amide-amide hydrogen bonding. It is proposed that in this way, the molecular structure of the spacer unit can directly control both the mesoscale morphology and the macroscopic properties of the gel. Solvent studies supported this hypothesis, with apolar, non-hydrogen-bonding solvents being the preferred solvent environment for the formation of these gels. Indeed, the thermal properties of the gel could be correlated with the polar solubility parameter δ_a and with the Kamlet–Taft hydrogen-bonding parameter α [73].

In addition, the effect of dendritic generation on the self-assembly was investigated [74]. Notably, in contrast to some other reports of dendrimer assembly, an optimum size of gelator unit was identified; second-generation branching gave a more thermally stable gel-phase material than first- and third-generation analogues. It was argued that in this case, optimal gelation conditions are reflected in a balance between the formation of enthalpically favourable hydrogen bonds and the steric and entropic cost of immobilising larger dendritic branches. It is notable that this observation is different from the results obtained with one-component dendritic systems based on L-lysine [56, 75]. This probably reflects differences in the mode of hierarchical self-assembly between one- and two-component gelators.

The effect of stereochemistry on the supramolecular chiral assembly was also studied [76]. The stereochemistry of the lysine groups in the dendritic building blocks played a key role in controlling the mode of self-assembly. Notably, the T_{gel} value, which reflects the macroscopic properties of the gel, was dependent on the stereochemistry. Racemic mixtures of L,L,L and D,D,D dendritic peptides possess lower T_{gel} values than their single- enantiomer analogues. Interestingly, SEM investigations indicated that the racemic gels possess a dramatically different morphology (Fig. 22). CD spectrometry indicated the helical nature of the homochiral assemblies and, furthermore, demonstrated that the presence of the "wrong" stereoisomer was able to disrupt the stacking process. This resulted in the break-up of well-ordered helical assemblies, giving rise to nanoscopic structures, which have less stereochemical and morphological definition.

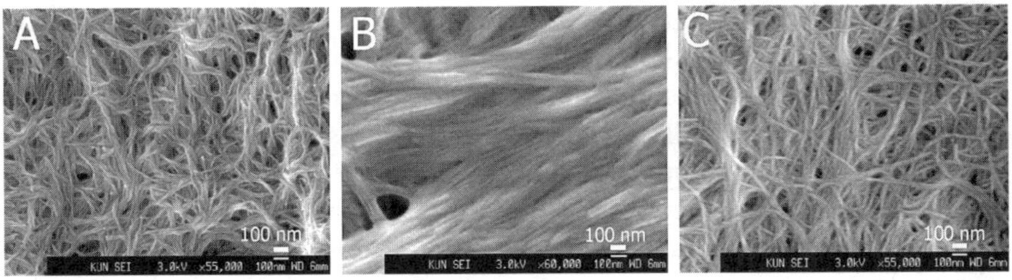

Fig. 22 Effect of chirality on the nanoscale morphology of the two-component gelation system. **a** = *l*, *l*, *l*. **b** = 50% D,D,D, 50% L,L,L. **c** = *d*, *d*, *d*

Stereoisomeric gels, in which one chiral centre of the dendritic peptide was changed, were also investigated. This important study proved that a subtle stereochemical change to a single chiral centre had a pronounced effect on the self-assembly process. The T_{gel} values of these gels were depressed and SEM indicated a less entangled fibrous network. SAXS data indicated that the mode of molecular packing was also modulated, as was the helicity of the fibres. It is particularly noteworthy, given that dendritic systems are capable of multiple interactions, that when one chiral centre is modified the gel formation is not completely switched off, but the nanostructure and strength of the material can be modulated. This is a clear example of the impact that subtle changes in chirality can have within an assembled dendritic architecture, and indicates that hydrogen-bonding interactions which yield dendritic gel-phase assemblies can enable exquisite levels of control over materials behaviour.

In a key study, Smith, Hirst and co-workers reported the effect of varying the ratio of the two components [77]. Remarkably, increasing the amount of diamine relative to the dendritic branch changed the propensity of this system to induce macroscopic gelation, and ultimately gave rise to a completely new morphology in which micrometer-sized platelets were observed

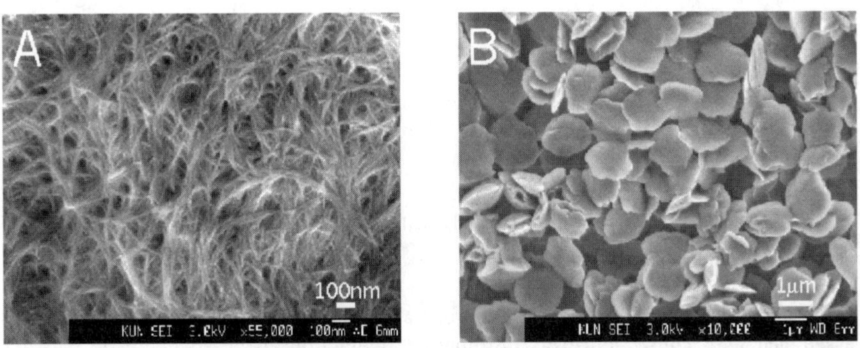

Fig. 23 The nanoscale morphology observed using SEM for the two-component gelation system. **a** = 2 : 1 dendron : diamine ratio. **b** = 1 : 4.5 dendron : diamine ratio

(Fig. 23). It is believed that as the amount of dendron was decreased relative to the diamine, it was less able to stabilise the extended fibrous morphology, and instead, microcrystalline chunks of diamine formed which were effectively capped by the small amounts of dendron present. This result illustrates a way in which two-component gels can be tuned that is not accessible for single-component gelators.

It is worth noting that Aida and co-workers have also reported what is effectively a two-component approach to a fibrous (columnar) assembled nanostructure [78]. They synthesised dendrons with a pyrazole unit at the focal point. Pyrazole is an exobidentate ligand capable of binding Group 11 univalent metal ions (Cu(I), Ag(I) and Au(I)), and appropriately functionalised pyrazoles are capable of forming metal pyrazolate coordination triangles (Fig. 24). Heating a paraffin suspension of the dendron–metal complexes at 200 °C and then cooling gave rise to a fibrous precipitate. In this case, the material formed did not show gel-phase properties due to its poor solubility in paraffin. It was argued that metal–metal interactions were responsible for holding the assembled fibrous superstructure together. The fibres were intensely luminescent, and the dendritic ligand was capable of acting as an antenna, transferring energy to the interior metal ion cluster. On dissolution in CH_2Cl_2, the characteristic luminescence disappeared as the fibre became dissociated into individual metallacycles. Interestingly, an optimum level of

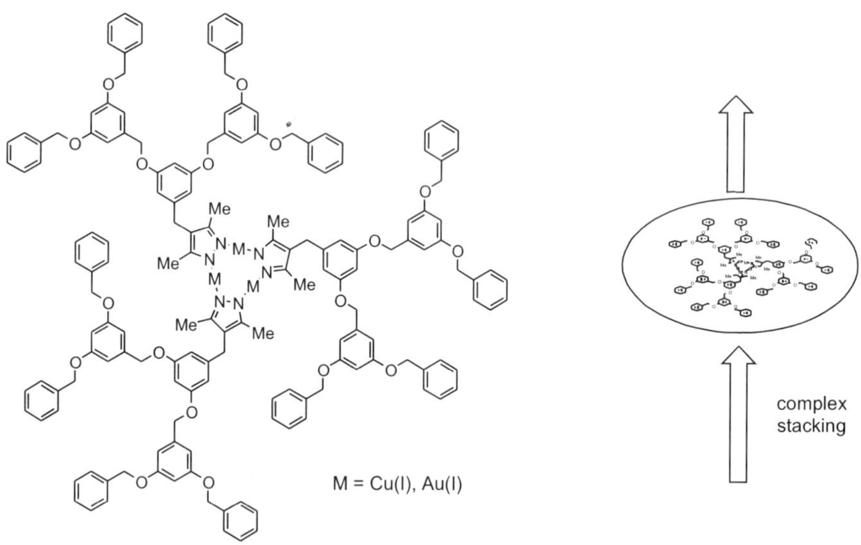

Fig. 24 Two-component approach to fibrous assemblies in which the initial complex is assembled as a consequence of coordination interactions between dendritic pyrazole ligands and metal ions. These complexes go on to assemble into columnar fibrous superstructures that precipitate from solution

dendritic branching was required for the assembly process to take place, indicating that unfavourable steric repulsion and favourable van der Waals forces between dendron units are in balance within the stacked superstructure—a common feature within dendritic assemblies [74]. On investigation by SEM it was found that, remarkably, even though the system does not contain any chiral centres, the fibres were helical and comprised a bundle of several loosely twisted fibrils.

4.3 Versatile One- and Two-Component Dendritic Gelators

In a recent publication, Smith and co-workers developed a dendritic system which is capable of acting as both a one-component and a two-component dendritic gelator (Fig. 25) [79]. By changing the "surface groups" on the L-lysine dendron from Boc protecting groups to long alkyl tails, the dendron becomes, in its own right, a one-component gelator. In fact, peptidic molecules with long alkyl tails are well-known to promote gelation as a consequence of their ability to form hydrogen bonds combined with a preference for solvophobic packing and their ability to form van der Waals interactions [80]. However, it had not previously been realised that such systems can also form a two-component complex on the addition of aliphatic diamine, and this was exploited to develop a very tunable gelation system.

When the dendrons were investigated as one-component gelators, second-generation dendron **B** formed effective gel-phase materials, with a T_{gel} value increasing up to a maximum of 87 °C. First-generation dendron **A**, on the other hand, only formed very soft, weak networks, with the T_{gel} values remaining at approximately 20 °C irrespective of concentration. This indicates

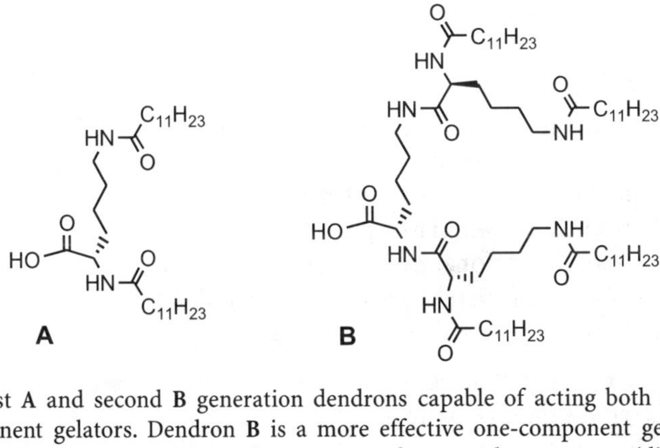

Fig. 25 First **A** and second **B** generation dendrons capable of acting both as one- *and* two-component gelators. Dendron **B** is a more effective one-component gelator, whilst dendron **A** forms stronger gels in the presence of a second component (diaminododecane)

that the additional dendritic branching of dendron **B** significantly assists in the assembly of gel-phase materials—presumably by allowing the formation of additional favourable non-covalent interactions. Interestingly, however, when investigated as two-component gelators, the mixture of dendron **A** (a poor gelator in its own right) with diaminododecane gave rise to a very effective gel, with a maximum T_{gel} value of 95 °C. In contrast, dendron **B** (a potent gelator in its own right) only formed very weak gels in the presence of diaminododecane, with a maximum T_{gel} value of 20 °C.

The addition of the second component therefore enhanced the gelation process for dendron **A**, but hindered the gelation process for dendron **B**. In summary, the gel-phase materials properties of the two dendrons responded in opposite ways to the presence of a diamine additive. This is a remarkable example in which dendritic generation and the use of a second (supramolecular) component can both control materials properties. This effect can be considered as orthogonal dendritic and supramolecular control. It was argued that, in the future, these systems may be exploited to develop switchable and highly controllable gel-phase materials.

5 Conclusions and Future Outlook

It is clear that the field of low molecular weight organogelators is one of the most exciting frontiers of supramolecular science. By controlling molecular recognition events, the self-assembly of molecular building blocks into nanoscale architectures is realised. This bottom-up fabrication approach will doubtlessly allow the generation of highly functional and tunable nanostructured materials of importance in the emerging field of nanochemistry. This article has focussed on the formation of soft materials, and in particular has illustrated the application of dendritic building blocks in the self-assembly of tunable gel-phase materials. The specific roles played by dendritic branching in the formation of gel-phase materials can be summarised as follows:

- Multiple interactions between dendritic branches enable the formation of strong fibrous assemblies as a consequence of multivalency (multiple non-covalent interactions are stronger than isolated ones).
- Dendritic branching can play a proactive steric role encouraging the formation of one-dimensional assemblies.
- For the formation of more traditional crosslinked polymer gels, dendritic building blocks act as ideal crosslinking units with multiple reactive sites.

It is also worth noting that very few dendritic gelators are the fully spherical systems traditionally associated with dendrimer chemistry. It is intuitively clear that spherical systems will have difficulty assembling into one-dimensional fibres, and consequently it is likely that the use of nonspherical dendritic systems to obtain gel-phase materials is a general design principle.

In terms of applications, it is expected that dendritic gels will have similar applications to other low molecular weight gelators. However, the branched framework of dendritic building blocks also offers a number of inherent advantages which may be of special interest. The following points offer some speculation about directions in which applications of dendritic gelators may develop in the future:

- The tunability of dendritic structures means that subtle modifications can easily be made to the dendritic architecture. Such modifications can modulate gel-phase properties without completely disrupting the gel formation process.
- The dendritic branching often forms the periphery of the fibrous assembly, and as such can be considered to play a role in the encapsulation of the centre of the fibre from the bulk solvent. This may lead to modification of, for example, electronic and optical properties within the assembled gel state.
- The dendritic branching provides multiple surface groups, and as such has potential for chemical modification to yield gelators with multiple covalently appended functionalities.
- The multiple surface groups present on the surface of the fibre (as a consequence of the dendritic branching) may offer multiple points for interaction with other species (e.g. other nanostructures or biological systems); this may give rise to gels with good adhesive properties.
- The repeating branched architecture can, in principle, be biodegraded into a relatively small number of similar building blocks. In many cases dendrimers are constructed from biocompatible branched motifs, such as L-lysine or L-glutamic acid derivatives; this may give rise to biocompatible gelators.

In summary, the unique architecture of dendritic molecules has proven to be amenable to the formation of gel-phase assemblies. We propose that the inherent branched architecture of dendritic macromolecules will be further exploited in the near future to yield a new generation of functional, dendritic, soft nanomaterials.

References

1. Graham T (1861) Phil Trans Roy Soc 151:183–224
2. (a) Lloyd DJ (1926) In: Alexander J (ed) Colloid chemistry, vol 1. Chemical Catalog Company, New York, p 767. (b) Keller A (1995) Faraday Discuss 101:1–49
3. Flory PJ (1953) Principles of polymer chemistry. Cornell University Press, Ithaca, pp 347–398
4. Corriu RJP, Leclercq D (1996) Angew Chem Int Ed Engl 35:1420–1436

5. See for example: (a) Burch W, Ross Murphy SB (eds) (1988) Physical networks — polymers and gels. Elsevier, London. (b) Guenet J-M (1992) Thermoreversible gelation of polymers and biopolymers. Academic, London. For a general overview of self-organising polymer systems including gels in a broader context see: (c) Förster S, Konrad M (2003) J Mater Chem 13:2671–2688
6. Fages F (ed) (2005) Top Curr Chem, chapters in this volume
7. Brunsveld L, Folmer BJB, Meijer EW, Sijbesma RP (2001) Chem Rev 101:4071–4097
8. (a) Whitesides GM, Grzybowski B (2002) Science 295:2418–2421. (b) Hamley IW (2003) Angew Chem Int Ed 42:1692–1712. (c) Elemans JAAW, Rowan AE, Nolte RJM (2003) J Mater Chem 13:2661–2670. (d) Schalley CA, Lützen A, Albrecht M (2004) Chem Eur J 10:1072–1080. (e) Lehn J-M (2002) Proc Natl Acad Sci USA 99:4763–4768
9. Estroff LA, Hamilton AD (2004) Chem Rev 104:1201–1217
10. (a) Terech P, Weiss RG (1997) Chem Rev 97:3133–3159. (b) Gronwald O, Snip E, Shinkai S (2002) Curr Opin Colloid Interface Sci 7:148–156. (c) van Esch JH, Feringa BL (2000) Angew Chem Int Ed 39:2263–2266. (d) Oda R, Huc I, Candau SJ (1998) Angew Chem Int Ed 37:2689–2691. (e) Abdallah DJ, Weiss RG (2000) Adv Mater 12:1237–1247. (f) Melendez RE, Carr AJ, Linton BR, Hamilton AD (2000) Struct Bond 96:31–61. (g) Shimizu T (2003) Polym J 25:1–22. (h) Pasini D, Kraft A (2004) Curr Opin Solid State Mater Sci 8:157–163
11. (a) Newkome GR, Moorefield CN, Vögtle F (2001) Dendrimers and dendrons—concepts, synthesis, applications. Wiley-VCH, Weinheim. (b) Fréchet JMJ, Tomalia DA (eds) (2002) Dendrimers and other dendritic polymers. Wiley, New York
12. (a) Smith DK, Hirst AR, Love CS, Hardy JG, Brignell SV, Huang B (2005) Prog Polym Sci 30:220–293. (b) Smith DK, Diederich F (2000) Top Curr Chem 210:183–227. (c) Zimmerman SC, Lawless LJ (2001) Top Curr Chem 217:95–120. (d) Fréchet JMJ (2002) Proc Natl Acad Sci USA 99:4782–4787. (e) Diederich F, Felber B (2002) Proc Natl Acad Sci USA 99:4778–4781. (f) Gittins PJ, Twyman LJ (2003) Supramol Chem 15:5–23
13. Lee RT, Lee YC (1994) In: Lee RT, Lee YC (eds) Neoglycoconjugates: preparation and applications. Academic, San Diego, p 23
14. See for example: (a) Roy R, Zanini D, Meunier SJ, Romanowska A (1993) J Chem Soc Chem Commun 1869–1872. (b) Zanini D, Roy R (1997) J Am Chem Soc 119:2088–2095
15. For an example in which dendritic branches play a steric buttressing role in controlling the self-assembly of a hexameric rosette-type aggregate see: (a) Zimmerman SC, Zeng F, Reichert DEC, Kolotuchin SV (1996) Science 271:1095–1098. (b) Zeng F, Zimmerman SC, Kolotuchin SV, Reichert DEC, Ma YG (2002) Tetrahedron 58:825–843
16. For reviews of dendrimer chemistry with a significant synthetic focus see: (a) Chow H-F, Mong TK-K, Nongrum MF, Wan C-W (1998) Tetrahedron 54:8543–8660. (b) Matthews OA, Shipway AN, Stoddart JF (1998) Prog Polym Sci 23:1–56
17. Smith DK, Diederich F (1998) Chem Eur J 4:1353–1361
18. Pyun J, Zhou X-Z, Drockenmuller E, Hawker CJ (2003) J Mater Chem 13:2653–2660
19. Buhleier E, Wehner W, Vögtle F (1978) Synthesis 155–158
20. (a) Denkewalter RG, Kolc JF, Lukasavage WJ (1979) US Patent 4,360,646. (b) Denkewalter RG, Kolc JF, Lukasavage WJ (1981) US Patent 4,289,872. (c) Denkewalter RG, Kolc JF, Lukasavage WJ (1983) US Patent 4,410,688
21. (a) Tomalia DA, Baker H, Dewald J, Hall M, Kallos G, Martin S, Roeck J, Ryder J, Smith P (1985) Polym J 17:117–132. (b) Tomalia DA, Baker H, Dewald J, Hall M, Kallos G, Martin S, Roeck J, Ryder J, Smith P (1986) Macromolecules 19:2466–2468
22. Newkome GR, Yao Z, Baker GR, Gupta VK (1985) J Org Chem 50:2003–2004

23. (a) de Brabander-van den Berg EMM, Meijer EW (1993) Angew Chem Int Ed 32:1308–1311. (b) Meijer EW, Bosman HJM, van den Booren HAMJ, de Brabander-van den Berg EMM, Castelijns AMCF, De Man HCJ, Reintjens RWEC, Stoelwinder CJC, Nijenhuis AJ (1996) US Patent 5,610,268
24. Wörner C, Mülhaupt R (1993) Angew Chem Int Ed 32:1306–1308
25. (a) Hawker C, Fréchet JMJ (1990) J Chem Soc Chem Commun 1010–1013. (b) Hawker CJ, Fréchet JMJ (1990) J Am Chem Soc 112:7638–7647
26. Miller TM, Neenan TX (1990) Chem Mater 2:346–349
27. Moore JS, Xu Z (1991) Macromolecules 24:5893–5894
28. Grayson SM, Fréchet JMJ (2001) Chem Rev 101:3819–3867
29. (a) Wooley KL, Hawker CJ, Fréchet JMJ (1991) J Am Chem Soc 113:4252–4261. (b) Kawaguchi T, Walker KL, Wilkins CL, Moore JS (1995) J Am Chem Soc 117:2159–2165. (c) Mattei S, Wallimann P, Kenda B, Amrein W, Diederich F (1997) Helv Chim Acta 80:2391–2417
30. (a) Inoue K (2000) Prog Polym Sci 25:453–571. (b) Yates CR, Hayes W (2004) Eur Polym J 40:1257–1281
31. Sunder A, Hanselmann R, Frey H, Mülhaupt R (1999) Macromolecules 32:4240–4246
32. (a) Kim MK, Jeon YM, Jeon WS, Kim HJ, Hong SG, Park CG, Kim K (2001) Chem Commun 667–668. (b) Wang RY, Yang J, Zheng ZP, Carducci MD, Jiao J, Seraphin S (2001) Angew Chem Int Ed 40:549–552. (c) Gopidas KR, Whitesell JK, Fox MA (2003) J Am Chem Soc 125:6491–6502. (d) Love CS, Chechik V, Smith DK, Brennan C (2004) J Mater Chem 14:919–924. (e) Daniel M-C, Ruiz J, Nlate S, Blais J-C, Astruc D (2003) J Am Chem Soc 125:2617–2628. (f) Wang YA, Li JJ, Chen HY, Peng XG (2002) J Am Chem Soc 124:2293–2298
33. Kim Y, Mayer MF, Zimmerman SC (2003) Angew Chem Int Ed 42:1121–1126
34. (a) van Hest JCM, Delnoye DAP, Baars MWPL, van Genderen MHP, Meijer EW (1995) Science 268:1592–1595. (b) van Hest JCM, Delnoye DAP, Baars MWPL, Elissen-Román C, van Genderen MHP, Meijer EW (1996) Chem Eur J 2:1616–1626
35. For reviews see: (a) Dennig J (2003) Top Curr Chem 228:227–236. (b) Stiriba SE, Frey H, Haag R (2002) Angew Chem Int Ed 41:1329–1334
36. For very selected references see: (a) Percec V, Chu PW, Ungar G, Zhou JP (1995) J Am Chem Soc 117:11 441–11 454. (b) Percec V, Cho W-D, Mosier PE, Ungar G, Yeardley DJP (1998) J Am Chem Soc 120:11 061–11 070. (c) Pesak DJ, Moore JS (1997) Angew Chem Int Ed Engl 36:1636–1639. (d) Lorenz K, Holter D, Stuhn B, Mülhaupt R, Frey H (1996) Adv Mater 8:414–416. (e) Baars MWPL, Sontjens SHM, Fischer HM, Peerlings HWI, Meijer EW (1998) Chem Eur J 4:2456–2466. (f) Saez IM, Goodby JW (2003) Chem Eur J 9:4869–4877
37. Newkome GR, Baker GR, Saunders MJ, Russo PS, Gupta VK, Yao Z-q, Miller JE, Bouillion K (1986) J Chem Soc Chem Commun 752–753
38. (a) Newkome GR, Baker GR, Arai S, Saunders MJ, Russo PS, Theriot KJ, Moorefield CN, Rogers LE, Miller JE, Lieux TR, Murray ME, Phillips B, Pascal L (1990) J Am Chem Soc 112:8458–8465. (b) Newkome GR, Moorefield CN, Baker GR, Behera RK, Escamilla GH, Saunders MJ (1992) Angew Chem Int Ed Engl 31:917–919. (c) Newkome GR, Lin X, Yaxiong C, Escamilla GH (1993) J Org Chem 58:3123–3129. (d) Yu KH, Russo PS, Younger L, Henk WG, Hua DW, Newkome GR, Baker G (1997) J Polym Sci Polym Phys 35:2787–2793
39. Jørgensen M, Bechgaard K, Bjørnholm T, Sommer-Larsen P, Hansen LG, Schaumburg K (1994) J Org Chem 59:5877–5882

40. Percec V, Glodde M, Bera TK, Miura Y, Shiyanovskaya I, Singer KD, Balagurusamy VSK, Heiney PA, Schnell I, Rapp A, Spiess HW, Hudson SD, Duan H (2002) Nature 419:384–387
41. Le Gall T, Pearson C, Bryce MR, Petty MC, Dahlgaard H, Becher J (2003) Eur J Org Chem 3562–3568
42. Marmillon C, Gauffre F, Gulik-Krzywicki T, Loup C, Caminade A-M, Majoral J-P, Vors J-P, Rump E (2001) Angew Chem Int Ed 40:2626–2629
43. (a) Launay N, Caminade A-M, Lahana R, Majoral J-P (1994) Angew Chem Int Ed Engl 33:1589–1592. (b) Chang JY, Ji HJ, Han MJ, Rhee SB, Cheong S, Yoon M (1994) Macromolecules 27:1376–1380
44. El Ghzaoui A, Gauffre F, Caminade A-M, Majoral J-P, Lannibois-Drean H (2004) Langmuir 20:9348–9353
45. Klok H-A, Hwang JJ, Hartgerink JD, Stupp SI (2002) Macromolecules 35:6101–6111
46. Stendahl JC, Li L, Claussen RC, Stupp SI (2004) Biomaterials 25:5847–5856
47. Namazi H, Adeli M (2003) Eur Poly J 39:1491–1500
48. (a) Carnahan MA, Grinstaff MW (2001) Macromolecules 34:7648–7655. (b) Carnahan MA, Middleton C, Kim J, Kim T, Grinstaff MW (2002) J Am Chem Soc 124:5291–5293. (c) Luman NR, Smeds KA, Grinstaff MW (2003) Chem Eur J 9:5618–5626
49. Gitsov I, Zhu C (2002) Macromolecules 35:8418–8427
50. Jang W-D, Jiang D-L, Aida T (2000) J Am Chem Soc 122:3232–3233
51. Jang W-D, Aida T (2003) Macromolecules 36:8461–8469
52. Jang W-D, Aida T (2004) Macromolecules 37:7325–7330
53. Zhang W, Gonzalez SO, Simanek EE (2002) Macromolecules 35:9015–9021
54. Kim C, Kim KT, Chang Y, Song HH, Cho T-Y, Jeon H-J (2001) J Am Chem Soc 123:5586–5587
55. Kim C, Lee SJ, Lee IH, Kim KT, Song HH, Jeon H-J (2003) Chem Mater 15:3638–3642
56. Love CS, Hirst AR, Chechik V, Smith DK, Brennan C, Ashworth I (2004) Langmuir 20:6580–6585
57. (a) Osterod F, Kraft A (1997) Chem Commun 1435–1436. (b) Kraft A, Osterod F (1998) J Chem Soc Perkin Trans 1 1019–1025
58. Mong TK-K, Niu A, Chow H-F, Wu C, Li L, Chen R (2001) Chem Eur J 7:686–699
59. Zubarev ER, Pralle MU, Sone ED, Stupp SI (2001) J Am Chem Soc 123:4105–4106
60. Zubarev ER, Stupp SI (2002) J Am Chem Soc 124:5762–5773
61. de Gans BJ, Wiegand S, Zubarev ER, Stupp SI (2002) J Phys Chem B 106:9730–9736
62. (a) Zubarev ER, Pralle MU, Sone ED, Stupp SI (2002) Adv Mater 14:198–203. (b) Stendahl JC, Li LM, Zubarev ER, Chen Y-R, Stupp SI (2002) Adv Mater 14:1540–1543
63. Sone ED, Zubarev ER, Stupp SI (2002) Angew Chem Int Ed 41:1705–1709
64. Li L, Beniash E, Zubarev ER, Xiang W, Rabatic BM, Zhang G, Stupp SI (2003) Nat Mater 2:689–694
65. Messmore BW, Hulvat JF, Sone ED, Stupp SI (2004) J Am Chem Soc 126:14452–14458
66. (a) Hanabusa K, Miki T, Taguchi Y, Koyama T, Shirai H (1993) J Chem Soc Chem Commun 1382–1384. (b) Jeong SW, Shinkai S (1997) Nanotechnology 8:179–185. (c) Inoue K, Ono Y, Kanekiyo Y, Ishi-i T, Yoshihara K, Shinkai S (1999) J Org Chem 64:2933–2937. (d) Tata M, John VT, Waguespack YY, McPherson GL (1994) J Am Chem Soc 116:9464–9470. (e) Simmons BA, Taylor CE, Landis FA, John VT, McPherson GL, Schwartz DK, Moore R (2001) J Am Chem Soc 123:2414–2421. (f) Willemen HM, Vermonden T, Marcelis ATM, Sudhölter EJR (2002) Langmuir 18:7102–7106. (g) de Loos M, van Esch J, Kellogg RM, Feringa BL (2001) Angew Chem Int Ed 40:613–616. (h) Nakano K, Hishikawa Y, Sada K, Miyata M, Hanabusa K (2000) Chem Lett 1170–1171

67. (a) Maitra U, Kumar PV, Chandra N, D'Sousa LJ, Prasanna MD, Raju AR (1999) Chem Commun 595–596. (b) Friggeri A, Gronwald O, van Bommel KJC, Shinkai S, Reinhoudt DN (2002) J Am Chem Soc 124:10754–10758. (c) Babu P, Sangeetha NM, Vijaykumar P, Maitra U, Rissanen K, Raju AR (2003) Chem Eur J 9:1922–1932
68. (a) Ihara H, Sakurai T, Yamada T, Hashimoto T, Takafuji M, Sagawa T, Hachisako H (2002) Langmuir 18:7120–7123 (b) Dukh M, Saman D, Kroulik J, Cerny I, Pouzar V, Kral V, Drasar P (2003) Tetrahedron 59:4069–4076
69. (a) George M, Weiss RG (2003) Langmuir 19:1017–1025. (b) George M, Weiss RG (2001) J Am Chem Soc 123:10393–10394
70. Partridge KS, Smith DK, Dykes GM, McGrail PT (2001) Chem Commun 319–320
71. Dykes GM, Smith DK (2003) Tetrahedron 59:3999–4009
72. Hirst AR, Smith DK, Feiters MC, Geurts HPM (2004) Langmuir 20:7070–7077
73. Hirst AR, Smith DK (2004) Langmuir 20:10851–10857
74. Hirst AR, Smith DK (2004) Org Biomol Chem 2:2965–2971
75. Huang B, Hirst AR, Smith DK Castelletto V, Hamley IW (2005) J Am Chem Soc 127:7130–7139
76. Hirst AR, Smith DK, Feiters MC, Geurts HPM (2004) Chem Eur J 10:5901–5910
77. Hirst AR, Smith DK, Feiters MC, Geurts HPM, Wright AC (2003) J Am Chem Soc 125:9010–9011
78. Enomoto M, Kishimura A, Aida T (2001) J Am Chem Soc 123:5608–5609
79. Hardy JG, Hirst AR, Smith DK, Ashworth I, Brennan C (2005) Chem Commun 385–387
80. (a) Luo X, Liu B, Liang Y (2001) Chem Commun 1556–1557. (b) Yamada N, Imai T, Koyama E (2001) Langmuir 17:961–963. (c) Bhattacharya S, Krishnan-Ghosh Y (2001) Chem Commun 185–186. (d) Estroff LA, Huang JS, Hamilton AD (2003) Chem Commun 2958–2959. (e) Suzuki M, Yumoto M, Kimura M, Shirai H, Hanabusa K (2004) Helv Chim Acta 87:1–10

Author Index Volumes 251–256

Author Index Vols. 26–50 see Vol. 50
Author Index Vols. 51–100 see Vol. 100
Author Index Vols. 101–150 see Vol. 150
Author Index Vols. 151–200 see Vol. 200
Author Index Vols. 201–250 see Vol. 250

The volume numbers are printed in italics

Subject Index

Printing: Krips bv, Meppel
Binding: Stürtz, Würzburg